T0202771

Universitext

Universitext

Universitext is a series of textbooks that presents material from a wide variety of mathematical disciplines at master's level and beyond. The books, often well class-tested by their author, may have an informal, personal, even experimental approach to their subject matter. Some of the most successful and established books in the series have evolved through several editions, always following the evolution of teaching curricula, into very polished texts.

Thus as research topics trickle down into graduate-level teaching, first textbooks written for new, cutting-edge courses may make their way into *Universitext*.

More information about this series at http://www.springer.com/series/223

Rabi Bhattacharya · Edward C. Waymire

A Basic Course in Probability Theory

Second Edition

 Springer

Rabi Bhattacharya
Department of Mathematics
University of Arizona
Tucson, AZ
USA

Edward C. Waymire
Department of Mathematics
Oregon State Univeristy
Corvallis, OR
USA

ISSN 0172-5939 ISSN 2191-6675 (electronic)
Universitext
ISBN 978-3-319-47972-9 ISBN 978-3-319-47974-3 (eBook)
DOI 10.1007/978-3-319-47974-3

Library of Congress Control Number: 2016955325

Mathematics Subject Classification (2010): 60-xx, 60Jxx

This Springer imprint is published by Springer Nature
The registered company is Springer International Publishing AG
The registered company address is: Gewerbestrasse 11, 6330 Cham, Switzerland

Preface to Second Edition

This second edition continues to serve primarily as a text for a lively two-quarter or one-semester course in probability theory for students from diverse disciplines, including mathematics and statistics. Exercises have been added and reorganized (i) for reinforcement in the use of techniques, and (ii) to complement some results. Sufficient material has been added so that in its entirety the book may also be used for a two-semester course in basic probability theory. The authors have reorganized material to make Chapters III–XIII as self-contained as possible. This will aid instructors of one semester (or two quarter) courses in picking and choosing some material, while omitting some other. Material from a former chapter on Laplace transforms has been redistributed to parts of the text where it is used.

The early introduction of conditional expectation and conditional probability maintains the pedagogic innovation of the first edition. This enables the student to quickly move to the fundamentally important notions of modern probability besides independence, namely, *martingale dependence* and *Markov dependence*, where new theory and examples have been added to the text. The former includes *Doob's upcrossing inequality*, the *submartingale convergence theorem*, and *reverse martingales*, and the (reverse) *martingale proof of the strong law of large numbers*, while retaining important earlier approaches such as those of Kolmogorov, Etemadi, and of Marcinkiewicz–Zygmund.

A theorem of *Polya* is added to the chapter on weak convergence to show that the convergence to the normal distribution function in the central limit theorem is uniform.

The *Cramér–Chernoff large deviation theory* in Chapter V is sharpened by the addition of a large deviation theorem of *Bahadur and Ranga Rao* using the Berry–Esseen convergence rate in the central limit theorem. Also added in Chapter V is a *concentration of measure* type inequality due to Hoeffding. The proof of the aforementioned *Berry–Esseen bound* is deferred to Chapter VI on Fourier series and Fourier transform. The *Chung–Fuchs transience/recurrence criteria* for random walk based on Fourier analysis is a new addition to the text.

Special examples of Markov processes such as *Brownian motion*, and *random walks* appear throughout the text to illustrate applications of (i) martingale theory and stopping times in computations of certain important probabilities. A culmination of the theory developed in the text occurs in Chapters XI and XII on Brownian motion. This continues to rank among the primary goals attainable for a course based on the text.

General *Markov dependent* sequences and their convergence to equilibrium is the subject matter of the entirely new Chapter XIII. Illustrative examples are provided, including some of historical importance to the development of the kinetic theory of matter in physics due to Boltzmann, Einstein, and Smoluchowski. The treatment centers on describing a prototypical framework, namely *Doeblin's theorem*, for existence and convergence to a unique *invariant probability* for Markov processes, together with illustrative examples for students with diverse interests ranging from mathematics and statistics to contemporary mathematical finance or biology. Examples include *iterated random maps*, the *Ehrenfest model*, and products of *random matrices*. The *Ornstein–Uhlenbeck process* is shown to be obtained as the unique solution to a *stochastic differential equation*, namely the *Langevin equation*, using *Picard iteration*. This provides students with a glimpse into the broad scope and utility of the probability that they have learned, while motivating continued study of stochastic processes.

Complete references to authors of books cited in footnotes are provided in a closing list of references. This also includes other textbook resources covering the same topics and/or further applications.

The authors are grateful to William Faris, University of Arizona, and to Enrique Thomann, Oregon State University, for providing comments and corrections to an earlier draft based on their teaching of the course. Partial support from the National Science Foundation under grants DMS 1406872 and DMS1408947, respectively, is gratefully acknowledged by the authors.

Tucson, AZ, USA   Rabi Bhattacharya
Corvallis, OR, USA   Edward C. Waymire
September 2016

Preface to First Edition

In 1937, A.N. Kolmogorov introduced a measure-theoretic mathematical framework for probability theory in response to David Hilbert's Sixth Problem. This text provides the basic elements of probability within this framework. It may be used for a one-semester course in probability, or as a reference to prerequisite material in a course on stochastic processes. Our pedagogical view is that the subsequent applications to stochastic processes provide a continued opportunity to motivate and reinforce these important mathematical foundations. The book is best suited for students with some prior, or at least concurrent, exposure to measure theory and analysis. But it also provides a fairly detailed overview, with proofs given in appendices, of the measure theory and analysis used.

The selection of material presented in this text grew out of our effort to provide a self-contained reference to foundational material that would facilitate a companion treatise on stochastic processes that *Theory and Applications of Stochastic Processes* we have been developing.[1] While there are many excellent textbooks available that provide the probability background for various continued studies of stochastic processes, the present treatment was designed with this as an explicit goal. This led to some unique features from the perspective of the ordering and selection of material.

We begin with Chapter I on various measure-theoretic concepts and results required for the proper mathematical formulation of a probability space, random maps, distributions, and expected values. Standard results from measure theory are motivated and explained with detailed proofs left to an appendix.

Chapter II is devoted to two of the most fundamental concepts in probability theory: independence and conditional expectation (and/or conditional probability). This continues to build upon, reinforce, and motivate basic ideas from real analysis and measure theory that are regularly employed in probability theory, such as Carathéodory constructions, the Radon–Nikodym theorem, and the Fubini–Tonelli

---

[1]Bhattacharya, R. and E. Waymire (2007): *Theory and Applications of Stochastic Processes*, Springer-Verlag, Graduate Texts in Mathematics.

theorem. A careful proof of the Markov property is given for discrete-parameter random walks on $\mathbb{R}^k$ to illustrate conditional probability calculations in some generality.

Chapter III provides some basic elements of martingale theory that have evolved to occupy a significant foundational role in probability theory. In particular, optional stopping and maximal inequalities are cornerstone elements. This chapter provides sufficient martingale background, for example, to take up a course in stochastic differential equations developed in a chapter of our text on stochastic processes. A more comprehensive treatment of martingale theory is deferred to stochastic processes with further applications there as well.

The various laws of large numbers and elements of large deviation theory are developed in Chapter IV. This includes the classical 0–1 laws of Kolmogorov and Hewitt–Savage. Some emphasis is given to size-biasing in large deviation calculations which are of contemporary interest.

Chapter V analyzes in detail the topology of weak convergence of probabilities defined on metric spaces, culminating in the notion of tightness and a proof of Prohorov's theorem.

The characteristic function is introduced in Chapter VI via a first principles development of Fourier series and the Fourier transform. In addition to the operational calculus and inversion theorem, Herglotz's theorem, Bochner's theorem, and the Cramér–Lévy continuity theorem are given. Probabilistic applications include the Chung–Fuchs criterion for recurrence of random walks on $\mathbb{R}^k$, and the classical central limit theorem for i.i.d. random vectors with finite second moments. The law of rare events (i.e., Poisson approximation to binomial) is also included as a simple illustration of the continuity theorem, although simple direct calculations are also possible.

In Chapter VII, central limit theorems of Lindeberg and Lyapounov are derived. Although there is some mention of stable and infinitely divisible laws, the full treatment of infinite divisibility and Lévy–Khinchine representation is more properly deferred to a study of stochastic processes with independent increments.

The Laplace transform is developed in Chapter VIII with Karamata's Tauberian theorem as the main goal. This includes a heavy dose of exponential size-biasing techniques to go from probabilistic considerations to general Radon measures. The standard operational calculus for the Laplace transform is developed along the way.

Random series of independent summands are treated in Chapter IX. This includes the mean square summability criterion and Kolmogorov's three series criteria based on Kolmogorov's maximal inequality. An alternative proof to that presented in Chapter IV for Kolmogorov's strong law of large numbers is given, together with the Marcinkiewicz and Zygmund extension, based on these criteria and Kronecker's lemma. The equivalence of a.s. convergence, convergence in probability, and convergence in distribution for series of independent summands is also included.

In Chapter X, Kolmogorov's consistency conditions lead to the construction of probability measures on the Cartesian product of infinitely many spaces. Applications include a construction of Gaussian random fields and discrete-parameter Markov processes. The deficiency of Kolmogorov's construction of a model for Brownian motion is described, and the Lévy–Ciesielski "wavelet" construction is provided.

Basic properties of Brownian motion are taken up in Chapter XI. Included are various rescalings and time inversion properties, together with the fine-scale structure embodied in the law of the iterated logarithm for Brownian motion.

In Chapter XII many of the basic notions introduced in the text are tied together via further considerations of Brownian motion. In particular, this chapter revisits conditional probabilities in terms of the Markov and strong Markov properties for Brownian motion, stopping times, and the optional stopping and/or sampling theorems for Brownian motion and related martingales, and leads to weak convergence of rescaled random walks with finite second moments to Brownian motion, i.e., Donsker's invariance principle or the functional central limit theorem, via the Skorokhod embedding theorem.

The text is concluded with a historical overview, Chapter XIII, on Brownian motion and its fundamental role in applications to physics, financial mathematics, and partial differential equations, which inspired its creation.

Most of the material in this book has been used by us in graduate probability courses taught at the University of Arizona, Indiana University, and Oregon State University. The authors are grateful to Virginia Jones for superb word processing skills that went into the preparation of this text. Also, two Oregon State University graduate students, Jorge Ramirez and David Wing, did an outstanding job in uncovering and reporting various bugs in earlier drafts of this text. Thanks go to the editorial staff at Springer and anonymous referees for their insightful remarks.

March 2007
Rabi Bhattacharya
Edward C. Waymire

NOTE: Some of the first edition chapter numbers have changed in the second edition. First edition Chapter VII was moved to Chapter IV, and the material in the first edition Chapter VIII has been redistributed into other chapters. An entirely new Chapter XIII was added to the second edition.

Contents

I   **Random Maps, Distribution, and Mathematical Expectation**   1
      Exercise Set I   18

II   **Independence, Conditional Expectation**   25
      Exercise Set II   47

III   **Martingales and Stopping Times**   53
      Exercise Set III   72

IV   **Classical Central Limit Theorems**   75
      Exercise Set IV   84

V   **Classical Zero–One Laws, Laws of Large Numbers**
      and Large Deviations   87
      Exercise Set V   100

VI   **Fourier Series, Fourier Transform, and Characteristic**
      Functions   103
      Exercise Set VI   129

VII   **Weak Convergence of Probability Measures on Metric**
      Spaces   135
      Exercise Set VII   155

VIII   **Random Series of Independent Summands**   159
      Exercise Set VIII   165

IX   **Kolmogorov's Extension Theorem and Brownian Motion**   167
      IX.1   A Wavelet Construction of Brownian Motion:
             The Lévy–Ciesielski Construction   173
      Exercise Set IX   176

X   **Brownian Motion: The LIL and Some Fine-Scale Properties**   179
      Exercise Set X   185

XI   Strong Markov Property, Skorokhod Embedding,
       and Donsker's Invariance Principle 187
       Exercise Set XI 203

XII   A Historical Note on Brownian Motion 207

XIII Some Elements of the Theory of Markov Processes
       and Their Convergence to Equilibrium 211
       Exercise Set XIII 221

Appendix A: Measure and Integration 225

Appendix B: Topology and Function Spaces 241

Appendix C: Hilbert Spaces and Applications in Measure Theory 247

References 255

Symbol Index 257

Index 259

Chapter I
Random Maps, Distribution, and Mathematical Expectation

In the spirit of a refresher, we begin with an overview of the measure–theoretic framework for probability. Readers for whom this is entirely new material may wish to consult the appendices for statements and proofs of basic theorems from analysis. A **measure space** is a triple $(S, \mathcal{S}, \mu)$, where S is a nonempty set; $\mathcal{S}$ is a collection of subsets of S, referred to as a σ-**field**, which includes $\emptyset$ and is closed under complements and countable unions; and $\mu : \mathcal{S} \to [0, \infty]$ satisfies (i) $\mu(\emptyset) = 0$, (ii) (**countable additivity**) $\mu(\cup_{n=1}^{\infty} A_n) = \sum_{n=1}^{\infty} \mu(A_n)$ if $A_1, A_2, \ldots$ is a sequence of disjoint sets in $\mathcal{S}$. Subsets of S belonging to $\mathcal{S}$ are called **measurable sets**. The pair $(S, \mathcal{S})$ is referred to as a **measurable space**, and the set function μ is called a **measure**. Familiar examples from real analysis are **Lebesgue measure** μ on $S = \mathbb{R}^k$, equipped with a σ-field $\mathcal{S}$ containing the class of all k-dimensional rectangles, say $R = (a_1, b_1] \times \cdots \times (a_k, b_k]$, of "volume" measure $\mu(R) = \prod_{j=1}^{k}(b_j - a_j)$; or **Dirac point mass measure** $\mu = \delta_x$ at $x \in S$ defined by $\delta_x(B) = 1$ if $x \in B$, $\delta_x(B) = 0$ if $x \in B^c$, for $B \in \mathcal{S}$. Such examples should suffice for the present, but see Appendix A for constructions of these and related measures based on the **Carathéodory extension theorem**. If $\mu(S) < \infty$ then μ is referred to as a **finite measure**. If one may write $S = \cup_{n=1}^{\infty} S_n$, where each $S_n \in \mathcal{S}(n \geq 1)$ and $\mu(S_n) < \infty, \forall n$, then μ is said to be a σ-**finite measure**.

A **probability space** is a triple $(\Omega, \mathcal{F}, P)$, where Ω is a nonempty set, $\mathcal{F}$ is a σ-field of subsets of Ω, and P is a finite measure on the measurable space $(\Omega, \mathcal{F})$ with $P(\Omega) = 1$. The measure P is referred to as a **probability**. Intuitively, Ω represents the set of all possible "outcomes" of a random experiment, real or conceptual, for some given coding of the results of the experiment. The set Ω is referred to as the **sample space** and the elements $\omega \in \Omega$ as **sample points** or **possible outcomes**. The σ-field $\mathcal{F}$ comprises "events" $A \subset \Omega$ whose probability $P(A)$ of occurrence is well defined.

The finite total probability and countable additivity of a probability have many important consequences, such as **finite additivity**, **finite** and **countable subadditivity**, **inclusion–exclusion**, **monotonicity**, and the formulas for both **relative**

© Springer International Publishing AG 2016
R. Bhattacharya and E.C. Waymire, *A Basic Course in Probability Theory*,
Universitext, DOI 10.1007/978-3-319-47974-3_I

complements and **universal complements**. Proofs of these properties are left to the reader and included among the exercises.

Example 1 *(Finite Sampling of a Fair Coin)* Consider m repeated tosses of a fair coin. Coding the individual outcomes as 1 or 0 (or, say, H, T), the possible outcomes may be represented as sequences of binary digits of length m. Let $\Omega = \{0, 1\}^m$ denote the set of all such sequences and $\mathcal{F} = 2^\Omega$, the power set of Ω. The condition that the coin be fair may be defined by the requirement that $P(\{\omega\})$ is the same for each sequence $\omega \in \Omega$. Since Ω has cardinality $|\Omega| = 2^m$, it follows from the finite additivity and total probability requirements that

$$P(\{\omega\}) = \frac{1}{2^m} = \frac{1}{|\Omega|}, \quad \omega \in \Omega.$$

Using finite additivity this completely and explicitly specifies the model $(\Omega, \mathcal{F}, P)$ with

$$P(A) = \sum_{\omega \in A} P(\{\omega\}) = \frac{|A|}{|\Omega|}, \quad A \subset \Omega.$$

The so-called **continuity properties** also follow from the definition as follows: A sequence of events $A_n, n \geq 1$, is said to be **increasing** (respectively, **decreasing**) with respect to set inclusion if $A_n \subset A_{n+1}, \forall n \geq 1$ (respectively $A_n \supset A_{n+1} \forall n \geq 1$). In the former case one defines $\lim_n A_n := \cup_n A_n$, while for decreasing measurable events $\lim_n A_n := \cap_n A_n$. In either case the continuity of a probability, from below or above, respectively, is the following consequence of countable additivity[1] (Exercise 1):

$$P(\lim_n A_n) = \lim_n P(A_n). \tag{1.1}$$

A bit more generally, if $\{A_n\}_{n=1}^\infty$ is a sequence of measurable events one defines

$$\limsup_n A_n := \cap_{n=1} \cup_{m \geq n} A_m \tag{1.2}$$

and

$$\liminf_n A_n := \cup_{n=1}^\infty \cap_{m \geq n} A_m. \tag{1.3}$$

The event $\limsup_n A_n$ denotes the collection of outcomes $\omega \in \Omega$ that correspond to the occurrences of A_n for infinitely many n; i.e., the events A_n occur **infinitely often** This event is also commonly denoted by $[A_n \ i.o.] := \limsup_n A_n$. On the other hand, $\liminf_n A_n$ is the set of outcomes ω that belong to A_n for all but finitely many n. Note that $[A_n \ i.o.]^c$ is the event that the complementary event A_n^c occurs for all but finitely many n and equals $\liminf_n A_n^c$.

---

[1] With the exception of properties for "complements" and "continuity from above," these and the aforementioned consequences can be checked to hold for any measure.

Lemma 1 (*Borel–Cantelli I*) Let $(\Omega, \mathcal{F}, P)$ be a probability space and $A_n \in \mathcal{F}, n = 1, 2, \ldots$. If $\sum_{n=1}^{\infty} P(A_n) < \infty$ then $P(A_n \ i.o.) = 0$.

Proof Apply (1.1) to the decreasing sequence of events

$$\cup_{m=1}^{\infty} A_m \supset \cup_{m=2}^{\infty} A_m \supset \cdots,$$

followed by subadditivity of the probability to get

$$P(\limsup_n A_n) = \lim_{n \to \infty} P(\cup_{m=n}^{\infty} A_m) \le \lim_{n \to \infty} \sum_{m=n}^{\infty} P(A_m) = 0.$$

∎

A partial converse (Borel–Cantelli II) will be given in the next chapter.

Example 2 Suppose that $T_1, T_2, \ldots$ is a sequence of positive random variables defined on a probability space $(\Omega, \mathcal{F}, P)$ such that for some constant $\lambda > 0$, $P(T_n > t) = e^{-\lambda t}, t \ge 0$, for $n = 1, 2, \ldots$. Then $P(T_n > n \ i.o.) = 0$. In fact, $P(T_n > \theta \log n \ i.o.) = 0$ for any value of $\theta > \frac{1}{\lambda}$. This may also be expressed as $P(T_n \le \theta \log n$ eventually for all $n) = 1$ if $\theta > \frac{1}{\lambda}$.

Example 3 (*Infinite Sampling of a Fair Coin*) The possible outcomes of nonterminated repeated coin tosses can be coded as infinite binary sequences of 1's and 0's. Thus the sample space is the infinite product space $\Omega = \{0, 1\}^{\infty}$. Observe that a sequence $\omega \in \Omega$ may be viewed as the digits in a binary expansion of a number x in the unit interval. The binary expansion $x = \sum_{n=1}^{\infty} \omega_n(x) 2^{-n}$, where $\omega_n(x) \in \{0, 1\}$, is not unique for binary rationals, e.g., $\frac{1}{2} = .1000000 \ldots = .011111 \ldots$. However it may be made unique by requiring that infinitely many 0's occur in the expansion. Thus, up to a subset of probability zero, Ω and $[0, 1)$ may be put in one-to-one correspondence. Observe that for a given specification $\varepsilon_n \in \{0, 1\}, n = 1, \ldots, m$, of the first m tosses, the event $A = \{\omega = (\omega_1, \omega_2, \ldots) \in \Omega : \omega_n = \varepsilon_n, n \le m\}$ corresponds to the subinterval $[\sum_{n=1}^{m} \varepsilon_n 2^{-n}, \sum_{n=1}^{m} \varepsilon_n 2^{-n} + 2^{-m})$ of $[0, 1)$ of length (Lebesgue measure) 2^{-m}. Again modeling the repeated tosses of a fair coin by the requirement that for each fixed m, $P(A)$ not depend on the specified values $\varepsilon_n \in \{0, 1\}, 1 \le n \le m$, it follows from finite additivity and total probability one that $P(A) = 2^{-m} = |A|$, where $|A|$ denotes the one-dimensional Lebesgue measure of A. Based on these considerations, one may use Lebesgue measure on $[0, 1)$ to define a probability model for infinitely many tosses of a fair coin. As we will see below, this is an essentially unique choice. For now, let us exploit the model with an illustration of the Borel–Cantelli Lemma 1. Fix a nondecreasing sequence r_n of positive integers and let $A_n = \{x \in [0, 1) : \omega_k(x) = 1, k = n, n+1, \ldots, n+r_n-1\}$ denote the event that a run of 1's occurs of length at least r_n starting at the nth toss. Note that this set is a union of length 2^{-r_n}. Thus, if r_n increases so quickly that $\sum_{n=1}^{\infty} 2^{-r_n} < \infty$ then the Borel–Cantelli Lemma 1 yields that $P(A_n \ i.o.) = 0$. For a concrete illustration, let $r_n = [\theta \log_2 n]$, for fixed $\theta > 0$, with $[\cdot]$ denoting the integer part. Then $P(A_n \ i.o.) = 0$ for $\theta > 1$. Analysis of the case $0 < \theta \le 1$ requires more detailed consideration of the fundamental notion of "statistical independence" of the outcomes of the individual tosses implicit to this model. This concept is among the most important in all of probability theory and will be precisely defined in the next chapter.

Remark 1.1 A detailed construction of Lebesgue measure is given in Example 1 of Appendix A. The *existence* of Lebsegue measure on $[0, 1)$ plays a fundamental role in providing the probability space for repeated unending tosses of a fair coin in the previous example. The existence of probability models corresponding to infinite sequences of experiments is as fundamentally important to probability as existence of Lebesgue measure is to analysis. A general existence theorem will be given in Chapter IX that will cover the theory developed in the chapters leading up to it. For now we generally take such existence theory for granted.

For a given collection C of subsets of Ω, the smallest σ-field that contains all of the events in C is called the σ-**field generated by** C and is denoted by $\sigma(C)$; if $\mathcal{G}$ is any σ-field containing C then $\sigma(C) \subset \mathcal{G}$. Note that, in general, if $\mathcal{F}_\lambda, \lambda \in \Lambda$, is an arbitrary collection of σ-fields of subsets of Ω, then $\bigcap_{\lambda \in \Lambda} \mathcal{F}_\lambda := \{F \subset \Omega : F \in \mathcal{F}_\lambda \forall \lambda \in \Lambda\}$ is a σ-field. On the other hand $\bigcup_{\lambda \in \Lambda} \mathcal{F}_\lambda := \{F \subset \Omega : F \in \mathcal{F}_\lambda \text{ for some } \lambda \in \Lambda\}$ is not generally a σ-field. Define the **join** σ-**field**, denoted by $\bigvee_{\lambda \in \Lambda} \mathcal{F}_\lambda$, to be the σ-field generated by $\bigcup_{\lambda \in \Lambda} \mathcal{F}_\lambda$.

It is not uncommon that $\mathcal{F} = \sigma(C)$ for a collection C *closed under finite intersections*; such a collection C is called a π-**system**, e.g., $\Omega = (-\infty, \infty)$, $C = \{(a, b] : -\infty \le a \le b < \infty\}$, or infinite sequence space $\Omega = \mathbb{R}^\infty$, and $C = \{(a_1, b_1] \times \cdots \times (a_k, b_k] \times \mathbb{R}^\infty : -\infty \le a_i \le b_i < \infty, i = 1, \ldots, k, k \ge 1\}$.

A λ-**system** is a collection $\mathcal{L}$ of subsets of Ω such that (i) $\Omega \in \mathcal{L}$, (ii) If $A \in \mathcal{L}$ then $A^c \in \mathcal{L}$, (iii) If $A_n \in \mathcal{L}, A_n \cap A_m = \emptyset, n \ne m, n, m = 1, 2, \ldots$, then $\cup_n A_n \in \mathcal{L}$. A σ-field is clearly also a λ-system. The following π-λ theorem provides a very useful tool for checking measurability.

Theorem 1.1 *(Dynkin's π-λ Theorem)* If $\mathcal{L}$ is a λ-system containing a π-system C, then $\sigma(C) \subset \mathcal{L}$.

Proof Let $\mathcal{L}(C) = \cap \mathcal{F}$, where the intersection is over all λ-systems $\mathcal{F}$ containing C. We will prove the theorem by showing (i) $\mathcal{L}(C)$ is a π-system, and (ii) $\mathcal{L}(C)$ is a λ-system. For then $\mathcal{L}(C)$ is a σ-field (see Exercise 15), and by its definition $\sigma(C) \subset \mathcal{L}(C) \subset \mathcal{L}$. Now (ii) is simple to check. For clearly $\Omega \in \mathcal{F}$ for all $\mathcal{F}$, and hence $\Omega \in \mathcal{L}(C)$. If $A \in \mathcal{L}(C)$, then $A \in \mathcal{F}$ for all $\mathcal{F}$, and since every $\mathcal{F}$ is a λ-system, $A^c \in \mathcal{F}$ for every $\mathcal{F}$. Thus $A^c \in \mathcal{L}(C)$. If $A_n \in \mathcal{L}(C), n \ge 1$, is a disjoint sequence, then for each $\mathcal{F}$, $A_n \in \mathcal{F}$, for all n and $A \equiv \cup_n A_n \in \mathcal{F}$ for all $\mathcal{F}$. Since this is true for every λ-system $\mathcal{F}$, one has $A \in \mathcal{L}(C)$. It remains to prove (i). For each set A, define the class $\mathcal{L}_A := \{B : A \cap B \in \mathcal{L}(C)\}$. It suffices to check that $\mathcal{L}_A \supset \mathcal{L}(C)$ for all $A \in \mathcal{L}(C)$. First note that if $A \in \mathcal{L}(C)$, then $\mathcal{L}_A$ is a λ-system, by arguments along the line of (ii) above (Exercise 15). In particular, if $A \in C$, then $A \cap B \in C$ for all $B \in C$, since C is closed under finite intersections. Thus $\mathcal{L}_A \supset C$. This implies, in turn that $\mathcal{L}(C) \subset \mathcal{L}_A$. This says that $A \cap B \in \mathcal{L}(C)$ for all $A \in C$ and for all $B \in \mathcal{L}(\mathcal{C})$. Thus, if we fix $B \in \mathcal{L}(\mathcal{C})$, then $\mathcal{L}_B \equiv \{A : B \cap A \in \mathcal{L}(\mathcal{C})\} \supset \mathcal{C}$. Therefore $\mathcal{L}_B \supset \mathcal{L}(\mathcal{C})$. In other words, for every $B \in \mathcal{L}(\mathcal{C})$ and $A \in \mathcal{L}(\mathcal{C})$, one has $A \cap B \in \mathcal{L}(\mathcal{C})$. ∎

In view of the additivity properties of a probability, the following is an immediate and important corollary to the π-λ theorem.

Corollary 1.2 *(Uniqueness)* If P_1, P_2 are two probability measures such that $P_1(C) = P_2(C)$ for all events C belonging to a π-system $\mathcal{C}$, then $P_1 = P_2$ on all of $\mathcal{F} = \sigma(\mathcal{C})$.

Proof Check that $\{A \in \mathcal{F} : P_1(A) = P_2(A)\} \supset \mathcal{C}$ is a λ-system. ∎

Remark 1.2 It is rather simple to construct examples of generating collections of sets $\mathcal{C}$ and probability measures P_1, P_2 such that $P_1 = P_2$ on $\mathcal{C}$, but $P_1 \neq P_2$ on $\sigma(\mathcal{C})$. For example take $\Omega = \{1, 2, 3, 4\}$, $\mathcal{C} = \{\{1, 2, 3\}, \{2, 3, 4\}\}$. Then $\sigma(\mathcal{C}) = \{\{1, 2, 3\}, \{2, 3, 4\}, \{1\}, \{4\}, \{1, 4\}, \{2, 3\}, \Omega, \emptyset\}$. Let $P_1(\{1\}) = P_2(\{1\}) = P_1(\{4\}) = P_2(\{4\}) = 1/8$, but $P_1(\{2\}) = P_2(\{3\}) = 1/8$, $P_1(\{3\}) = P_2(\{2\}) = 5/8$.

For a related application suppose that (S, ρ) is a metric space. The **Borel σ-field** of S, denoted by $\mathcal{B}(S)$, is defined as the σ-field generated by the collection $\mathcal{C} = \mathcal{T}$ of open subsets of S, the collection $\mathcal{T}$ being referred to as the topology on S specified by the metric ρ. More generally, one may specify a **topology** for a set S by a collection $\mathcal{T}$ of subsets of S that includes both $\emptyset$ and S, and is closed under arbitrary unions and finite intersections. Then $(S, \mathcal{T})$ is called a **topological space** and members of $\mathcal{T}$ define the open subsets of S. The topology is said to be **metrizable** when it may be specified by a metric ρ as above. In any case, one defines the Borel σ-field by $\mathcal{B}(S) := \sigma(\mathcal{T})$.

Definition 1.1 A class $\mathcal{C} \subset \mathcal{B}(S)$ is said to be **measure-determining** if for any two finite measures μ, ν such that $\mu(C) = \nu(C) \ \forall C \in \mathcal{C}$, it follows that $\mu = \nu$ on $\mathcal{B}(S)$.

One may directly apply the π-λ theorem, noting that S is both open and closed, to see that the class $\mathcal{T}$ of all open sets is measure-determining, as is the class $\mathcal{K}$ of all closed sets.

If $(S_i, \mathcal{S}_i)$, $i = 1, 2$, is a pair of measurable spaces then a function $f : S_1 \to S_2$ is said to be a **measurable map** if $f^{-1}(B) := \{x \in S_1 : f(x) \in B\} \in \mathcal{S}_1$ for all $B \in \mathcal{S}_2$. In usual mathematical discourse the σ-fields required for this definition may not be explicitly mentioned and will need to be inferred from the context. For example, if $(S, \mathcal{S})$ is a measurable space, by a **Borel-measurable function** $f : S \to \mathbb{R}$ is meant measurability when $\mathbb{R}$ is given its Borel σ-field. A **random variable,** or a **random map**, X is a measurable map on a probability space $(\Omega, \mathcal{F}, P)$ into a measurable space $(S, \mathcal{S})$. Measurability of X means that each event[2] $[X \in B] := X^{-1}(B)$ belongs to $\mathcal{F} \ \forall B \in \mathcal{S}$. The σ**-field generated by X**, denoted $\sigma(X)$, is the smallest σ-field of subsets of Ω for which $X : \Omega \to S$ is measurable. In

---

[2]Throughout, this square-bracket notation will be used to denote events defined by inverse images.

particular, therefore, $\sigma(X) = \{[X \in A] : A \in \mathcal{S}\}$ (Exercise 11). The term **random
variable** is most often used to denote a real-valued random variable, i.e., where
$S = \mathbb{R}$, $\mathcal{S} = \mathcal{B}(\mathbb{R})$. When $S = \mathbb{R}^k$, $\mathcal{S} = \mathcal{B}(\mathbb{R}^k)$, $k > 1$, one uses the term **random
vector**.

A common alternative to the use of a metric to define a metric space topology, is
to indirectly characterize the topology by specifying what it means for a sequence to
converge in the metric. That is, if $\mathcal{T}$ is a topology on S, then a sequence $\{x_n\}_{n=1}^{\infty}$ in S
converges to $x \in S$ **with respect to the topology** $\mathcal{T}$ if for arbitrary $U \in \mathcal{T}$ such that
$x \in U$, there is an N such that $x_n \in U$ for all $n \geq N$. A topological space $(S, \mathcal{T})$,
or a topology $\mathcal{T}$, is said to be **metrizable** if $\mathcal{T}$ coincides with the class of open sets
defined by a metric ρ on S. Alternatively, by specifying the meaning of convergence
in the metric, one has that closed sets, and therefore open sets via complements,
can also be defined. Using this notion, other commonly occurring measurable image
spaces may be described as follows: (i) $S = \mathbb{R}^{\infty}$—the space of all sequences of
reals with the (metrizable) **topology of pointwise convergence**, and $\mathcal{S} = \mathcal{B}(\mathbb{R}^{\infty})$,
(ii) $S = C[0, 1]$—the space of all real-valued continuous functions on the interval
$[0, 1]$ with the (metrizable) **topology of uniform convergence**, and $\mathcal{S} = \mathcal{B}(C[0, 1])$,
and (iii) $S = C([0, \infty): \mathbb{R}^k)$—the space of all continuous functions on $[0, \infty)$ into
$\mathbb{R}^k$, with the (metrizable) **topology of uniform convergence on compact subsets of**
$[0, \infty)$, $\mathcal{S} = \mathcal{B}(S)$ (see Exercise 10).

The relevant quantities for a random map X on a probability space $(\Omega, \mathcal{F}, P)$ are
the probabilities with which X takes sets of values. In this regard, P determines the
most important aspect of X, namely, its **distribution** $Q \equiv P \circ X^{-1}$ defined on the
image space $(S, \mathcal{S})$ by

$$Q(B) := P(X^{-1}(B)) \equiv P(X \in B), \quad B \in \mathcal{S}. \tag{1.4}$$

The distribution is sometimes referred to as the **induced measure** of X under P.
For random vectors $\mathbf{X} = (X_1, \ldots, X_k)$ with values in $\mathbb{R}^k$, it is often convenient to
restrict consideration to the (multivariate) **distribution function** defined by $F(\mathbf{x}) =
P(\mathbf{X} \leq \mathbf{x}) \equiv P(X_1 \leq x_1, \ldots, X_k \leq x_k)$, $\mathbf{x} = (x_1, \ldots, x_k) \in \mathbb{R}^k$; see Exercise 16. A
familiar and important special case is that of an **absolutely continuous distribution
function** given by

$$F(\mathbf{x}) = \int_{-\infty}^{x_k} \cdots \int_{-\infty}^{x_1} g(\mathbf{u}) d\mathbf{u}, \quad \mathbf{x} \in \mathbb{R}^k,$$

for a nonnegative density function g with respect to Lebesgue measure on $\mathbb{R}^k$; here
we have used the convention of representing Lebesgue measure as $d\mathbf{u}$. In an abuse of
terminology, a random variable with an absolutely continuous distribution is often
referred to as a **continuous random variable**.

If a real-valued random variable X has the distribution function F, then $P(X \in
(a, b]) = P(a < X \leq b) = F(b) - F(a)$. Moreover, $P(X \in (a, b)) = P(a < X < b)
= F(b^-) - F(a)$. Since the collection $\mathcal{C}$ of all open intervals (a, b), $-\infty < a \leq b <$

∞ is closed under finite intersections, and every open subset of $\mathbb{R}$ can be expressed as a countable disjoint union of sets in $\mathcal{C}$, the collection $\mathcal{C}$ is measure-determining (cf. Exercise 19). One may similarly check that the (multivariate) distribution function of a probability Q on the Borel sigma-field of $\mathbb{R}^k$ uniquely determines Q; cf. Exercise 19.

In general, let us also note that given any probability measure Q on a measurable space $(S, \mathcal{S})$ one can construct a probability space $(\Omega, \mathcal{F}, P)$ and a random map X on $(\Omega, \mathcal{F})$ with distribution Q. The simplest such construction is given by letting $\Omega = S$, $\mathcal{F} = \mathcal{S}$, $P = Q$, and X the **identity map**, $X(\omega) = \omega$, $\omega \in S$. This is often called a **canonical construction**, and $(S, \mathcal{S}, Q)$ with the identity map X is called a **canonical model**. Note that any canonical model for X will generally be a noncanonical model for a function of X. So it would not be prudent to restrict the theoretical development to canonical models alone!

Before proceeding, it is of value to review the manner in which abstract Lebesgue integration and, more specifically, **mathematical expectation** is defined. Throughout $\mathbf{1}_A$ denotes the *indicator function* of the set A, i.e., $\mathbf{1}_A(x) = 1$ if $x \in A$, and is zero otherwise. If $X = \sum_{j=1}^{m} a_j \mathbf{1}_{A_j}$, $A_j \in \mathcal{F}$, $A_i \cap A_j = \emptyset (i \neq j)$, is a **discrete random variable** or, equivalently, a **simple random variable**, then $\mathbb{E}X \equiv \int_\Omega X dP := \sum_{j=1}^{m} a_j P(A_j)$. If $X : \Omega \to [0, \infty)$ is a random variable, then $\mathbb{E}X$, expected is defined by the "simple function approximation" $\mathbb{E}X \equiv \int_\Omega X dP := \sup\{\mathbb{E}Y : 0 \leq Y \leq X, Y \text{ simple}\}$. In particular, one may apply the standard simple function approximations $X = \lim_{n \to \infty} X_n$ given by the nondecreasing sequence

$$X_n := \sum_{j=0}^{n2^n - 1} \frac{j}{2^n} \mathbf{1}_{[j2^{-n} \leq X < (j+1)2^{-n}]} + n\mathbf{1}_{[X \geq n]}, \quad n = 1, 2, \ldots, \tag{1.5}$$

to write

$$\mathbb{E}X = \lim_{n \to \infty} \mathbb{E}X_n = \lim_{n \to \infty} \left\{ \sum_{j=0}^{n2^n - 1} \frac{j}{2^n} P(j2^{-n} \leq X < (j+1)2^{-n}) + nP(X \geq n) \right\}. \tag{1.6}$$

Note that if $\mathbb{E}X < \infty$, then $nP(X > n) \to 0$ as $n \to \infty$ (Exercise 30). Now, more generally, if X is a real-valued random variable, then the **expected value** (or, **mean, first moment**) of X is defined as

$$\mathbb{E}(X) \equiv \int_\Omega X dP := \mathbb{E}X^+ - \mathbb{E}X^-, \tag{1.7}$$

provided at least one of $\mathbb{E}(X^+)$ and $\mathbb{E}(X^-)$ is finite, where $X^+ = X\mathbf{1}_{[X \geq 0]}$ and $X^- = -X\mathbf{1}_{[X \leq 0]}$. If both $\mathbb{E}X^+ < \infty$ and $\mathbb{E}X^- < \infty$, or equivalently, $\mathbb{E}|X| = \mathbb{E}X^+ + \mathbb{E}X^- < \infty$, then X is said to be **integrable** with respect to the probability P. Note that if X is bounded a.s., then applying (1.5) to X^+ and X^-, one obtains a sequence

$X_n (n \geq 1)$ of simple functions that converge uniformly to X, outside a P-null set (Exercise 1.5(i)).

If X is a random variable with values in $(S, \mathcal{S})$ and if h is a real-valued Borel-measurable function on S, then using simple function approximations to h, one may obtain the following basic **change of variables formula**

$$\mathbb{E}(h(X)) \equiv \int_\Omega h(X(\omega)) P(d\omega) = \int_S h(x) Q(dx), \qquad (1.8)$$

where Q is the distribution of X, provided one of the two indicated integrals may be shown to exist.

For arbitrary $p \geq 1$, the **order p-moment** of a random variable X on $(\Omega, \mathcal{F}, P)$ having distribution Q is defined by

$$\mu_p := \mathbb{E}X^p = \int_\Omega X^p(\omega) P(d\omega) = \int_\mathbb{R} x^p Q(dx), \qquad (1.9)$$

provided that X^p is integrable, or nonnegative. Moments of lower order p than one, including negative order ($p < 0$) moments, may be defined similarly so long as X^p is real-valued random variable. Moments of absolute values $|X|$ are referred to as **absolute moments** of X. Let us record a useful formula for the moments of a random variable derived from the Fubini–Tonelli theorem before proceeding. Namely,

Proposition 1.3 If X is a random variable on $(\Omega, \mathcal{F}, P)$, then for any $p > 0$,

$$\mathbb{E}|X|^p = p \int_0^\infty y^{p-1} P(|X| > y) dy. \qquad (1.10)$$

Proof For $x \geq 0$, simply use $x^p = p \int_0^x y^{p-1} dy$ in the formula

$$\mathbb{E}|X|^p = \int_\Omega |X(\omega)|^p P(d\omega) = \int_\Omega \left(p \int_0^{|X(\omega)|} y^{p-1} dy \right) P(d\omega)$$

and apply the Tonelli part (a) to reverse the order of integration. The assertion follows. $\blacksquare$

Example 4 As a generalization of Example 2, suppose that $X_1, X_2, \ldots$ is a sequence of positive random variables, each having distribution Q, with a finite moment of order $p > 0$. Then an application of Borel–Cantelli I together with Proposition 1.3 shows that $P(X_n > n^{\frac{1}{p}} i.o.) = 0$ (Use Exercise 29 applied to X_1^p.).

If $X = (X_1, X_2, \ldots, X_k)$ is a random vector whose components are integrable real-valued random variables, then define $\mathbb{E}(X) = (\mathbb{E}(X_1), \ldots, \mathbb{E}(X_k))$. Similarly for complex valued random variables $X = U + iV$, where U, V are integrable real-valued random variables, one defines $\mathbb{E}X = \mathbb{E}U + i\mathbb{E}V$. In particular, for complex valued random variables $\mathbb{E}X$ exists if and only if $\mathbb{E}|X| = \mathbb{E}\sqrt{U^2 + V^2} < \infty$; Exercise 36.

This definition of expectation as an *integral in the sense of Lebesgue* is precisely the same as that used in real analysis to define $\int_S f(x)\mu(dx)$ for a real-valued Borel-measurable function f on an arbitrary measure space $(S, \mathcal{S}, \mu)$; see Appendix A. **Almost sure** convergence of a sequence of random maps $X_n, n \geq 1$, to X, each defined on $(\Omega, \mathcal{F}, P)$, is defined by $X_n(\omega) \to X(\omega)$ as $n \to \infty$ for all $\omega \in \Omega$ up to a subset of probability zero; i.e., convergence almost everywhere with respect to P. One may exploit standard tools of real analysis (see Appendices A and C), such as **Lebesgue's dominated convergence theorem, Lebesgue's monotone convergence theorem, Fatou's lemma, Fubini–Tonelli theorem, Radon–Nykodym theorem**, for estimates and computations involving expected values.

The following lemma and proposition illustrate the often used exchange in the order of integration.

Lemma 2 *(Integration by parts)* Let μ_1, μ_2 be signed measures on $\mathbb{R}$, which are finite on finite intervals. Let

$$F_i(y) = \mu_i(0, y], \quad i = 1, 2, \quad -\infty < y < \infty.$$

Then for any $-\infty < a < b < \infty$, one has

$$\int_{(a,b]} F_1(y)\mu_2(dy) = F_1(b)F_2(b) - F_1(a)F_2(a) - \int_{(a,b]} F_2(y-)\mu_1(dy).$$

Proof Since a signed measure may be expressed as the difference of two measures, without loss of generality it is sufficient to let both μ_1 and μ_2 be measures that are finite on finite intervals. Then, using the Fubini–Tonelli theorem, one has

$$\int_{a<u\leq v, a<v\leq b} \mu_1(du)\mu_2(dv) = \int_{a<v\leq b} [F_1(v) - F_1(a)]\mu_2(dv)$$

$$= -F_1(a)[F_2(b) - F_2(a)] + \int_{a<v\leq b} F_1(v)\mu_2(dv).$$

Also,

$$\int_{a<u\leq v, a<v\leq b} \mu_1(du)\mu_2(dv) = \int_{a<u\leq b} [F_2(b) - F_2(u-)]\mu_1(du)$$

$$= F_2(b)[F_1(b) - F_1(a)] - \int_{a<u\leq b} F_2(u-)\mu_1(du).$$

Comparing these two iterations yields the asserted formula.   ∎

Remark 1.3 The "distribution functions" $F_i, i = 1, 2$, can be defined as $F_i(y) = \mu_i((c, y]), i = 1, 2, y \in \mathbb{R}$, for any real number c in place of zero, and the lemma still holds. This formula has special utility when applied to a nondecreasing function, or more generally a function of bounded variation, as an integrand.

The following is a useful version for expected values. A clever application is given in Theorem 3.4.

Proposition 1.4 Let μ_1 be an arbitrary measure on $(0, \infty]$ which is finite on finite intervals, and such that $\mu_1(\{0\}) = 0$. Suppose that μ_2 is a probability measure on $[0, \infty)$, and let Y be a random variable with distribution μ_2. Then, with $F_i(y) = \mu_i((0, y])$, $i = 1, 2$, $y \geq 0$, one has

$$\mathbb{E} F_1(Y) = \int_{[0,\infty)} P(Y \geq y)\mu_1(dy)$$

Proof By the lemma, for any $b > 0$ one has

$$\int_{(0,b]} F_1(y)\mu_2(dy) = F_1(b)F_2(b) - F_1(0)F_2(0) - \int_{(0,b]} F_2(y-)\mu_1(dy)$$

$$= F_1(b)F_2(b) - \int_{(0,b]} F_2(y-)\mu_1(dy)$$

$$= \int_{(0,b]} [F_2(b) - F_2(y-)]\mu_1(dy). \tag{1.11}$$

The assertion follows by letting $b \uparrow \infty$ on both sides, and using Lebesgue's monotone convergence theorem to obtain

$$\int_{(0,\infty)} F_1(y)\mu_2(dy) = \int_{(0,\infty)} [1 - F_2(y-)]\mu_1(dy).$$

∎

Definition 1.2 A sequence $\{X_n\}_{n=1}^{\infty}$ of random variables on a probability space $(\Omega, \mathcal{F}, P)$ is said to **converge in probability** to a random variable X if for each $\varepsilon > 0$, $\lim_{n\to\infty} P(|X_n - X| > \varepsilon) = 0$. The convergence is said to be **almost sure** (a.s.) if the event $[X_n \not\to X] \equiv \{\omega \in \Omega : X_n(\omega) \not\to X(\omega)\}$ has P-measure zero.

Convergence in probability is referred to as "convergence in measure" in analysis; see Appendix A. Note that almost sure convergence always implies convergence in probability, since for arbitrary $\varepsilon > 0$ one has $0 = P(\cap_{n=1}^{\infty} \cup_{m=n}^{\infty} [|X_m - X| > \varepsilon]) = \lim_{n\to\infty} P(\cup_{m=n}^{\infty} [|X_m - X| > \varepsilon]) \geq \lim\sup_{n\to\infty} P(|X_n - X| > \varepsilon)$; also see Exercise 5. An equivalent formulation of convergence in probability can be cast in terms of almost sure convergence as follows.

Proposition 1.5 A sequence of random variables $\{X_n\}_{n=1}^{\infty}$ on $(\Omega, \mathcal{F}, P)$ converges in probability to a random variable X on $(\Omega, \mathcal{F}, P)$ if and only if every subsequence has an a.s. convergent subsequence to X.

Proof Suppose that $X_n \to X$ in probability as $n \to \infty$. Let $\{X_{n_k}\}_{k=1}^{\infty}$ be a subsequence, and for each $m \geq 1$ recursively choose $n_{k(0)} = 1$, $n_{k(m)} = \min\{n_k > n_{k(m-1)}:$

$P(|X_{n_k} - X| > 1/m) \leq 2^{-m}\}$. Then it follows from the Borel–Cantelli lemma (Part I) that $X_{n_{k(m)}} \to X$ a.s. as $m \to \infty$. For the converse suppose that X_n does not converge to X in probability. Then there exists $\varepsilon > 0$ and a sequence $n_1, n_2, \ldots$ such that $\lim_k P(|X_{n_k} - X| > \varepsilon) = \alpha > 0$. Since a.s. convergence implies convergence in probability (see Appendix A, Proposition 2.4), there cannot be an a.s. convergent subsequence of $\{X_{n_k}\}_{k=1}^{\infty}$. ∎

The utility of Proposition 1.5 can be seen, for example, in demonstrating that if a sequence of random variables $X_n, n \geq 1$, say, converges in probability to X, then X_n^2 will converge in probability to X^2 by virtue of continuity of the map $x \to x^2$, and considerations of almost sure convergence; see Exercise 6.

The notion of measure-determining classes of sets extends to classes of functions as follows. Let μ, ν be arbitrary finite measures on the Borel σ-field of a metric space S. A class Γ of real-valued bounded Borel-measurable functions on S is **measure-determining** if $\int_S g \, d\mu = \int_S g \, d\nu \; \forall g \in \Gamma$ implies $\mu = \nu$.

Proposition 1.6 The class $C_b(S)$ of real-valued bounded continuous functions on S is measure-determining.

Proof To prove this, it is enough to show that for each (closed) $F \in \mathcal{K}$ there exists a sequence of nonnegative functions $\{f_n\} \subset C_b(S)$ such that $f_n \downarrow \mathbf{1}_F$ as $n \uparrow \infty$. Since F is closed, one may view $x \in F$ in terms of the equivalent condition that $\rho(x, F) = 0$, where $\rho(x, F) := \inf\{\rho(x, y) : y \in F\}$. Let $h_n(r) = 1 - nr$ for $0 \leq r \leq 1/n, h_n(r) = 0$ for $r \geq 1/n$. Then take $f_n(x) = h_n(\rho(x, F))$. In particular, $\mathbf{1}_F(x) = \lim_n f_n(x), x \in S$, and Lebesgue's dominated convergence theorem applies. ∎

Note that the functions f_n in the proof of Proposition 1.6 are uniformly continuous, since $|f_n(x) - f_n(y)| \leq (n\rho(x, y)) \wedge (2 \sup_x |f_n(x)|)$. It follows that the **set** $UC_b(S)$ **of bounded uniformly continuous real-valued functions on** S is measure-determining. Measure-determining classes of functions are generally actually quite extensive. For example, since the Borel σ-field on $\mathbb{R}$ can be generated by classes of open intervals, closed intervals, half-lines etc., each of the corresponding class of indicator functions $\mathbf{1}_{(a,b)}, -\infty \leq a < b < \infty, \mathbf{1}_{[a,b]}, -\infty < a < b < \infty$, $\mathbf{1}_{(-\infty,x]}, x \in \mathbb{R}$, is measure-determining (see Exercises 9, 16).

Consider the L^p-space $L^p(\Omega, \mathcal{F}, P)$ of (real-valued) random variables X such that $\mathbb{E}|X|^p < \infty$. When random variables that differ only on a P-null set are identified, then for $p \geq 1$, it follows from Theorem 1.7(e) below that $L^p(\Omega, \mathcal{F}, P)$ is a normed linear space with norm $\|X\|_p := (\int_{\Omega} |X|^p dP)^p)^{\frac{1}{p}} \equiv (\mathbb{E}|X|^p)^{\frac{1}{p}}$. It is in this sense that elements of $L^p(\Omega, \mathcal{F}, P)$ are, strictly speaking, represented by equivalence classes of random variables that are equal almost surely. It may be shown that with this norm (and distance $\|X - Y\|_p$), it is a complete metric space, and therefore a **Banach space** (Exercise 35). In particular, $L^2(\Omega, \mathcal{F}, P)$ is a **Hilbert space** with inner product (see Appendix C)

$$\langle X, Y \rangle = \mathbb{E}XY \equiv \int_\Omega XY dP, \quad ||X||_2 = \langle X, X \rangle^{\frac{1}{2}} . \qquad (1.12)$$

The $L^2(S, \mathcal{S}, \mu)$ spaces are the only Hilbert spaces that are required in this text, where $(S, \mathcal{S}, \mu)$ is a σ-finite measure space; see Appendix C for an exposition of the essential structure of such spaces. Note that by taking S to be a countable set with counting measure μ, this includes the l^2 sequence space. Unlike the case of a measure space $(\Omega, \mathcal{F}, \mu)$ with an infinite measure μ, for finite measures it is always true that

$$L^r(\Omega, \mathcal{F}, P) \subset L^s(\Omega, \mathcal{F}, P) \quad \text{if } r > s \geq 1, \qquad (1.13)$$

as can be checked using $|x|^s < |x|^r$ for $|x| > 1$. The basic inequalities in the following Theorem 1.7 are consequences of *convexity* at some level. So let us be precise about this notion.

Definition 1.3 A function φ defined on an open interval J is said to be a **convex function** if $\varphi(ta + (1 - t)b) \leq t\varphi(a) + (1 - t)\varphi(b)$, for all $a, b \in J, 0 \leq t \leq 1$.

If the function φ is sufficiently smooth, one may use calculus to check convexity, see Exercise 24. The following lemma is required to establish a geometrically obvious "line of support property" of convex functions.

Lemma 3 (*Line of Support*) Suppose φ is convex on an interval J. (a) If J is open, then (i) the left-hand and right-hand derivatives φ^- and φ^+ exist and are finite and nondecreasing on J, and $\varphi^- \leq \varphi^+$. Also (ii) for each $x_0 \in J$ there is a constant $m = m(x_0)$ such that $\varphi(x) \geq \varphi(x_0) + m(x - x_0), \forall x \in J$. (b) If J has a left (or right) endpoint and the right-hand (left-hand) derivative is finite, then the line of support property holds at this endpoint x_0.

Proof (a) In the definition of convexity, one may take $a < b, 0 < t < 1$. Thus convexity is equivalent to the following inequality with the identification $a = x$, $b = z, t = (z - y)/(z - x)$: For any $x, y, z \in J$ with $x < y < z$,

$$\frac{\varphi(y) - \varphi(x)}{y - x} \leq \frac{\varphi(z) - \varphi(y)}{z - y}. \qquad (1.14)$$

More generally, use the definition of convexity to analyze monotonicity and bounds on the Newton quotients (slopes of secant lines) from the right and left to see that (1.14) implies $\frac{\varphi(y)-\varphi(x)}{y-x} \leq \frac{\varphi(z)-\varphi(x)}{z-x} \leq \frac{\varphi(z)-\varphi(y)}{z-y}$ (use the fact that $c/d \leq e/f$ for $d, f > 0$ implies $c/d \leq (c + e)/(d + f) \leq e/f$). The first of these inequalities shows that $\frac{\varphi(y)-\varphi(x)}{y-x}$ decreases as y decreases, so that the right-hand derivative $\varphi^+(x)$ exists and $\frac{\varphi(y)-\varphi(x)}{y-x} \geq \varphi^+(x)$. Letting $z \downarrow y$ in (1.14), one gets $\frac{\varphi(y)-\varphi(x)}{y-x} \leq \varphi^+(y)$ for all $y \in J$. Hence φ^+ is finite and nondecreasing on J. Now fix $x_0 \in J$. By taking $x = x_0$ and $y = x_0$ in turn in these two inequalities for φ^+, it follows that $\varphi(y) - \varphi(x_0) \geq \varphi^+(x_0)(y - x_0)$ for all $y \geq x_0$, and $\varphi(x_0) - \varphi(x) \leq \varphi^+(x_0)(x_0 - x)$ for all

$x \leq x_0$. Thus the "line of support" property holds with $m = \varphi^+(x_0)$. (b) If J has a left (right) endpoint x_0, and $\varphi^+(x_0)$ ($\varphi^-(x_0)$) is finite, then the above argument remains valid with $m = \varphi^+(x_0)$ ($\varphi^-(x_0)$).

A similar proof applies to the left-hand derivative $\varphi^-(x)$ (Exercise 24). On letting $x \uparrow y$ and $z \downarrow y$ in (1.14), one obtains $\varphi^-(y) \leq \varphi^+(y)$ for all y. In particular, the line of support property now follows for $\varphi^-(x_0) \leq m \leq \varphi^+(x_0)$.   ∎

Theorem 1.7 *(Basic Inequalities)* Let X, Y be random variables on $(\Omega, \mathcal{F}, P)$.

(a) *(Jensen's Inequality)* If φ is a convex function on the interval J and $P(X \in J) = 1$, then $\varphi(\mathbb{E}X) \leq \mathbb{E}(\varphi(X))$ provided that the indicated expectations exist. Moreover, if φ is strictly convex, then equality holds if and only if X is a.s. constant.

(b) *(Lyapounov Inequality)* If $0 < r < s$ then $(\mathbb{E}|X|^r)^{\frac{1}{r}} \leq (\mathbb{E}|X|^s)^{\frac{1}{s}}$.

(c) *(Hölder Inequality)* Let $p \geq 1$. If $X \in L^p$, $Y \in L^q$, $\frac{1}{p} + \frac{1}{q} = 1$, then $XY \in L^1$ and $\mathbb{E}|XY| \leq (\mathbb{E}|X|^p)^{\frac{1}{p}} (\mathbb{E}|Y|^q)^{\frac{1}{q}}$.

(d) *(Cauchy–Schwarz Inequality)* If $X, Y \in L^2$ then $XY \in L^1$ and one has $|\mathbb{E}(XY)| \leq \sqrt{\mathbb{E}X^2}\sqrt{\mathbb{E}Y^2}$.

(e) *(Minkowski Triangle Inequality)* Let $p \geq 1$. If $X, Y \in L^p$ then $\|X + Y\|_p \leq \|X\|_p + \|Y\|_p$.

(f) *(Markov and Chebyshev-type Inequalities)* Let $p \geq 1$. If $X \in L^p$ then $P(|X| \geq \lambda) \leq \frac{\mathbb{E}(|X|^p \mathbf{1}_{(|X|\geq\lambda)})}{\lambda^p} \leq \frac{\mathbb{E}|X|^p}{\lambda^p}$, $\lambda > 0$. More generally, if h is a nonnegative increasing function on an interval containing the range of X, then $P(X \geq \lambda) \leq \mathbb{E}(h(X)\mathbf{1}_{[X\geq\lambda]})/h(\lambda)$.

Proof The proof of Jensen's inequality hinges on the line of support property of convex functions in Lemma 3 by taking $x = X(\omega), \omega \in \Omega, x_0 = \mathbb{E}X$. The Lyapounov inequality follows from Jensen's inequality by writing $|X|^s = (|X|^r)^{\frac{s}{r}}$, for $0 < r < s$, since $\varphi(x) = x^{\frac{s}{r}}$ is convex on $[0, \infty)$. For the Hölder inequality, let $p, q > 1$ be **conjugate exponents** in the sense that $\frac{1}{p} + \frac{1}{q} = 1$. Using convexity of the function $\exp(x)$ one sees that $|ab| = \exp(\ln(|a|^p)/p + \ln(|b|^q)/q)) \leq \frac{1}{p}|a|^p + \frac{1}{q}|b|^q$. Applying this to $a = \frac{|X|}{\|X\|_p}, b = \frac{|Y|}{\|Y\|_q}$ and integrating, it follows that $\mathbb{E}|XY| \leq (\mathbb{E}|X|^p)^{\frac{1}{p}} (\mathbb{E}|Y|^q)^{\frac{1}{q}}$. The Cauchy–Schwarz inequality is the Hölder inequality with $p = q = 2$. For the proof of Minkowski's inequality, first use the inequality (1.27) to see that $|X + Y|^p$ is integrable from the integrability of $|X|^p$ and $|Y|^p$. Applying Hölder's inequality to each term of the expansion $\mathbb{E}(|X| + |Y|)^p = \mathbb{E}|X|(|X| + |Y|)^{p-1} + \mathbb{E}|Y|(|X| + |Y|)^{p-1}$, and solving the resulting inequality for $\mathbb{E}(|X| + |Y|)^p$ (using conjugacy of exponents), it follows that $\|X + Y\|_p \leq \|X\|_p + \|Y\|_p$. Finally, for the Markov and Chebyshev-type inequalities simply observe that since $\mathbf{1}_{(|X|\geq\lambda)} \leq \frac{|X|^p\mathbf{1}_{(|X|\geq\lambda)}}{\lambda^p} \leq \frac{|X|^p}{\lambda^p}$ on Ω, taking expectations yields $P(|X| \geq \lambda) \leq \frac{\mathbb{E}(|X|^p\mathbf{1}_{(|X|\geq\lambda)})}{\lambda^p} \leq \frac{\mathbb{E}|X|^p}{\lambda^p}$, $\lambda > 0$. More generally, for increasing h with $h(\lambda) > 0$, one has $\mathbb{E}(h(X)\mathbf{1}_{[X\geq\lambda]}) \geq h(\lambda)P(X \geq \lambda)$.   ∎

Remark 1.4 One may note that the same proof may be used to check that corresponding formulations of both the Hölder and the Minkowski inequality for functions on arbitrary measure spaces can be verified with the same proof as above.

The **Markov inequality** refers to the case $p = 1$ in (f). Observe from the proofs that (c–e) hold with the random variables X, Y replaced by measurable functions, in fact complex valued, on an arbitrary (not necessarily finite) measure space $(S, \mathcal{S}, \mu)$; see Exercise 36.

Example 5 *(Chebyshev Estimation of a Distribution Function)* Suppose that X is a nonnegative random variable with distribution function F, and having finite moments μ_p of order $p < s$ for some $s > 1$. According to the Chebyshev inequality one has

$$1 - F(x) = P(X > x) \le \frac{\mu_p}{x^p}. \tag{1.15}$$

By Liapounov's inequality one also has $\mu_p^{\frac{1}{p}} \le \mu_{p+1}^{\frac{1}{p+1}}, p = 1, 2, \dots$. Since $p \to \log \mu_p$ is convex on $[0, s]$ (Exercise 23), it follows that $\frac{\mu_p}{\mu_{p-1}} \le \frac{\mu_{p+1}}{\mu_p}$. So

$$\frac{\mu_p}{x^p} \le \frac{\mu_{p+1}}{x^{p+1}} \iff \frac{x^{p+1}}{x^p} \le \frac{\mu_{p+1}}{\mu_p} \iff x \le \frac{\mu_{p+1}}{\mu_p}. \tag{1.16}$$

Thus upper bound estimates of $1 - F(x)$ may be obtained as follows:

$$\overline{F}(x) := 1 - F(x) \le \begin{cases} 1 & \text{if } x \le \mu_1, \\ \frac{\mu_2}{x^2} & \text{if } \mu_1 < x \le \frac{\mu_2}{\mu_1}, \\ \dots & \\ \frac{\mu_p}{x^p} & \text{if } \frac{\mu_p}{\mu_{p-1}} < x \le \frac{\mu_{p+1}}{\mu_p}, 1 \le p \le s - 1. \end{cases} \tag{1.17}$$

From here one also has Chebyshev estimated lower bounds on $F(x)$ in terms of the moments μ_p of $\frac{1}{X}$ (Exercise 2).[3]

Suppose that $(S, \mathcal{S}, \mu)$ is an arbitrary measure space and $g : S \to [0, \infty)$ a Borel-measurable function, though not necessarily integrable. One may use g as a **density** with respect to μ to define another measure ν on $(S, \mathcal{S})$, i.e., with g as its Radon–Nykodym derivative $d\nu/d\mu = g$, also commonly denoted by $d\nu = g\,d\mu$, and meaning that $\nu(A) = \int_A g\,d\mu$, $A \in \mathcal{S}$; see Appendix C for a full treatment of the Radon–Nikodym theorem.

Recall that a sequence of measurable functions $\{g_n\}_{n=1}^{\infty}$ on S is said to **converge** μ-**a.e.** to a measurable function g on S if and only if $\mu(\{x \in S : \lim_n g_n(x) \ne g(x)\}) = 0$. The following simple theorem finds diverse uses in the framework of probability theory.

Theorem 1.8 *(Scheffé)* Let $(S, \mathcal{S}, \mu)$ be a measure space and suppose that ν, $\{\nu_n\}_{n=1}^{\infty}$ are measures on $(S, \mathcal{S})$ with respective nonnegative densities g, $\{g_n\}_{n=1}^{\infty}$ with respect to μ, such that

---

[3]For an application see Bhattacharya, R.N., Kim, H., Majumdar, M.K. (2015): Sustainability in the Stochastic Ramsey Model, *J. Quant. Econ.* **13**, 169–184.

$$\int_S g_n \, d\mu = \int_S g \, d\mu < \infty, \quad \forall n = 1, 2, \ldots.$$

If $g_n \to g$ as $n \to \infty$, μ-a.e., then

$$\sup_{A \in \mathcal{S}} \left| \int_A g \, d\mu - \int_A g_n \, d\mu \right| \le \int_S |g - g_n| \, d\mu \to 0, \text{ as } n \to \infty.$$

Proof The indicated bound on the supremum follows from the triangle inequality for integrals. Since $\int_S (g - g_n) \, d\mu = 0$ for each n, $\int_S (g - g_n)^+ \, d\mu = \int_\Omega (g - g_n)^- \, d\mu$. In particular, since $|g - g_n| = (g - g_n)^+ + (g - g_n)^-$,

$$\int_S |g - g_n| \, d\mu = 2 \int_S (g - g_n)^+ \, d\mu.$$

But $0 \le (g - g_n)^+ \le g$. Since g is μ-integrable, one obtains $\int_S (g - g_n)^+ \, d\mu \to 0$ as $n \to \infty$ from Lebesgue's dominated convergence theorem. ∎

Remark 1.5 Suppose g_n, g are probability densities (with respect to a σ-finite measure μ) and $g_n \to g$ in μ-measure. Then the conclusion of Theorem 1.8 holds.

For a measurable space $(S, \mathcal{S})$, a useful metric (see Exercise 33) defined on the space $\mathcal{P}(S)$ of probabilities on $\mathcal{S} = \mathcal{B}(S)$ is furnished by the **total variation distance** defined by

$$d_v(\mu, \nu) := \sup\{|\mu(A) - \nu(A)| : A \in \mathcal{B}(S)\}, \quad \mu, \nu \in \mathcal{P}(S). \tag{1.18}$$

Proposition 1.9 Suppose that $(S, \mathcal{S})$ is a measurable space. Then

$$d_v(\mu, \nu) = \frac{1}{2} \sup \left\{ \left| \int_S f \, d\mu - \int_S f \, d\nu \right| : f \in B(S), |f| \le 1 \right\},$$

where $B(S)$ denotes the space of bounded Borel-measurable functions on S. Moreover, $(\mathcal{P}(S), d_v)$ is a complete metric space.

Proof Let us first establish the formula for the total variation distance. By standard simple function approximation it suffices to consider bounded simple functions in the supremum. Fix arbitrary $\mu, \nu \in \mathcal{P}(S)$. Let $f = \sum_{i=1}^k a_i \mathbf{1}_{A_i} \in B(S)$ with $|a_i| \le 1, i = 1, \ldots, k$ and disjoint sets $A_i \in \mathcal{S}, 1 \le i \le k$. Let $I^+ := \{i \le k : \mu(A_i) \ge \nu(A_i)\}$. Let I^- denote the complementary set of indices. Then by definition of the integral of a simple function and splitting the sum over $I^\pm$ one has upon twice using the triangle inequality that

$$\left|\int_S f d\mu - \int_S f d\nu\right| \leq \sum_{i \in I^+} |a_i|(\mu(A_i) - \nu(A_i)) + \sum_{i \in I^-} |a_i|(\nu(A_i) - \mu(A_i))$$

$$\leq \sum_{i \in I^+} (\mu(A_i) - \nu(A_i)) + \sum_{i \in I^-} (\nu(A_i) - \mu(A_i))$$

$$= \mu(\cup_{i \in I^+} A_i) - \nu(\cup_{i \in I^+} A_i) + \nu(\cup_{i \in I^-} A_i) - \mu(\cup_{i \in I^-} A_i)$$

$$\leq 2 \sup\{|\mu(A) - \nu(A)| : A \in \mathcal{S}\}. \tag{1.19}$$

On the other hand, taking $f = \mathbf{1}_A - \mathbf{1}_{A^c}$, $A \in \mathcal{S}$, one has

$$\left|\int_S f d\mu - \int_S f d\nu\right| = |\mu(A) - \mu(A^c) - \nu(A) + \nu(A^c)|$$

$$= |\mu(A) - \nu(A) - 1 + \mu(A) + 1 - \nu(A)|$$

$$= 2|\mu(A) - \nu(A)|. \tag{1.20}$$

Thus, taking the supremum over sets $A \in \mathcal{S}$ establishes the asserted formula for the total variation distance. Next, to prove that the space $\mathcal{P}(S)$ of probabilities is complete for this metric, let $\{\mu_n\}_{n=1}^\infty$ be a Cauchy sequence in $\mathcal{P}(S)$. Since the closed interval $[0, 1]$ of real numbers is complete, one may define $\mu(A) := \lim_n \mu_n(A)$, $A \in \mathcal{S}$. Because this convergence is uniform over $\mathcal{S}$, it is simple to check that $\mu \in \mathcal{P}(S)$ and $\mu_n \to \mu$ in the metric d_v; see Exercise 33. ∎

For real-valued random variables X_n, $n \geq 1$, and X, having absolutely continuous distribution functions with densities g_n, g, say, with respect to Lebesgue measure, a notion of **convergence in distribution** can be defined as $F_n(x) \equiv P(X_n \leq x) \to F(x) \equiv P(X \leq x)$ for all $x \in \mathbb{R}$. It follows from Scheffé's theorem that pointwise convergence a.e. of the densities implies uniform convergence of the distribution functions; Exercise 25. For contrast in the absence of a density see Exercise 26. A treatment of convergence in distribution is the subject of Chapter V.

One may also note that Scheffé's theorem provides conditions under which convergence in measure implies $L^1(S, \mathcal{S}, \mu)$-convergence of the densities g_n to g, and convergence in the total variation metric of the probabilities ν_n to ν.

We will conclude this chapter with some additional basic convergence theorems for probability spaces. Namely, we consider L^p-convergence ($p \geq 1$) of a sequence of random variables X_n, $n \geq 1$ in $L^p(\Omega, \mathcal{F}, P)$, to X, i.e., $\mathbb{E}|X_n - X|^p \to 0$ as $n \to \infty$. We start with $p = 1$. Of course $|\mathbb{E}X_n - \mathbb{E}X| \leq \mathbb{E}|X_n - X|$, $n \geq 1$, so that L^1-convergence implies convergence of the expected values.

For this purpose we require a definition.

Definition 1.4 A sequence $\{X_n\}_{n=1}^\infty$ of random variables on a probability space $(\Omega, \mathcal{F}, P)$ is said to be **uniformly integrable** if

$$\lim_{\lambda \to \infty} \sup_n \mathbb{E}\{|X_n|\mathbf{1}_{[|X_n| \geq \lambda]}\} = 0.$$

Theorem 1.10 (*L^1-Convergence Criterion*) Let $\{X_n\}_{n=1}^{\infty}$ be a sequence of random variables on a probability space $(\Omega, \mathcal{F}, P)$, $X_n \in L^1$ $(n \geq 1)$. Then $\{X_n\}_{n=1}^{\infty}$ converges in L^1 to a random variable X if and only if (i) $X_n \to X$ in probability as $n \to \infty$, and (ii) $\{X_n\}_{n=1}^{\infty}$ is uniformly integrable.

Proof (Necessity) If $X_n \to X$ in L^1 then convergence in probability (i) follows from the Markov inequality. Also

$$
\begin{aligned}
\int_{[|X_n| \geq \lambda]} |X_n| dP &\leq \int_{[|X_n| \geq \lambda]} |X_n - X| dP + \int_{[|X_n| \geq \lambda]} |X| dP \\
&\leq \int_{\Omega} |X_n - X| dP + \int_{[|X| \geq \lambda/2]} |X| dP \\
&\quad + \int_{[|X| < \lambda/2, |X_n - X| \geq \lambda/2]} |X| dP.
\end{aligned}
\tag{1.21}
$$

The first term of the last sum goes to zero as $n \to \infty$ by hypothesis. For each $\lambda > 0$ the third term goes to zero by the dominated convergence theorem as $n \to \infty$. The second term goes to zero as $\lambda \to \infty$ by the dominated convergence theorem too. Thus, there are numbers $n(\varepsilon)$ and $\lambda(\varepsilon)$ such that for all $\lambda \geq \lambda(\varepsilon)$,

$$
\sup_{n \geq n(\varepsilon)} \int_{[|X_n| \geq \lambda]} |X_n| dP \leq \varepsilon.
\tag{1.22}
$$

Since a *finite* sequence of integrable random variables $\{X_n : 1 \leq n \leq n(\varepsilon)\}$ is always uniformly integrable, it follows that the full sequence $\{X_n\}$ is uniformly integrable.

(Sufficiency) Under the hypotheses (i), (ii), given $\varepsilon > 0$ one has for all n that

$$
\int_{\Omega} |X_n| dP \leq \int_{[|X_n| \geq \lambda]} |X_n| dP + \lambda \leq \varepsilon + \lambda(\varepsilon)
\tag{1.23}
$$

for sufficiently large $\lambda(\varepsilon)$. In particular, $\{\int_{\Omega} |X_n| dP\}_{n=1}^{\infty}$ is a bounded sequence. Thus $\int_{\Omega} |X| dP < \infty$ since, using Fatou's lemma, one has $\int_{\Omega} |X| dP = \int_{\Omega} \lim_n |X_n| dP \leq \liminf_n \int_{\Omega} |X_n| dP < \infty$. Now

$$
\begin{aligned}
\int_{[|X_n - X| \geq \lambda]} |X_n - X| dP &= \int_{[|X_n - X| \geq \lambda, |X_n| \geq \lambda/2]} |X_n - X| dP \\
&\quad + \int_{[|X_n| < \lambda/2, |X_n - X| \geq \lambda]} |X_n - X| dP \\
&\leq \int_{[|X_n| \geq \lambda/2]} |X_n| dP + \int_{[|X_n - X| \geq \lambda]} |X| dP \\
&\quad + \int_{[|X_n| < \lambda/2, |X_n - X| \geq \lambda]} (\frac{\lambda}{2} + |X|) dP.
\end{aligned}
\tag{1.24}
$$

Now, using (ii), given $\varepsilon > 0$, choose $\lambda = \lambda(\varepsilon) > 0$ so large that the first term of the last sum is smaller than ε. With this value of $\lambda = \lambda(\varepsilon)$ the second and third terms go to zero as $n \to \infty$ by Lebesgue's dominated convergence theorem, using (i). Thus,

$$\limsup_{n\to\infty} \int_{[|X_n-X|\geq\lambda(\varepsilon)]} |X_n - X| dP \leq \varepsilon. \tag{1.25}$$

But again applying the dominated convergence theorem one also has

$$\limsup_{n\to\infty} \int_{[|X_n-X|<\lambda(\varepsilon)]} |X_n - X| dP = 0. \tag{1.26}$$

Thus, the conditions are also sufficient for L^1 convergence to X. ∎

The next result follows as a corollary.

Theorem 1.11 (L^p-*Convergence Criterion*) Let $p \geq 1$. Let $\{X_n\}_{n=1}^\infty$ be a sequence of random variables on a probability space $(\Omega, \mathcal{F}, P)$, $X_n \in L^p (n \geq 1)$. Then $\{X_n\}_{n=1}^\infty$ converges in L^p to a random variable X if and only if (i) $X_n \to X$ in probability as $n \to \infty$, and (ii) $\{|X_n|^p\}_{n=1}^\infty$ is uniformly integrable.

Proof Apply the preceding result to the sequence $\{|X_n - X|^p\}_{n=1}^\infty$. The proof of necessity is analogous to (1.21) and (1.22) using the following elementary inequalities:

$$|a + b|^p \leq (|a| + |b|)^p \leq (2 \max\{|a|, |b|\})^p \leq 2^p(|a|^p + |b|^p). \tag{1.27}$$

For sufficiency, note as in (1.23) that (i), (ii) imply $X \in L^p$, and then argue as in (1.24) that the uniform integrability of $\{|X_n|^p : n \geq 1\}$ implies that of $\{|X_n - X|^p : n \geq 1\}$. ∎

Chebyshev-type inequalities often provide useful ways to check uniform integrability of $\{|X_n|^p\}_{n=1}^\infty$ in the case that $\{\mathbb{E}|X_n|^m\}$ can be shown to be a bounded sequence for some $m > p$ (see Exercise 28).

Exercise Set I

1. Let $(\Omega, \mathcal{F}, P)$ be an arbitrary probability space and let $A_1, A_2, \ldots$ be measurable events. Prove each of the following.

    (i) (Finite Additivity). If $A_1, \ldots, A_m$ are disjoint then $P(\cup_{j=1}^m A_j) = \sum_{j=1}^m P(A_j)$.
    (ii) (Monotonicity). If $A_1 \subset A_2$ then $P(A_1) \leq P(A_2)$.
    (iii) (Inclusion–Exclusion). $P(\cup_{j=1}^m A_j) = \sum_{k=1}^m (-1)^{k+1} \sum_{1\leq j_1 < \cdots < j_k \leq m} P(A_{j_1} \cap \cdots \cap A_{j_k})$.
    (iv) (Subadditivity). $P(\cup_j A_j) \leq \sum_j P(A_j)$.

(v) Show that the property $\mu(A_n) \uparrow \mu(A)$ if $A_n \uparrow A$, holds for all measures μ. [*Hint*: $A = \cup_n B_n$, $B_1 = A_1$, $B_2 = A_1^c \cap A_2, \ldots, B_n = A_1^c \cap \cdots \cap A_{n-1}^c \cap A_n$, so that $A_n = \cup_{j=1}^n B_j$.]

(vi) Show that the property: $\mu(A_n) \downarrow \mu(A)$ if $A_n \downarrow A$ holds for *finite* measures. Show by counterexample that it does not, in general, hold for measures μ that are not finite.

2. (a) Write out the corresponding lower bounds on the distribution function F in Example 5. (b) Compute the Chebyshev bounds on the exponential distribution $F(x) = 1 - e^{-x}$, $x \geq 0$, $F(x) = 0$, $x \leq 0$.

3. (*Bonferroni Inequalities*) Show that for odd $m \in \{1, 2, \ldots, n\}$, (a) $P(\cup_{j=1}^n A_j) \leq \sum_{k=1}^m \sum_{1 \leq j_1 \leq j_2 \leq \cdots \leq j_k \leq n} (-1)^{k+1} P(A_{j_1} \cap \cdots \cap A_{j_k})$, and for even $m \in \{2, \ldots, n\}$, (b) $P(\cup_{j=1}^n A_j) \geq \sum_{k=1}^m \sum_{1 \leq j_1 \leq j_2 \leq \cdots \leq j_k \leq n} (-1)^{k+1} P(A_{j_1} \cap \cdots \cap A_{j_k})$.

4. Let $(\Omega, \mathcal{F}, P)$ be an arbitrary probability space and suppose $A, B \in \mathcal{F}$ are *independent* events, i.e., $P(A \cap B) = P(A)P(B)$, with both $P(A) \geq \frac{1}{2}$ and $P(B) \geq \frac{1}{2}$. Show that $P(A \cup B) \geq \frac{1}{4}$.

5. Suppose that X_n, $n \geq 1$, is a sequence of random variables that converge to X in probability as $n \to \infty$, "sufficiently fast" that for any $\varepsilon > 0$, one has $\sum_{n=1}^\infty P(|X_n - X| > \varepsilon) < \infty$. Show that $X_n \to X$ a.s. as $n \to \infty$.

6. Suppose that X_n, $n \geq 1$, is a sequence of random variables that converge to X in probability as $n \to \infty$, and g is a continuous function. Show that $g(X_n)$, $n \geq 1$, converges in probability to $g(X)$.

7. Suppse that X_n, $n \geq 1$ and Y_n, $n \geq 1$, converge in probability to X and Y, and $X_n - Y_n \to 0$ in probability as $n \to \infty$, respectively. Show that $X = Y$ a.s.

8. Suppose that X_n, $n \geq 1$, is a sequence of real-valued random variables such that $|X_n| \leq Y$ on Ω with $\mathbb{E}Y < \infty$. Show that if $X_n \to X$ in probability as $n \to \infty$, then $\mathbb{E}X_n \to \mathbb{E}X$ as $n \to \infty$. [*Hint*: Use Proposition 1.5 and the dominated convergence theorem for almost surely convergent sequences.]

9. Show that the Borel σ-field of $\mathbb{R}$ is generated by any one of the following classes of sets: (i) $\mathcal{C} = \{(a, b) : -\infty \leq a \leq b \leq \infty\}$; (ii) $\mathcal{C} = \{(a, b] : -\infty \leq a \leq b < \infty\}$; (iii) $\mathcal{C} = \{(-\infty, x] : x \in \mathbb{R}\}$.

10. In each case below, show that ρ is a metric for the indicated topology.

(i) For $S = \mathbb{R}^\infty$, $\rho(x, y) = \sum_{k=1}^\infty 2^{-k} |x_k - y_k| / (1 + |x_k - y_k|)$, for $x = (x_1, x_2, \ldots)$, $y = (y_1, y_2, \ldots) \in \mathbb{R}^\infty$ metrizes the topology of pointwise convergence: $x^{(n)} \to x$ if and only if $x_k^{(n)} \to x_k$ for each k, as $n \to \infty$.

(ii) For $S = C[0, 1]$, $\rho(f, g) = \max\{|f(x) - g(x)| : x \in [0, 1]\}$ metrizes the topology of uniform convergence of continuous functions on $[0, 1]$.

(iii) For $S = C([0, \infty) \to \mathbb{R}^k)$, $\rho(f, g) = \sum_{n=1}^\infty 2^{-n} \|f - g\|_n / (1 + \|f - g\|_n)$, where $\|f - g\|_n := \max\{\|f(x) - g(x)\| : x \in [0, n]\}$, $\| \cdot \|$ denoting the Euclidean norm on $\mathbb{R}^k$, metrizes the topology of uniform convergence on compacts.

11. Let X be a random map on $(\Omega, \mathcal{F}, P)$ with values in a measurable space $(S, \mathcal{S})$. Show that $\mathcal{G} := \{[X \in A] : A \in \mathcal{S}\}$ is the smallest sub-σ-field of $\mathcal{F}$ such that

$X : \Omega \rightarrow S$ is a random map on $(\Omega, \mathcal{G})$, i.e., such that $[X \in A] \in \mathcal{G}$ for all $A \in \mathcal{S}$.

12. Let $\Omega = \{(1, 1), (2, 2), (1, 2), (2, 1)\}$ equipped with the power set $\mathcal{F}$. Define a simple random variable by $X(\omega) = \omega_1 + \omega_2$, $\omega = (\omega_1, \omega_2) \in \Omega$. Give an explicit description of $\sigma(X)$ as a subcollection of sets in $\mathcal{F}$ and give an example of a set in $\mathcal{F}$ that is not in $\sigma(X)$.

13. (i) Let $(\Omega, \mathcal{F}, P)$ be a probability space and let $\mathcal{P} = \{A_1, A_2, \dots, A_m\}$, $\emptyset \neq A_j \in \mathcal{F}$, $1 \leq j \leq m$, be a disjoint partition of Ω. Let $(S, \mathcal{S})$ be an arbitrary measurable space such that $\mathcal{S}$ contains all of the singleton sets $\{x\}$ for $x \in S$. Show that a random map $X : \Omega \rightarrow S$ is $\sigma(\mathcal{P})$-measurable if and only if X is a $\sigma(\mathcal{P})$-measurable simple function; i.e., simple in the sense of finitely many values. Give a counterexample in the case that $\mathcal{S}$ does not contain singletons.
    (ii) Let $A_1, \dots, A_k$ be nonempty subsets of Ω. Describe the smallest σ-field containing $\{A_1, \dots, A_k\}$ and show that its cardinality is at most 2^{k+1}.

14. Give a proof of the change of variables formula. [*Hint*: (Method of simple function approximation) Begin with h an indicator function, then h a simple function, then $h \geq 0$, and finally write $h = h^+ - h^-$.]

15. Show that if $\mathcal{L}$ is a π-system and a λ-system, then it is a σ-field. In the proof of Dynkin's π-λ theorem, show that if $A \in \mathcal{L}(\mathcal{C})$, then $\mathcal{L}_A$ is a λ-system.

16. Let X_1, X_2 be real-valued random variables on $(\Omega, \mathcal{F}, P)$. Suppose that $F_i(x) = P(X_i \leq x)$, $x \in \mathbb{R}(i = 1, 2)$ are two distribution functions on $(\mathbb{R}, \mathcal{B})$ and $F_1 = F_2$. Show that X_1 and X_2 have the same distribution. Extend this to random vectors $\mathbf{X}_1, \mathbf{X}_2$ with values in $\mathbb{R}^k$.

17. Suppose that X_1 and X_2 are two bounded real-valued random variables on $(\Omega, \mathcal{F}, P)$ such that $\mathbb{E}X_1^m = \mathbb{E}X_2^m$, $m = 1, 2, \dots$. Show that X_1 and X_2 must have the same distribution. [*Hint*: According to the Weierstrass approximation theorem, a continuous function on a closed and bounded interval may be approximated by polynomials uniformly over the interval (see Appendix B).]

18. Let S be a metric space. (a) Then any family $\mathcal{F} \subset \mathcal{B}(S)$ with the following two properties is measure-determining: (i) $\mathcal{F}$ is closed under finite intersections, (ii) every open set in S is the union of a finite or a countable number of sets in $\mathcal{F}$. [*Hint*: Let P, Q be two probability measures which agree on $\mathcal{F}$. If an open set G is a finite union of sets in $\mathcal{F}$, then using the inclusion–exclusion formula, $P(G) = Q(G)$. If $G = \cup_{i \geq 1} G_i (G_i \in F, i \geq 1)$, then given any $\varepsilon > 0$, there exists k such that $P(\cup_{1 \leq i \leq k} G_i) \geq P(G) - \varepsilon$, which implies $Q(G) \geq P(G)$; similarly, $P(G) \geq Q(G)$]. (b) Prove that the distribution function of a real-valued random variable determines its distribution. [*Hint*: Apply (a) to the class of intervals $(a, b], -\infty < a < b < \infty$.] (c) Prove that the distribution function of a random vector $\mathbf{X}$ determines its distribution. [*Hint*: Let $F(\mathbf{x}) = P(X_i \leq x_i, 1 \leq i \leq k)$, $\mathbf{x} = (x_1, \dots, x_k)$. Let $a_i < b_i$, $1 \leq i \leq k$. Then, by induction one may see that $P(\mathbf{X} \in (\mathbf{a}, \mathbf{b}]) = \sum (-1)^{k-j} F(c_1, \dots, c_k)$, where the sum is over all $c_i \in \{a_i, b_i\}$, and $j = \#\{i : c_i = b_i\}$.]

19. (*Finite-dimensional distributions are measure-determining on* $\mathbb{R}^\infty$) The space $\mathbb{R}^\infty$ of all sequences $x = (x_1, x_2, \dots)$ of real numbers is a metric space with the topology of pointwise convergence metrized by $\rho(x, y) = \sum_{n \geq 1} \frac{|x_n - y_n|}{1 + |x_n - y_n|} 2^{-n}$.

Prove the following. (a) $\mathbb{R}^\infty$ is separable. (b) The class $\mathcal{F}$ of all finite-dimensional open sets in $\mathbb{R}^\infty$ is measure-determining, so that finite-dimensional Borel sets form a measure-determining class [*Hint*: Apply Exercise 18].

20. Let (S, d) be a separable metric space, and S^∞ the space of sequences $x = (x_1, x_2, ...)$ with $x_j \in S$ for all $j \geq 1$. On S^∞ define the metric $\rho(x, y) = \sum_{n \geq 1} \frac{d(x_n, y_n)}{1 + d(x_n, y_n)} 2^{-n}$. Prove that S^∞ is separable and the finite-dimensional distributions are measure-determining on S^∞.

21. (*Finite-dimensional distributions are measure-determining on* $C((0, T) : \mathbb{R}^k)$)
Let $S = C([0, T] : \mathbb{R}^k)$ be the class of $\mathbb{R}^k$-valued continuous functions on the interval $[0, T]$, with the topology of uniform convergence, metrized by $\rho(f, g) = \|f - g\|_T \equiv \sup\{|f(s) - g(s)| : 0 \leq s \leq T\}$. Note that, by Corollary 1.5 in Appendix B, $C([0, T] : \mathbb{R})$ is a separable metric space, and it is complete because $\mathbb{R}$ is complete. (a) Give a direct proof of the separability of $C([0, T] : \mathbb{R}^k)$. [*Hint*: For each $m = 2, 3, \ldots$, choose $m + 1$ equidistant points in $[0, T]$ with the first point 0 and the last point T. At each of these points assign a point from $\mathbb{R}^k$ with rational coordinates, and define a (continuous) function on $[0, T]$ by linear interpolation. The family of such functions forms a countable dense subset of $C([0, T] : \mathbb{R}^k)$.] (b) For each m and points $0 = t_0 < t_1 < \cdots < t_m = T$, consider the projection $\pi = \pi(t_0, t_1, \ldots, t_m)$ of $f = \{f(t) : 0 \leq t \leq T\}$ in $C([0, T] : \mathbb{R}^k)$ onto the vector $(f(t_0), f(t_1), \ldots, f(t_m))$. For a probability P on $C([0, T] : \mathbb{R}^k)$, $P \circ \pi^{-1}$ is a probability measure on $(\mathbb{R}^k)^{m+1}$, called a *finite-dimensional distribution* of P. Prove that the collection of all finite-dimensional distributions is measure- determining.

22. Let $S = C([0, \infty) : \mathbb{R}^k)$ be the class of $\mathbb{R}^k$-valued continuous functions on $[0, \infty)$ with the topology of uniform convergence on compact subsets of $[0, \infty)$ metrized by $\rho(f, g) = \sum_{N \geq 1} 2^{-N} \frac{\|f - g\|_N}{1 + \|f - g\|_N}$. Prove the following. (a) S is separable and complete. (b) Finite-dimensional distributions of S are measure-determining.

23. Let X be a nonnegative random variable which is not degenerate at 0. (a) Prove that $r \to \ln \mu_r$ is convex on $[0, s]$, where $\mu_r = \mathbb{E}X^r < \infty$ for $r \leq s$. [*Hint*: Let $0 < \alpha < 1$ and $0 \leq r_1, r_2 \leq s$. Then $\mu_{\alpha r_1 + (1 - \alpha) r_2} = \mathbb{E}YZ$, where $Y = X^{\alpha r_1}$, and $Z = X^{(1 - \alpha) r_2}$. Apply Hölder's inequality with $p = \frac{1}{\alpha}$.] (b) Prove that $\frac{\mu_r}{\mu_{r-1}} \leq \frac{\mu_{r+1}}{\mu_r}$, for $0 \leq r \leq s - 1$ if $\mu_r < \infty$ for $0 \leq r \leq s$. [*Hint*: Use $\ln \mu_r \leq \frac{1}{2} \ln \mu_{r-1} + \frac{1}{2} \ln \mu_{r+1}$.]

24. (i) Show that for a convex function φ on an open interval J, φ^- is finite and nondecreasing, and the "line of support" property holds with $m = \varphi^-(x_0)$, as well as with any $m \in [\varphi^-(x_0), \varphi^+(x_0)]$. (ii) Show that while a convex φ is continuous on an open interval, it need not be so on an interval with left-hand and/or right-hand endpoints. (iii) Show that if φ has a continuous, nondecreasing derivative φ' on J, then φ is convex. In particular, if φ is twice differentiable and $\varphi'' \geq 0$ on J, then φ is convex. [*Hint*: Use the mean value theorem from calculus.]

25. Assume that X_n, $n \geq 1$, and X are real-valued random variables with absolutely continuous distribution functions having densities $g_n, n \geq 1$, g, with respect

to Lebesgue measure. Show that if $g_n \to g$ pointwise a.e., then the distribution functions $F_n(x) = P(X_n \leq x), x \in \mathbb{R}$, converge uniformly on $\mathbb{R}$ to $F(x) = P(X \leq x), x \in \mathbb{R}$.

26. Suppose that $X = 0$, and $X_n = 1/n$ with probability one for $n = 1, 2, \dots$. Let $F_n(x) = P(X_n \leq x), F(x) = P(X \leq x), x \in \mathbb{R}, n \geq 1$. Show that $F_n(x) \to F(x)$ as $n \to \infty$ for each $x \neq 0$. Is the convergence uniform?

27. Suppose that X_n has a Binomial distribution with parameters n, $p_n = \frac{1}{n}, n = 1, 2, \dots$, i.e., $P(X_n = k) = g_n(k) = \binom{n}{k} p_n^k (1-p_n)^{n-k}, k = 0, 1, \dots n$. Suppose that X has a Poisson distribution with parameter $\lambda = 1$, i.e., $P(X = k) = g(k) = \frac{1}{k!} e^{-1}, k = 0, 1, 2, \dots$. Show that the distribution functions $F_n(x) = P(X_n \leq x), x \in \mathbb{R}$ converge uniformly to the distribution function $F(x) = P(X \leq x), x \in \mathbb{R}$. [*Hint:* The distributions of X_n, X have respective densities g_n, g with respect to counting measure $\mu(dx) = \sum_{k=0}^{\infty} \delta_{\{k\}}(dx)$ concentrated on $\mathbb{Z}^+ \subset S = \mathbb{R}, \mathcal{S} = \mathcal{B}$.]

28. Let $p \geq 1, X_n \in L^m(\Omega, \mathcal{F}, P)$ for some $m > p$. Suppose there is an M such that $\mathbb{E}|X_n|^m \leq M, \forall n \geq 1$. Show that $\{|X_n|^p\}_{n=1}^{\infty}$ is uniformly integrable. [*Hint:* Use Holder inequality with $||\cdot||_{\frac{m}{p}}$ and its conjugate, followed by Chebyshev. Alternatively, use a Chebyshev-type inequality after inserting a factor $(|X_n|/\lambda)^{m-p} > 1$ on $[|X_n| > \lambda]$.]

29. Suppose that $X_1, X_2, \dots$ is a sequence of identically distributed random variables defined on a probability space $(\Omega, \mathcal{F}, P)$. Show that if $\mathbb{E}e^{|X_1|} < \infty$, then a.s. $\limsup_{n\to\infty} \frac{|X_n|}{\ln n} \leq 1$.

30. Let X be a nonnegative random variable. (i) Show that $nP(X > n) \to 0$ as $n \to \infty$ if $\mathbb{E}X < \infty$. [*Hint:* $nP(X > n) \leq \mathbb{E}X\mathbf{1}_{[X>n]}$.] (ii) Prove that $\sum_{n=1}^{\infty} P(X > n) \leq \mathbb{E}X \leq \sum_{n=0}^{\infty} P(X > n)$. [*Hint:* $\sum_{n=1}^{\infty}(n-1)P(n-1 < X \leq n) \leq \mathbb{E}X \leq \sum_{n=1}^{\infty} nP(n-1 < X \leq n)$. Write $P(n-1 < X \leq n) = P(X > n-1) - P(X > n), n = \sum_{k=1}^{n} 1$, and reverse the order of summation.]

31. Let $\mathbf{U}_n = (U_{n,1}, \dots, U_{n,n})$, be uniformly distributed over the n-dimensional cube $C_n = [0,2]^n$ for each $n = 1, 2, \dots$. That is, the distribution of $\mathbf{U}_n$ is $2^{-n}\mathbf{1}_{C_n}(\mathbf{x})m_n(d\mathbf{x})$, where m_n is n-dimensional Lebesgue measure. Define $X_n = U_{n,1} \cdots U_{n,n}, n \geq 1$. Show that (a) $X_n \to 0$ in probability as $n \to \infty$, [*Hint:* Compute $\mathbb{E}X_n^t$ as an iterated integral for strategic choices of $t > 0$], and (b) $\{X_n : n \geq 1\}$ is *not* uniformly integrable;

32. Suppose U is uniformly distributed on the unit interval $[0,1]$. That is, the distribution of U is Lebesgue measure on $[0,1]$. Define $X_n = n\mathbf{1}_{[0,\frac{1}{n}]}(U), n \geq 1$. Show that $X_n \to 0$ in probability as $n \to \infty$, but $\mathbb{E}X_n = 1$ for all n. Show from definition that $\{X_1, X_2, \dots\}$ is not uniformly integrable.

33. Let $(S, \mathcal{S})$ be a measurable space. (i) Show that if f is a real-valued bounded measurable function, $|f(x)| \leq c$ for all x, then the standard simple function approximations (1.5) to f^+ and f^- provide a sequence of simple functions f_n converging to f *uniformly* on S, and satisfying $|f_n(x)| \leq c$ for all x and for all n. (ii) Show that d_v defines a metric on $\mathcal{P}(S)$ i.e., is a well-defined nonnegative symmetric function on $\mathcal{P}(S) \times \mathcal{P}(S)$ satisfying the triangle inequality with $d_v(\mu, \nu) = 0$ if and only if $\mu = \nu$. Also show for a Cauchy sequence $\{\mu_n\}_{n=1}^{\infty}$ in

$\mathcal{P}(S)$ that the set function defined by $\mu(A) := \lim_n \mu_n(A) \in [0, 1]$, $A \in \mathcal{S}$ is a probability measure. [*Hint*: The convergence of the real numbers $\mu_n(A) \to \mu(A)$ is uniform for $A \in \mathcal{S}$.]

34. Let $\{f_n : n \geq 1\}$ be a Cauchy sequence in measure: $\mu(|f_n - f_m| > \varepsilon) \to 0$ as $n, m \to \infty$. Prove that there exists a measurable function f such that $f_n \to f$ in measure. [*Hint*: Find a sequence $n_1 < n_2 < \cdots$ such that $\mu(|f_{n_k} - f_{n_{k+1}}| > 2^{-k}) < 2^{-k}, k = 1, 2, \ldots$. Let $B = [|f_{n_k} - f_{n_{k+1}}| > 2^{-k} \ i.o.]$, and show that $\mu(B) = 0$. On B^c, $\{f_{n_k}\}_{k=1}^{\infty}$ is a Cauchy sequence, converging to some function f. Also for every $\varepsilon > 0$, $\mu(|f_n - f| > \varepsilon) \leq \mu(|f_n - f_{n_k}| > \varepsilon/2) + \mu(|f_{n_k} - f| > \varepsilon/2)$. The first term on the right of this inequality is $o(1)$ as $k \to \infty, n \to \infty$. Also, outside $B_k := \cup_{m=k}^{\infty}[|f_{n_m} - f_{n_{m+1}}| > 2^{-m}]$, one has $|f_{n_k} - f| \leq \sum_{m=k}^{\infty} 2^{-m} = 2^{-(k-1)}$. By choosing k_0 such that $2^{-(k_0-1)} < \varepsilon/2$, one gets $\mu(|f_{n_k} - f| > \varepsilon/2) \leq \mu(B_{k_0}) \leq \varepsilon/2$ for all $k \geq k_0$.]

35. Show that for every $p \geq 1$, $L^p(S, \mathcal{S}, \mu)$ is a complete metric space.

36. (*Integration of Complex-Valued Functions*) A Borel measurable function $f = g + ih$ on a measure space $(S, \mathcal{S}, \mu)$ into $\mathbb{C}$, set, (i.e., g, h are real-valued Borel-measurable), is said to be *integrable* if its real and imaginary parts g and h are both integrable. Since $2^{-\frac{1}{2}}(|g| + |h|) \leq |f| \equiv \sqrt{g^2 + h^2} \leq |g| + |h|$, f is integrable if and only if $|f|$ is integrable. The following extend a number of standard results for measurable real-valued functions to measurable complex-valued functions.

   (a) Extend Lebesgue's dominated convergence theorem (Appendix A) to complex-valued functions.

   (b) Extend the inequalities of Lyapounov, Hölder, Minkowski, and Markov–Chebyshev (Theorem 1.7(b), (c), (e), (f)) to complex-valued functions.

   (c) For $p \geq 1$, let the L^p-space of complex-valued functions be defined by equivalence classes of complex-valued functions f induced by equality a.e. such that $|f|^p$ is integrable. Show that this L^p-space is a Banach space over the field of complex numbers with norm $\|f\|_p = (\int_S |f|^p d\mu)^{\frac{1}{p}}$.

   (d) Show that the L^2-space of complex-valued square-integrable functions is a Hilbert space with inner product $\langle f_1, f_2 \rangle = \int_S f_1 \overline{f_2} \, d\mu$, where $\overline{f_2}$ is the complex conjugate of f_2.

   (e) Show that for the special case of real-valued functions, the L^p-norm defined above reduces to that introduced in the text.

Chapter II
Independence, Conditional Expectation

The notions of statistical independence, conditional expectation, and conditional probability are the cornerstones of probability theory. Since probabilities may be expressed as expected values (integrals) of random variables, i.e., $P(A) = \mathbb{E}\mathbf{1}_A$, $A \in \mathcal{F}$, much can be gained by beginning with a formulation of independence of random maps, and conditional expectations of random variables.

Consider a finite set of random variables (maps) $X_1, X_2, \ldots, X_n$, where each X_i is a measurable map on $(\Omega, \mathcal{F}, P)$ into $(S_i, \mathcal{S}_i)$ $(1 \leq i \leq k)$. The **product σ-field**, denoted by $\mathcal{S}_1 \otimes \cdots \otimes \mathcal{S}_n$ is defined as the σ-field generated by the collection $\mathcal{R}$ of **measurable rectangles** of the form $R = \{\mathbf{x} \in S_1 \times \cdots \times S_n : (x_1, \ldots, x_n) \in B_1 \times \cdots \times B_n\}$, for $B_i \in \mathcal{S}_i$, $1 \leq i \leq n$. Alternatively, the product σ-field may be viewed, equivalently, as the smallest σ-field of subsets of $S_1 \times \cdots \times S_n$ which makes each of the **coordinate projections**, say $\pi_k(\mathbf{x}) = x_k$, $\mathbf{x} \in S_1 \times \cdots \times S_n$, a measurable map. In particular, if one gives $S_1 \times \cdots \times S_n$ the product σ-field $\mathcal{S}_1 \otimes \cdots \otimes \mathcal{S}_n$, then the vector $X := (X_1, \ldots, X_n) : \Omega \to S_1 \times \cdots \times S_n$ is a measurable map. This makes $\mathcal{S}_1 \otimes \cdots \otimes \mathcal{S}_n$ a natural choice for a σ-field on $S_1 \times \cdots \times S_n$.

The essential idea of the definition of independence of $X_1, \ldots, X_n$ below is embodied in the extension of the formula

$$P(\cap_{j=1}^n [X_j \in A_j]) = \prod_{j=1}^n P(X_j \in A_j), \, A_j \in \mathcal{S}_j, 1 \leq j \leq n,$$

or equivalently

$$P((X_1, \ldots, X_n) \in A_1 \times \cdots \times A_n) = P(X_1 \in A_1) \cdots P(X_n \in A_n),$$

for $A_j \in \mathcal{S}_j$, $1 \leq j \leq n$, to the full distribution of $(X_1, \ldots, X_n)$. This is readily obtained via the notion of **product measure** (see Appendix A).

© Springer International Publishing AG 2016
R. Bhattacharya and E.C. Waymire, *A Basic Course in Probability Theory*,
Universitext, DOI 10.1007/978-3-319-47974-3_II

Definition 2.1 Finitely many random variables (maps) $X_1, X_2, \ldots, X_n$, with X_i a measurable map on $(\Omega, \mathcal{F}, P)$ into $(S_j, \mathcal{S}_j)$ $(1 \leq j \leq k)$, are said to be **independent** if the distribution Q of $(X_1, X_2, \ldots, X_n)$ on the product space $(S = S_1 \times S_2 \times \cdots \times S_n, S = S_1 \otimes S_2 \otimes \cdots \otimes S_n)$ is a product measure $Q_1 \times Q_2 \times \cdots \times Q_n$, where Q_j is a probability measure on $(S_j, \mathcal{S}_j)$, $1 \leq i \leq n$. Events $A_j \in \mathcal{S}_j$, $1 \leq j \leq n$ are said to be **independent events** if the corresponding indicator random variables $\mathbf{1}_{A_j}$, $1 \leq j \leq n$, are independent.

Notice that if $X_1, \ldots, X_n$ are independent then

$$P(X_i \in B_i) = P((X_1, \ldots, X_n) \in S_1 \times \cdots \times B_i \times \cdots \times S_n)$$
$$= Q_1(S_1) \cdots Q_i(B_i) \cdots Q_n(S_n)$$
$$= Q_i(B_i), \quad B_i \in \mathcal{S}_i, 1 \leq i \leq n.$$

Moreover since, by the $\pi - \lambda$ theorem, product measure is uniquely determined by its values on the π-system $\mathcal{R}$ of measurable rectangles, $X_1, X_2, \ldots, X_n$ are independent if and only if $Q(B_1 \times B_2 \times \cdots \times B_n) = \prod_{i=1}^n Q_i(B_i)$, $\forall B_i \in \mathcal{S}_i$, $1 \leq i \leq n$, or equivalently

$$P(X_i \in B_i, 1 \leq i \leq n) = \prod_{i=1}^n P(X_i \in B_i), \quad B_i \in \mathcal{S}_i, 1 \leq i \leq n. \qquad (2.1)$$

Observe that any subcollection of finitely many independent random variables will be independent. In particular, pairs of random variables will be independent. The converse is not true (Exercise 15).

The following important application of Fubini–Tonelli readily extends to any finite sum of independent random variables, Exercise 2. Also see Exercise 8 for applications to sums of independent exponentially distributed random variables and Gaussian random variables.

Theorem 2.1 Suppose $\mathbf{X}_1, \mathbf{X}_2$ are independent k-dimensional random vectors having distributions Q_1, Q_2, respectively. The distribution of $\mathbf{X}_1 + \mathbf{X}_2$ is given by the **convolution** of Q_1 and Q_2:

$$Q_1 * Q_2(B) = \int_{\mathbb{R}^k} Q_1(B - y) Q_2(dy), \quad B \in \mathcal{B}^k,$$

where $B - y := \{x - y : x \in B\}$.

Proof Since sums of measurable functions are measurable, $\mathbf{S} = \mathbf{X}_1 + \mathbf{X}_2$ is a random vector, and for $B \in \mathcal{B}^k$

$$P_S(B) = P(S \in B) = P((\mathbf{X}_1, \mathbf{X}_2) \in C) = P_{X_1} \times P_{X_2}(C),$$

where $C = \{(x, y) : x + y \in B\}$. Now simply observe that $C^y = B - y$ and apply Fubini–Tonelli to $P_{X_1} \times P_{X_2}(C)$. ∎

Corollary 2.2 Suppose $\mathbf{X}_1, \mathbf{X}_2$ are independent k-dimensional random vectors having distributions Q_1, Q_2, respectively. Assume that at least one of Q_1 or Q_2 is absolutely continuous with respect to k-dimensional Lebesgue measure with pdf g. Then the distribution of $S = \mathbf{X}_1 + \mathbf{X}_2$ is absolutely continuous with density

$$f_S(s) = \int_{\mathbb{R}^k} g(s - y) Q_2(dy).$$

Proof Without loss of generality, assume Q_1 has pdf g. Then, using Theorem 2.1, change of variable, and Fubini-Tonelli, one has for any Borel set B

$$P(S \in B) = \int_{\mathbb{R}^k} Q_1(B - y) Q_2(dy) = \int_{\mathbb{R}^k} \int_{B-y} g(z) dz \, Q_2(dy)$$

$$= \int_B \left\{ \int_{\mathbb{R}^k} g(s - y) Q_2(dy) \right\} ds. \tag{2.2}$$

This establishes the assertion.   ∎

From the Corollary 2.2, one sees that if both Q_1 and Q_2 have pdf's g_1, g_2, say, then the pdf of $Q_1 * Q_2$ may be expressed as a convolution of densities $g_1 * g_2$ as given by

$$f_S(s) = \int_{\mathbb{R}^k} g_1(s - y) g_2(y) dy. \tag{2.3}$$

As given in Appendix A, the notion of product measure $\mu_1 \times \cdots \times \mu_n$ can be established for a finite number of σ-finite measure spaces $(S_1, \mathcal{S}_1, \mu_1), \ldots, (S_n, \mathcal{S}_n, \mu_n)$. The σ-finiteness is essential for such important properties as **associativity** of the product, i.e., $(\mu_1 \times \mu_2) \times \mu_3 = \mu_1 \times (\mu_2 \times \mu_3)$, the Fubini-Tonelli theorem, and other such useful properties. In practice, to determine integrability of a measurable function $f : S_1 \times \cdots \times S_n \to \mathbb{R}$, one typically applies the Tonelli part (a) of the Fubini-Tonelli theorem (requiring nonnegativity) to $|f|$ in order to determine whether the Fubini part (b) is applicable to f; cf. Appendix A.

The following result is an often used consequence of independence.

Theorem 2.3 If $X_1, \ldots, X_n$ are independent random variables on $(\Omega, \mathcal{F}, P)$ such that $\mathbb{E}|X_j| < \infty, 1 \leq j \leq n$, then $\mathbb{E}|X_1 \cdots X_n| < \infty$ and

$$\mathbb{E}(X_1 \cdots X_n) = \mathbb{E}(X_1) \cdots \mathbb{E}(X_n).$$

Proof Let $Q_j = P \circ X_j^{-1}, j \geq 1$. Since by independence, $(X_1, \ldots, X_n)$ has product measure as joint distribution, one may apply a change of variables and the Tonelli part to obtain

$$\mathbb{E}|X_1 \cdots X_n| = \int_\Omega |X_1 \cdots X_n| dP$$

$$= \int_{\mathbb{R}^n} |x_1 \cdots x_n| Q_1 \times \cdots \times Q_n(dx_1 \times \cdots \times dx_n)$$

$$= \prod_{j=1}^n \int_{\mathbb{R}} |x_j| Q_j(dx_j) = \prod_{j=1}^n \mathbb{E}|X_j| < \infty.$$

With the integrability established one may apply the Fubini part to do the same thing for $\mathbb{E}(X_1 \cdots X_n)$ and the product measure distribution $P \circ X_1^{-1} \times \cdots \times P \circ X_n^{-1}$ of $(X_1, \ldots, X_n)$. ∎

The role of independence in modeling occurs either as an assumption about various random variables defining a particular model or, alternatively, as a property that one may check within a specific model.

Example 1 *(Finitely Many Repeated Coin Tosses)* As a model of n-repeated tosses of a fair coin, one might assume that the successive binary outcomes are represented by a sequence of independent of $0 - 1$ valued random variables $X_1, X_2, \ldots X_n$ such that $P(X_j = 0) = P(X_j = 1) = 1/2, j = 1, \ldots, n$, defined on a probability space $(\Omega, \mathcal{F}, P)$. Alternatively, as described in Example 1 in Chapter I, one may define a probability space $\Omega = \{0, 1\}^n$, with sigma-field $\mathcal{F} = 2^\Omega$, and probability $P(\{\omega\}) = 2^{-n}$, for all $\omega \in \Omega$. Within the framework of this model one may then check that the Bernoulli variables $X_j(\omega) = \omega_j, \omega = (\omega_1, \ldots, \omega_n) \in \Omega, 1 \le j \le n$, define a sequence of independent random variables with $P(X_j = 0) = P(X_j = 1) = 1/2, 1 \le j \le n$. For a parameter $p \in [0, 1]$, the model of n independent repeated tosses of a (possibly biased) coin is naturally defined as a sequence of independent Bernoulli $0 - 1$ valued random variables $X_1, \ldots, X_n$ with $P(X_j = 1) = p = 1 - P(X_j = 0), 1 \le j \le n$.

Example 2 *(Percolation on Binary Trees)* The set $T_n = \cup_{j=0}^n \{1, 2\}^n$ may be viewed as a *rooted binary tree graph* in which, for $1 \le j \le n$, $v = (v_1, \ldots, v_j) \in T_n$ is a *vertex* of length $|v| = j$, with the added convention that $\{1, 2\}^0 = \{\theta\}$ and $v = \theta$ has length $|\theta| = 0$. The special vertex θ is also designated the root. The *parent vertex* of $v = (v_1, \ldots, v_j), j \ge 2$, is defined by $\overleftarrow{v} = (v_1, \ldots, v_{j-1})$, with $\theta = \overleftarrow{(1)} = \overleftarrow{(2)}$. A pair of vertices v, w are connected by an *edge* if either $\overleftarrow{v} = w$ or $\overleftarrow{w} = v$. The infinite tree graph is defined by $T = \cup_{n=0}^\infty T_n = \cup_{j=0}^\infty \{1, 2\}^j$, with the corresponding definitions of vertices and edges. For $v = (v_1, v_2, \ldots, v_n) \in \partial T_n := \{1, 2\}^n$, or $v = (v_1, v_2, \ldots) \in \partial T := \{1, 2\}^\infty$, denote the restriction to the first j generations by $v|j = (v_1, \ldots, v_j)$, with $v|0 = \theta$. Then $\theta = v|0, v|1, v|2, \ldots$ may be viewed as a *path* of nearest neighboring vertices to the root. Now, let $\{X_v : v \in T\}$ be a family of independent and identically distributed (i.i.d.) Bernoulli $0 - 1$-valued random variables with $p = P(X_v = 1)$. Define $X_\theta \equiv 1$ with probability one. A path $v \in \partial T$ is said to be *open* if $X_{v|j} = 1 \, \forall j = 0, 1, 2 \ldots$. The tree graph is said to *percolate* if there is at least one open nearest neighbor path from θ to ∂T. Let B denote the event that the graph percolates. The problem is to find p such that $P(B) > 0$. Let N_n be the number of open nearest neighbor paths from the root to ∂T_n, and N the number of open nearest neighbor paths to ∂T. Then

$$B = [N > 0] = \cap_{n=1}^{\infty} [N_n > 0], \quad [N_{n+1} > 0] \subset [N_n > 0], n = 1, 2 \ldots.$$

So it suffices to investigate p for which there is a positive lower bound on $\lim_{n \to \infty} P(N_n > 0)$. First observe that $\mathbb{E}N_n = \mathbb{E} \sum_{|v|=n} \prod_{j=0}^{n} X_{v|j} = (2p)^n$. Since $P(N_n > 0) \leq \mathbb{E}N_n$, it follows that $P(N = 0) = 1$ for $p < \frac{1}{2}$, i.e., $P(B) = 0$ and the graph does not percolate. For larger p we use the second moment bound $(\mathbb{E}N_n)^2 \leq P(N_n > 0)\mathbb{E}N_n^2$ by Cauchy-Schwarz. In particular,

$$\mathbb{E}N_n^2 = \sum_{|u|=n, |v|=n} \mathbb{E}\prod_{j=1}^{n} X_{u|j} X_{v|j}$$

$$= \sum_{k=1}^{n} \sum_{|w|=k, |u|=n-k, |v|=n-k, u_1 \neq v_1} \prod_{i=1}^{k} \mathbb{E}X_{w|i}^2$$

$$\prod_{j=k+1}^{n} \mathbb{E}X_{w*u|j}\mathbb{E}X_{w*v|j}, \tag{2.4}$$

where $w * v = (w_1, \ldots, w_k, v_1, \ldots, v_{n-k})$ denotes concatenation of the paths. It follows for $2p > 1$ that $\mathbb{E}N_n^2 \leq \frac{1}{2p-1}(2p)^{2n}$. In particular, from the second moment bound one finds that $P(B) \geq \inf_n (\mathbb{E}N_n)^2 / \mathbb{E}N_n^2 \geq 2p - 1 \geq 0 \ for \ p > \frac{1}{2}$. The parameter value $p_c = \frac{1}{2}$ is thus the *critical* probability for percolation on the tree graph.[1] It should be clear that the methods used for the binary tree carry over to the b-ary tree for $b = 3, 4, \ldots$ by precisely the same method, Exercise 13.

Two random variables X_1, X_2 in $L^2 = L^2(\Omega, \mathcal{F}, P)$ are said to be **uncorrelated** if their **covariance** $\text{Cov}(X_1, X_2)$ is zero, where

$$\text{Cov}(X_1, X_2) := \mathbb{E}\left[(X_1 - \mathbb{E}(X_1))(X_2 - \mathbb{E}(X_2))\right] = \mathbb{E}(X_1 X_2) - \mathbb{E}(X_1)\mathbb{E}(X_2). \tag{2.5}$$

The **variance** $\text{Var}(Y)$ of a random variable $Y \in L^2$ is defined by the average squared deviation of Y from its mean $\mathbb{E}Y$. That is,

$$\text{Var}(Y) = \text{cov}(Y, Y) = \mathbb{E}(Y - \mathbb{E}Y)^2 = \mathbb{E}Y^2 - (\mathbb{E}Y)^2.$$

The covariance term naturally appears in consideration of the variance of sums of random variables $X_j \in L^2(\Omega, \mathcal{F}, P), 1 \leq j \leq n$, i.e.,

---

[1] Criteria for percolation on the d-dimensional integer lattice is a much deeper and technically challenging problem. In the case $d = 2$ the precise identification of the critical probability for (bond) percolation as $p_c = \frac{1}{2}$ is a highly regarded mathematical achievement of Harry Kesten, see Kesten, H. (1982). For $d \geq 3$ the best known results for p_c are expressed in terms of bounds.

$$\text{Var}\left(\sum_{j=1}^{n} X_j\right) = \sum_{j=1}^{n} \text{Var}(X_j) + 2 \sum_{1 \le i < j \le n} \text{Cov}(X_i, X_j).$$

Note that if X_1 and X_2 are independent, then it follows from Theorem 2.3 that they are uncorrelated; but the converse is easily shown to be false. For the record one has the following important corollary to Theorem 2.3.

Corollary 2.4 If $X_1, \ldots, X_n$ are independent random variables with finite second moment, then

$$\text{Var}(X_1 + \cdots + X_n) = \text{Var}(X_1) + \cdots + \text{Var}(X_n).$$

Example 3 *(Chebyshev Sample Size)* Suppose that $X_1, \ldots, X_n$ is a sequence of independent and identically distributed (i.i.d.) Bernoulli $0 - 1$ valued random variables with $P(X_j = 1) = p, 1 \le j \le n$. As often happens in random polls, for example, the parameter p is unknown and one seeks an estimate based on the sample proportion $\hat{p}_n := \frac{X_1 + \cdots + X_n}{n}$. Observe that $n\hat{p}_n$ has the Binomial distribution with parameters n, p obtained by

$$P(n\hat{p}_n = k) = \sum_{(\varepsilon_1, \ldots, \varepsilon_n) \in \{0,1\}^n : \sum_{j=1}^{n} \varepsilon_j = k} P(X_1 = \varepsilon_1, \ldots, X_n = \varepsilon_n)$$

$$= \binom{n}{k} p^k (1 - p)^k, \tag{2.6}$$

for $k = 0, 1, \ldots, n$. In a typical polling application one seeks a sample size n such that

$$P(|\hat{p}_n - p| > .03) \le .05.$$

Since $\mathbb{E}|\hat{p}_n - p|^2 = var(\hat{p}_n) = np(1-p)/n^2 \le 1/4n$, the second moment Chebyshev bound yields $n = 5,556$. This is obviously much larger than that used in a standard poll ! To improve on this one may consider a fourth moment Chebyshev bound. Rather tedious calculation yields (Exercise 3)

$$\mathbb{E}|\hat{p}_n - p|^4 = n(p(1-p)^4 + p^4(1-p) + 3(n-1)p^2(1-p)^2)/n^4$$
$$\le \frac{(3n-2)}{16n^3} \le \frac{3}{16n^2}.$$

Thus the fourth moment Chebyshev bound yields a reduction to see that $n = 2,154$ is a sufficient sample size. This example will be used in subsequent chapters to explore various other inequalities involving deviations from the mean.

Definition 2.2 Let $\{X_t : t \in \Lambda\}$ be a possibly infinite family of random maps on $(\Omega, \mathcal{F}, P)$, with X_t a measurable map into $(S_t, \mathcal{S}_t)$, $t \in \Lambda$. We will say that $\{X_t : t \in \Lambda\}$ is a **family of independent maps** if every finite subfamily is a family of independent maps. That is, for all $n \geq 1$ and for every n-tuple $(t_1, t_2, \ldots, t_n)$ of distinct points in Λ, the maps $X_{t_1}, X_{t_2}, \ldots, X_{t_n}$ are independent (in the sense of (2.1)).

Definition 2.3 A sequence of independent random maps $X_1, X_2, \ldots$ is said to be **independent and identically distributed**, denoted **i.i.d.**, if the distribution of X_n does not depend on n, i.e., is the same for each $n = 1, 2, \ldots$.

Remark 2.1 The general problem of establishing the existence of infinite families $\{X_t : t \in \Lambda\}$ of random maps, including that of infinite sequences, defined on a common probability space $(\Omega, \mathcal{F}, P)$ and having consistently specified distributions of finitely many variables $(X_{t_1}, \ldots, X_{t_n})$, say, for $t_j \in \Lambda, j = 1, \ldots, n$, is treated in Chapter VIII under the guise of **Kolmogorov's extension theorem**. This is a nontrivial problem of constructing probability measures with prescribed properties on an infinite product space, also elaborated upon at the close of this chapter. Kolmogorov provided a frequently useful solution, especially for countable Λ.

Let us consider the notions of uncorrelated and independent random variables a bit more fully. Before stating the main result in this regard the following proposition provides a very useful perspective on the meaning of measurabililty with respect to a σ-field generated by random variables.

Proposition 2.5 Let $Z, Y_1, \ldots, Y_k$ be real-valued random variables on a measurable space $(\Omega, \mathcal{F})$. A random variable $Z : \Omega \to \mathbb{R}$ is $\sigma(Y_1, \ldots, Y_k)$-measurable iff there is a Borel measurable function $g : \mathbb{R}^k \to \mathbb{R}$ such that $Z = g(Y_1, \ldots, Y_k)$.

Proof If $Z = g(Y_1, \ldots, Y_k)$, then $\sigma(Y_1, \ldots, Y_k)$-measurability is clear, since for $B \in \mathcal{B}(\mathbb{R})$, $[Z \in B] = [(Y_1, \ldots, Y_k) \in g^{-1}(B)]$ and $g^{-1}(B) \in \mathcal{B}(\mathbb{R}^k)$ for Borel measurable g.

For the converse, suppose that Z is a simple $\sigma(Y_1, \ldots, Y_k)$-measurable random variable with *distinct* values $z_1, \ldots, z_m$. Then $[Z = z_j] \in \sigma(Y_1, \ldots, Y_k)$ implies that there is a $B_j \in \mathcal{B}(\mathbb{R}^k)$ such that $[Z = z_j] = [(Y_1, \ldots, Y_k) \in B_j]$, since the class of all sets of the form $[(Y_1, \ldots, Y_k) \in B], B \in B(\mathbb{R}^k), is \sigma(Y_1, \ldots, Y_k)$. $Z = \sum_{j=1}^{k} f_j(Y_1, \ldots, Y_k)$, where $f_j(y_1, \ldots, y_k) = z_j \mathbf{1}_{B_j}(y_1, \ldots, y_k)$, so that $Z = g(Y_1, \ldots, Y_k)$ with $g = \sum_{j=1}^{k} f_j$. More generally, one may use approximation by simple functions to write $Z(\omega) = \lim_{n \to \infty} Z_n(\omega)$, for each $\omega \in \Omega$, where Z_n is a $\sigma(Y_1, \ldots, Y_k)$-measurable simple function, $Z_n(\omega) = g_n(Y_1(\omega), \ldots, Y_k(\omega)), n \geq 1, \omega \in \Omega$. Let $\tilde{B} = \{(y_1, \ldots, y_k) \in \mathbb{R}^k : \lim_{n \to \infty} g_n(y_1, \ldots, y_k) \text{ exists}\}$. Then $\tilde{B} \in \mathcal{B}(\mathbb{R}^k)$ (Exercise 38). Denoting this limit by g on $\tilde{B}$ and letting g = 0 on $(\tilde{B})^c$, one has $Z(\omega) = g(Y_1(\omega), \ldots, Y_k(\omega))$. ∎

Corollary 2.6 Suppose that X_1, X_2 are two independent random maps with values in $(S_1, \mathcal{S}_1), (S_2, \mathcal{S}_2)$, respectively. Then for Borel measurable functions $g_i : S_i \to \mathbb{R}, i = 1, 2$, the random variables $Z_1 = g_1(X_1)$ and $Z_2 = g_2(X_2)$ are independent.

Proof For Borel sets B_1, B_2 one has $g_i^{-1}(B_i) \in \mathcal{S}_i, i = 1, 2$, and

$$P(Z_1 \in B_1, Z_2 \in B_2) = P(X_1 \in g_1^{-1}(B_1), X_2 \in g_2^{-1}(B_2))$$
$$= P(X_1 \in g_1^{-1}(B_1))P(X_2 \in g_2^{-1}(B_2)). \qquad (2.7)$$

Thus (2.1) follows.   ∎

Now let us return to the formulation of independence in terms of correlations. Although zero correlation is a weaker notion than statistical independence, if a sufficiently large class of functions of disjoint finite sets of random variables are uncorrelated then independence will follow. More specifically

Proposition 2.7 A family of random maps $\{X_t : t \in \Lambda\}$ (with X_t a measurable map into $(S_t, \mathcal{S}_t)$) is an independent family if and only if for every pair of disjoint finite subsets Λ_1, Λ_2 of Λ, any random variable $V_1 \in L^2(\sigma\{X_t : t \in \Lambda_1\})$ is uncorrelated with any random variable $V_2 \in L^2(\sigma\{X_t : t \in \Lambda_2\})$

Proof Observe that the content of the uncorrelation condition may be expressed multiplicatively as

$$\mathbb{E}V_1 V_2 = \mathbb{E}V_1\mathbb{E}V_2, \ V_i \in L^2(\sigma\{X_t : t \in \Lambda_i\}), i = 1, 2.$$

Suppose the uncorrelation condition holds as stated. Let $\{t_1, \ldots, t_n\} \subset \Lambda$ be an arbitrary finite subset. To establish (2.1) proceed inductively by first selecting $\Lambda_1 = \{t_1\}$, $\Lambda_2 = \{t_2, \ldots, t_n\}$, with $V_1 = \mathbf{1}_{B_1}(X_{t_1})$, $V_2 = \prod_{j=2}^n \mathbf{1}_{B_j}(X_{t_j})$, for arbitrary $B_j \in \mathcal{S}_{t_j}, 1 \leq j \leq n$. Then

$$P(X_{t_j} \in B_j, 1 \leq j \leq n) = \mathbb{E}V_1 V_2 = \mathbb{E}V_1\mathbb{E}V_2$$
$$= P(X_{t_1} \in B_1)P(X_{t_j} \in B_j, 2 \leq j \leq n).$$
$$(2.8)$$

Iterating this process establishes (2.1). For the converse one may simply apply Proposition 2.5 and its Corollary to see V_1, V_2 are independent and therefore uncorrelated.   ∎

Definition 2.4 A collection $\mathcal{C}$ of events $A \in \mathcal{F}$ is defined to be a set of **independent events** if the set of indicator random variables $\{\mathbf{1}_A : A \in \mathcal{C}\}$ is an independent collection.

The notion of independence of events may also be equivalently defined in terms of sub-σ-fields of $\mathcal{F}$.

Definition 2.5 Given $(\Omega, \mathcal{F}, P)$, a family $\{\mathcal{F}_t : t \in \Lambda\}$ of σ-fields (contained in $\mathcal{F}$) is a **family of independent σ-fields** if for every n-tuple of distinct indices $(t_1, t_2, \ldots, t_n)$ in Λ one has $P(F_{t_1} \cap F_{t_2} \cap \cdots \cap F_{t_n}) = P(F_{t_1})P(F_{t_2}) \cdots P(F_{t_n})$ for all $F_{t_i} \in \mathcal{F}_{t_i}$ $(1 \leq i \leq n)$; here n is an arbitrary finite integer, $n \leq$ cardinality of Λ.

It is straightforward to check that the independence of a family $\{X_t : t \in \Lambda\}$ of random maps is equivalent to the independence of the family $\{\sigma(X_t) : t \in \Lambda\}$ of

σ-fields $\sigma(X_t) \equiv \{[X_t \in B] : B \in \mathcal{S}_t\}$ generated by $X_t(t \in \Lambda)$, where $(S_t, \mathcal{S}_t)$ is the image space of X_t. One may also check that $\sigma(X_t)$ is the smallest sub-σ-field of measurable subsets of Ω that makes $X_t : \Omega \to S$ measurable. More generally

Definition 2.6 Suppose $\{X_t : t \in \Lambda\}$ is a collection of random maps defined on $(\Omega, \mathcal{F})$. The smallest sub-σ-field of $\mathcal{F}$ such that every $X_t, t \in \Lambda$, is measurable, denoted $\sigma(X_t : t \in \Lambda)$, is referred to as the **σ-field generated by** $\{X_t : t \in \Lambda\}$.

Proposition 2.8 Let $X_1, X_2, \ldots$ be a sequence of independent random maps with values in measurable spaces $(S_1, \mathcal{S}_1), (S_2, \mathcal{S}_2), \ldots$, respectively, and let $n_1 < n_2 < \cdots$ be a nondecreasing sequence of positive integers. Suppose that $Y_1 = f_1(X_1, \ldots, X_{n_1}), Y_2 = f_2(X_{n_1+1}, \ldots, X_{n_2}), \ldots$, where $f_1, f_2, \ldots$ are Borel-measurable functions on the respective product measure spaces $S_1 \times \cdots \times S_{n_1}, S_{n_1+1} \times \cdots \times S_{n_2}, \ldots$. Then $Y_1, Y_2, \ldots$ is a sequence of independent random variables.

Proof It suffices to check that the distribution of an arbitrary finite subset of random variables $(Y_{n_1}, \ldots Y_{n_m}), 1 \leq n_1 < n_2 < \cdots < n_m$, is product measure. But this follows readily from the distribution of $(X_1, \ldots, X_{n_m})$ being product measure, by observing for any $k \geq 2$,

$$
\begin{aligned}
&P(Y_1 \in B_1, \ldots, Y_k \in B_k) \\
&= P((X_1, \ldots, X_{n_k}) \in f_1^{-1}(B_1), \ldots, (X_{n_{k-1}+1}, \ldots, X_{n_k}) \in f_k^{-1}(B_k)) \\
&= \prod_{j=1}^{k} P(X_{n_{j-1}+1}, \ldots, X_{n_j}) \in f_j^{-1}(B_j)).
\end{aligned}
$$

Taking $k = n_m$ and $B_i = S_i$ for $n_{j-1} < i < n_j$, the assertion follows. ∎

The σ-field formulation of independence can be especially helpful in tracking independence, as illustrated by the following consequence of the $\pi - \lambda$ theorem.

Proposition 2.9 If $\{\mathcal{C}_t\}_{t \in \Lambda}$ is a family of π-systems such that $P(C_{t_1} \cap \cdots \cap C_{t_n}) = \prod_{i=1}^{n} P(C_{t_i}), C_{t_i} \in \mathcal{C}_{t_i}$, for any distinct $t_i \in \Lambda, n \geq 2$, then $\{\sigma(\mathcal{C}_t)\}_{t \in \Lambda}$ is a family of independent σ-fields.

Proof Fix a finite set $\{t_1, \ldots, t_n\} \subset \Lambda$. Fix arbitrary $C_{t_i} \in \mathcal{C}_{t_i}, 2 \leq i \leq n$. Then $\{F_1 \in \sigma(\mathcal{C}_{t_1}) : P(F_1 \cap C_{t_2} \cap \cdots \cap C_{t_n}) = P(F_1)P(C_{t_2}) \cdots P(C_{t_n})\}$ is a λ-system containing the π-system $\mathcal{C}_{t_1}$. Thus, by the $\pi - \lambda$ theorem, one has

$$
P(F_1 \cap C_{t_2} \cap \cdots \cap C_{t_n}) = P(F_1)P(C_{t_2}) \cdots P(C_{t_n}), \forall F_1 \in \sigma(\mathcal{C}_{t_1}).
$$

Now proceed inductively to obtain

$$
P(F_1 \cap F_2 \cap \cdots \cap F_n) = P(F_1)P(F_2) \cdots P(F_n), \forall F_i \in \sigma(\mathcal{C}_{t_i})\ 1 \leq i \leq n.
$$

∎

The simple example in which $\Omega = \{a, b, c, d\}$ consists of four equally probable outcomes and $C_1 = \{\{a, b\}\}$, $C_2 = \{\{a, c\}, \{a, d\}\}$, shows that the π-system requirement is indispensable. For a more positive perspective, note that if $A, B \in \mathcal{F}$ are independent events then it follows immediately that A, B^c and A^c, B^c are respective pairs of independent events, since $\sigma(\{A\}) = \{A, A^c, \emptyset, \Omega\}$ and similarly for $\sigma(\{B\})$, and each of the singletons $\{A\}$ and $\{B\}$ is a π-system. More generally, a collection of events $\mathcal{C} \subset \mathcal{F}$ is a collection of independent events if and only if $\{\sigma(\{C\}) : C \in \mathcal{C}\}$ is a collection of independent σ-fields. As a result, for example, $C_1, \ldots, C_n$ are independent events in $\mathcal{F}$ if and only if the 2^n equations

$$P(A_1 \cap A_2 \cap \cdots \cap A_n) = \prod_{i=1}^{n} P(A_i),$$

where $A_i \in \{C_i, C_i^c\}$, $1 \leq i \leq n$, are satisfied; also see Exercise 15.

Often one also needs the notion of independence of (among) several families of σ-fields or random maps.

Definition 2.7 Let Λ_i, $i \in \mathcal{I}$, be a family of index sets and, for each $i \in \mathcal{I}$, $\{\mathcal{F}_t : t \in \Lambda_i\}$ a collection of (sub) σ-fields of $\mathcal{F}$. The **families** $\{\mathcal{F}_t : t \in \Lambda_i\}_{i \in \mathcal{I}}$ are said to be **independent** (of each other) if the σ-fields $\mathcal{G}_i := \sigma(\{\mathcal{F}_t : t \in \Lambda_i\})$ generated by $\{\mathcal{F}_t : t \in \Lambda_i\}$ (i.e., $\mathcal{G}_i$ is the smallest σ-field containing $\cup_{t \in \Lambda_i} \mathcal{F}_t, i \in \mathcal{I}$),[2] are independent in the sense of Definition 2.5.

The corresponding definition of **independence of (among) families of random maps** $\{X_t : t \in \Lambda_i\}_{i \in \mathcal{I}}$ can now be expressed in terms of independence of the σ-fields $\mathcal{F}_t := \sigma(X_t), t \in \Lambda_i, i \in \mathcal{I}$.

We will conclude the discussion of independence with a return to considerations of a converse to the Borel–Cantelli lemma I. Clearly, by taking $A_n = A_1, \forall n$, then $P(A_n \ i.o.) = P(A_1) \in [0, 1]$. So there is no general theorem without some restriction on how much dependence exists among the events in the sequence. Write A_n **eventually for all** n to denote the event $[A_n^c \ i.o.]^c$, i.e., "A_n occurs for all but finitely many n."

Lemma 1 (Borel–Cantelli II) Let $\{A_n\}_{n=1}^{\infty}$ be a sequence of independent events in a probability space $(\Omega, \mathcal{F}, P)$. If $\sum_{n=1}^{\infty} P(A_n) = \infty$ then $P(A_n \ i.o.) = 1$.

Proof Consider the complementary event to get from continuity properties of P, independence of complements, and the simple bound $1 - x \leq e^{-x}, x \geq 0$, that
$1 \geq P(A_n \ i.o.) = 1 - P(A_n^c \ \text{eventually for all } n) = 1 - P(\cup_{n=1}^{\infty} \cap_{m=n}^{\infty} A_m^c) = 1 - \lim_{n \to \infty} \prod_{m=n}^{\infty} P(A_m^c) \geq 1 - \lim_{n \to \infty} \exp\{-\sum_{m=n}^{\infty} P(A_m)\} = 1$. ∎

Example 4 Suppose that $\{X_n\}_{n=1}^{\infty}$ is a sequence of independent and identically distributed (i.i.d.) Bernoulli 0 or 1-valued random variables with $P(X_1 = 1) = p > 0$. Then $P(X_n = 1 \ i.o.) = 1$ is a quick and easy consequence of Borel–Cantelli II.

---

[2]Recall that the σ-field $\mathcal{G}_i$ generated by $\cup_{t \in \Lambda_i} \mathcal{F}_t$ is referred to as the *join* σ-field and denoted $\bigvee_{t \in \Lambda_i} \mathcal{F}_t$.

Example 5 *(Random Power Series)* Consider the random formal power series

$$S(x) = \sum_{n=0}^{\infty} \varepsilon_n x^n, \tag{2.9}$$

where $\varepsilon_0, \varepsilon_1, \ldots,$ is an i.i.d. sequence of random variables, *not* a.s. zero, with $\mathbb{E} \log^+ |\varepsilon_1| < \infty$. Let us see how the Borel–Cantelli lemmas can be used to determine those values of x for which $S(x)$ is actually the almost sure limit of the sequence of partial sums, i.e., almost surely a power series in x. As a warm-up notice that if for some positive constants b, c, one has $|\varepsilon_n| \leq cb^n$ a.s. eventually for all n then a.s. convergence holds for $|x| < \frac{1}{b}$. More generally, if $|x| < 1$, then there is a number $1 < b < \frac{1}{|x|}$,

$$\sum_{n=0}^{\infty} P(|\varepsilon_n| \geq b^n) = \sum_{n=0}^{\infty} P(|\varepsilon_1| > b^n)$$

$$= \sum_{n=0}^{\infty} P(\log^+ |\varepsilon_1| > n \log^+ b)$$

$$\leq \frac{1}{\log^+ b} \int_0^{\infty} P(\log^+ |\varepsilon_1| > x) dx < \infty. \tag{2.10}$$

It follows from Borel-Cantelli I that almost surely $|\varepsilon_n| \leq b^n$ eventually for all n. In particular the series is almost surely convergent for $|x| < 1$. Conversely, if $|x| > 1$ then there is a $1 < b < |x|$ such that

$$\sum_{n=0}^{\infty} P(|\varepsilon_n x^n| \geq b^n) = \sum_{n=0}^{\infty} P(|\varepsilon_1| > (\frac{b}{|x|})^n) = \infty, \tag{2.11}$$

since $\lim_{n\to\infty} P(|\varepsilon_1| > (\frac{b}{|x|})^n) = P(|\varepsilon_1| > 0) > 0$. Thus, using Borel–Cantelli II, one sees that $P(|\varepsilon_n x^n| > b^n \text{ i.o.}) = 1$. In particular, the series is almost surely divergent for $|x| > 1$. In the case $|x| = 1$, one may choose $\delta > 0$ such that $P(|\varepsilon_1| > \delta) > 0$. Again it follows from Borel–Cantelli II that $|\varepsilon_n x^n| \equiv |\varepsilon_n| > \delta$ i.o. with probability one. Thus the series is a.s. divergent for $|x| = 1$. Analysis of the case $\mathbb{E} \log^+ \varepsilon_1 = \infty$ is left to Exercise 20.

We now come to another basic notion of fundamental importance in probability—the notion of **conditional probability** and **conditional expectation**. It is useful to consider the spaces $L^p(\Omega, \mathcal{G}, P)$, where $\mathcal{G}$ is a sub-σ-field of $\mathcal{F}$. A little thought reveals that an element of this last (Banach) space is not in general an element of $L^p(\Omega, \mathcal{F}, P)$. For if Z is $\mathcal{G}$-measurable, then the set (equivalence class) $\tilde{Z}$ of all $\mathcal{F}$-measurable random variables each of which differs from Z on at most a P-null set may contain random variables that are not $\mathcal{G}$-measurable. However, if we denote by $L^p(\mathcal{G})$ the set of all elements of $L^p(\Omega, \mathcal{F}, P)$, each equivalent to some

$\mathcal{G}$-measurable Z with $\mathbb{E}|Z|^p < \infty$, then $L^p(\mathcal{G})$ becomes a closed linear subspace of $L^p(\Omega, \mathcal{F}, P)$. In particular, under this convention, $L^2(\mathcal{G})$ is a closed linear subspace of $L^2 \equiv L^2(\mathcal{F}) \equiv L^2(\Omega, \mathcal{F}, P)$, for every σ-field $\mathcal{G} \subset \mathcal{F}$. The first definition below exploits the Hilbert space structure of L^2 through the projection theorem (see Appendix C) to obtain the conditional expectation of X, given $\mathcal{G}$, as the orthogonal projection of X onto $L^2(\mathcal{G})$. Since this projection, as an element of $L^2(\Omega, \mathcal{F}, P)$, is an equivalence class which contains in general $\mathcal{F}$-measurable functions which may not be $\mathcal{G}$-measurable, one needs to select from it a $\mathcal{G}$-measurable version, i.e., an equivalent element of $L^2(\Omega, \mathcal{G}, P)$. If $\mathcal{F}$ is P-complete, and so are all its sub-sigma fields, then such a modification is not necessary.

Definition 2.8 (*First Definition of Conditional Expectation (on L^2)*). Let $X \in L^2$ and $\mathcal{G}$ be a sub-σ-field of $\mathcal{F}$. Then a **conditional expectation of X given $\mathcal{G}$**, denoted by $\mathbb{E}(X|\mathcal{G})$, is a $\mathcal{G}$-measurable version of the orthogonal projection of X onto $L^2(\mathcal{G})$.

Intuitively, $\mathbb{E}(X|\mathcal{G})$ is the best prediction of X (in the sense of least mean square error), given information about the experiment coded by events that constitute $\mathcal{G}$. In the case $\mathcal{G} = \sigma\{Y\}$ is a random map with values in a measurable space $(S, \mathcal{S})$, this makes $\mathbb{E}(X|\mathcal{G})$ a version of a Borel measurable function of Y. This is because of the Proposition 2.5.

As simple examples, consider the sub-σ-fields $\mathcal{G}_0 = \{\Omega, \emptyset\}$, $\sigma(X)$, and $\mathcal{F}$. (The σ-field $\mathcal{G}_0$, or the one comprising only P-null sets and their complements, is called the **trivial σ-field**). One has for all $X \in L^2$,

$$\mathbb{E}(X|\mathcal{G}_0) = \mathbb{E}(X), \qquad \mathbb{E}(X|\sigma(X)) = X, \qquad \mathbb{E}(X|\mathcal{F}) = X. \qquad (2.12)$$

The first of these follows from the facts that (i) the only $\mathcal{G}_0$-measurable functions are constants, and (ii) $\mathbb{E}(X - c)^2$ is minimized, uniquely, by the constant $c = \mathbb{E}X$. The other two relations in (2.12) are obvious from the definition.

For another perspective, if $X \in L^2$, then the orthogonal projection of X onto $1^\perp \equiv \{Y \in L^2 : Y \perp 1\} = \{Y \in L^2 : \mathbb{E}Y = 0\}$ is given by $X - \mathbb{E}(X)$, or equivalently, the projection of X onto the space of (equivalence classes of) constants is $\mathbb{E}(X)$.

In addition to the intuitive interpretation of $\mathbb{E}(X|\mathcal{G})$ as a best predictor of X, there is also an interpretation based on *smoothing* in the sense of averages that extends beyond L^2. For example, as noted above, $\mathbb{E}(X|\{\emptyset, \Omega\}) = \mathbb{E}X = \int_\Omega X(\omega)P(d\omega)$. In particular, this may be viewed as a *smoothing* of the function X over all sample points $\omega \in \Omega$. Similarly, and $B \in \mathcal{F}$, $0 < P(B) < 1$, and $X \in L^2$, one may check that (Exercise 24)

$$\mathbb{E}(X|\{\emptyset, B, B^c, \Omega\}) = \left(\frac{1}{P(B)} \int_B X dP\right) \mathbf{1}_B + \left(\frac{1}{P(B^c)} \int_{B^c} X dP\right) \mathbf{1}_{B^c}. \quad (2.13)$$

It is worth repeating that the conditional expectation is well defined only up to a $\mathcal{G}$-measurable P-null set. That is, if X is a version of $\mathbb{E}(X|\mathcal{G})$, then so is any

$\mathcal{G}$-measurable Y such that $P(Y \neq X) = 0$. Thus the conditional expectation $\mathbb{E}(X|\mathcal{G})$ is uniquely defined only as an element of $L^2(\Omega, \mathcal{G}, P)$. We will, however, continue to regard $\mathbb{E}(X|\mathcal{G})$ as a $\mathcal{G}$-measurable version of the orthogonal projection of X onto $L^2(\mathcal{G})$. The orthogonality condition is expressed by

$$\int_{\Omega} (X - \mathbb{E}(X|\mathcal{G}))Z dP = 0 \quad \forall Z \in L^2(\Omega, \mathcal{G}, P), \tag{2.14}$$

or

$$\int_{\Omega} XZdP = \int_{\Omega} \mathbb{E}(X|\mathcal{G})Z dP \quad \forall Z \in L^2(\Omega, \mathcal{G}, P). \tag{2.15}$$

In particular, with $Z = \mathbf{1}_G$ for $G \in \mathcal{G}$ in (2.15), one has

$$\int_G X dP = \int_G \mathbb{E}(X|\mathcal{G})dP \quad \forall G \in \mathcal{G}. \tag{2.16}$$

It is simple to check that for $X \in L^2(\Omega, \mathcal{F}, P)$, (2.16) is equivalent to (2.14) or (2.15). But (2.16) makes sense for all $X \in L^1 \equiv L^1(\Omega, \mathcal{F}, P)$, which leads to the second, more general, definition.

Definition 2.9 (*Second Definition of Conditional Expectation (on L^1)*). Let $X \in L^1(\Omega, \mathcal{F}, P)$, and let $\mathcal{G}$ be a sub-σ-field of $\mathcal{F}$. A $\mathcal{G}$-measurable random variable is said to be a **conditional expectation of X given $\mathcal{G}$**, denoted by $\mathbb{E}(X|\mathcal{G})$, if (2.16) holds.

That $\mathbb{E}(X|\mathcal{G})$ exists for $X \in L^1$, and is well defined a.e., may be proved by letting $X_n \in L^2$ converge to X in L^1 (i.e., $\|X_n - X\|_1 \to 0$ as $n \to \infty$), applying (2.16) to X_n, and letting $n \to \infty$. Note that L^2 is dense in L^1 (Exercise 39). Alternatively, one may apply the Radon–Nikodym theorem to the finite (signed) measure $\nu(G) := \int_G X dP$ on $(\Omega, \mathcal{G})$, which is absolutely continuous with respect to P (restricted to $(\omega, \mathcal{G})$), i.e., if $P(G) = 0$, then $\nu(G) = 0$. Hence there exists a $\mathcal{G}$-measurable function, denoted $\mathbb{E}(X|\mathcal{G})$, such that (2.16) holds. Viewed as an element of $L^1(\Omega, \mathcal{G}, P)$, $\mathbb{E}(X|\mathcal{G})$ is unique.

There are variations on the requirement (2.16) in the definition of conditional expectation that may be noted. In particular, a version of $\mathbb{E}(X|\mathcal{G})$ is uniquely determined by the condition that (1) it be a $\mathcal{G}$-measurable random variable on Ω and, (2) it satisfy the equivalent version

$$\mathbb{E}\{XZ\} = \mathbb{E}\{\mathbb{E}(X|\mathcal{G})Z\} \quad \forall Z \in \Gamma, \tag{2.17}$$

of (2.16), where, in view of the Radon–Nikodym theorem, Γ may be taken as the set of indicator random variables $\{\mathbf{1}_G : G \in \mathcal{G}\}$. There is some flexibility in the choice of Γ as a set of *test functions* when verifying and/or using the definition. Depending on the context, Γ may more generally be selected as (i) the collection of all bounded nonnegative $\mathcal{G}$-measurable random variables Z on Ω or (ii) the collection of all bounded $\mathcal{G}$ measurable random variables Z on Ω, or even the collection of all nonnegative, bounded $\mathcal{G}$ measurable random variables Z on Ω, for example, as convenient. Certainly all of these would include the indicator random variables and therefore more than sufficient. Moreover, by simple function approximation, the defining condition extends to such choices for Γ and may be applied accordingly in computations involving conditional expectations.

The following properties of $\mathbb{E}(X|\mathcal{G})$ are important and, for the most part, immediate consequences of the definitions. As illustrated by the proofs, the basic approach to the determination of $\mathbb{E}(X|\mathcal{G})$ may be by "guessing"a $\mathcal{G}$-measurable random variable and then checking that it satisfies (2.17), or by starting from the left side of (2.17) and making calculations/deductions that lead to the right side, with an explicit $\mathcal{G}$-measurable random variable that reveals $\mathbb{E}(X|\mathcal{G})$. To check almost sure properties, on the other hand, an approach is to show that the event G, say, for which the desired property fails, has probability zero. These alternative approaches are illustrated in the proofs of the properties given in the following theorem.

Theorem 2.10 Let $(\Omega, \mathcal{F}, P)$ be a probability space, $L^1 = L^1(\Omega, \mathcal{F}, P)$, $\mathcal{G}, \mathcal{D}$ sub-σ-fields of $\mathcal{F}$, $X, Y \in L^1$. Then the following holds, almost surely (P):

(a) $\mathbb{E}(X|\{\Omega, \phi\}) = \mathbb{E}(X)$.
(b) $\mathbb{E}[\mathbb{E}(X|\mathcal{G})] = \mathbb{E}(X)$.
(c) If X is $\mathcal{G}$-measurable, then $\mathbb{E}(X|\mathcal{G}) = X$.
(d) *(Linearity)*. $\mathbb{E}(cX + dY|\mathcal{G}) = c\mathbb{E}(X|\mathcal{G}) + d\mathbb{E}(Y|\mathcal{G})$ for all constants c, d.
(e) *(Order)*. If $X \le Y$ a.s., then $\mathbb{E}(X|\mathcal{G}) \le \mathbb{E}(Y|\mathcal{G})$.
(f) *(Smoothing)*. If $\mathcal{D} \subset \mathcal{G}$, then $\mathbb{E}[\mathbb{E}(X|\mathcal{G})|\mathcal{D}] = \mathbb{E}(X|\mathcal{D})$.
(g) If $XY \in L^1$ and X is $\mathcal{G}$-measurable, then $\mathbb{E}(XY|\mathcal{G}) = X\mathbb{E}(Y|\mathcal{G})$.
(h) If $\sigma(X)$ and $\mathcal{G}$ are independent then $\mathbb{E}(X|\mathcal{G}) = \mathbb{E}(X)$.
(i) *(Conditional Jensen's Inequality)*. Let ψ be a convex function on an interval J such that ψ has finite right- (or left-)hand derivative(s) at left (or right) endpoint(s) of J if J is not open. If $P(X \in J) = 1$, and if $\psi(X) \in L^1$, then

$$\psi(\mathbb{E}(X|\mathcal{G})) \le \mathbb{E}(\psi(X)|\mathcal{G}).$$

(j) *(Contraction)*. For $X \in L^p(\Omega, \mathcal{F}, P)$, $p \ge 1$, $\|\mathbb{E}(X|\mathcal{G})\|_p \le \|X\|_p \, \forall \, p \ge 1$.
(k) *(Convergences)*.

(k1) If $X_n \to X$ in L^p then $\mathbb{E}(X_n|\mathcal{G}) \to \mathbb{E}(X|\mathcal{G})$ in L^p $(p \ge 1)$.
(k2) *(Conditional Monotone Convergence)* If $0 \le X_n \uparrow X$ a.s., X_n and $X \in L^1$ $(n \ge 1)$, then $\mathbb{E}(X_n|\mathcal{G}) \uparrow \mathbb{E}(X|\mathcal{G})$ a.s. and $\mathbb{E}(X_n|\mathcal{G}) \to \mathbb{E}(X|\mathcal{G})$ in L^1.
(k3) *(Conditional Dominated Convergence)* If $X_n \to X$ a.s. and $|X_n| \le Y \in L^1$, then $\mathbb{E}(X_n|\mathcal{G}) \to \mathbb{E}(X|\mathcal{G})$ *a.s.*

(ℓ) *(Substitution Property)* Let U, V be random maps into $(S_1, \mathcal{S}_1)$ and $(S_2, \mathcal{S}_2)$, respectively. Let ψ be a measurable real-valued function on $(S_1 \times S_2, \mathcal{S}_1 \otimes \mathcal{S}_2)$. If U is $\mathcal{G}$-measurable, $\sigma(V)$ and $\mathcal{G}$ are independent, and $\mathbb{E}|\psi(U, V)| < \infty$, then one has that $\mathbb{E}[\psi(U, V)|\mathcal{G}] = h(U)$, where $h(u) := E\psi(u, V)$.

(m) $\mathbb{E}(X|\sigma(Y, Z)) = \mathbb{E}(X|\sigma(Y))$ if (X, Y) and Z are independent.

Proof (a–h) follow easily from the definition. In the case of (e) take $G = [\mathbb{E}(X|\mathcal{G}) > \mathbb{E}(Y|\mathcal{G})] \in \mathcal{G}$ in the definition (2.16) of conditional expectation with X replaced by $Y - X$. To prove (g), let $Z \in \Gamma$, the set of bounded, $\mathcal{G}$-measurable random variables. Then, since X and XY are both integrable, $XZ \in L^1(\Omega, \mathcal{G}, P)$ and $XYZ \in L^1(\Omega, \mathcal{F}, P)$. Thus, $X\mathbb{E}(Y|\mathcal{G})$ is $\mathcal{G}$-measurable, and $\mathbb{E}(ZX\mathbb{E}(Y|\mathcal{G})) = \mathbb{E}(ZXY) = \mathbb{E}(Z\mathbb{E}(XY|\mathcal{G}))$. To prove (h), again let $Z \in \Gamma$ be a bounded, $\mathcal{G}$-measurable random variable. By independence of $\sigma(X)$ and $\mathcal{G}$, one has that X and Z are independent random variables; namely since $[Z \in B] \in \mathcal{G}$ for all Borel sets B, one has $P(X \in A, Z \in B) = P(X \in A)P(Z \in B)$ for all Borel sets A, B. Thus one has, using Theorem 2.3, $\mathbb{E}(ZX) = \mathbb{E}(Z)\mathbb{E}(X) = \mathbb{E}(Z\mathbb{E}(X))$. Since the constant $\mathbb{E}(X)$ is $\mathcal{G}$-measurable, indeed, constant random variables are measurable with respect to any σ-field, (h) follows by the defining property (2.17).

For (i) use the line of support Lemma 3 from Chapter I. If J does not have a right endpoint, take $x_0 = \mathbb{E}(X|\mathcal{G})$, and $m = \psi^+(\mathbb{E}(X|\mathcal{G}))$, where ψ^+ is the right-hand derivative of ψ, to get $\psi(X) \geq \psi(\mathbb{E}(X|\mathcal{G})) + \psi^+(\mathbb{E}(X|\mathcal{G}))(X - \mathbb{E}(X|\mathcal{G}))$. Now take the conditional expectation, given $\mathcal{G}$, and use (e) to get (i). Similarly, if J does not have a left endpoint, take $m = \psi^-(\mathbb{E}(X|\mathcal{G}))$. If J has both right and left endpoints, say $a < b$, let $m = \psi^+(\mathbb{E}(X|\mathcal{G}))$ on $[\mathbb{E}(X|\mathcal{G}) \neq b]$ and $m = \psi^-(\mathbb{E}(X|\mathcal{G}))$ on $[\mathbb{E}(X|\mathcal{G}) \neq a]$.

The contraction property (j) follows from this by taking $\psi(x) = |x|^p$ in the conditional Jensen inequality (i), and then taking expectations on both sides. The first convergence in (k) follows from (j) applied to $X_n - X$. The second convergence in (k) follows from the order property (e), and the monotone convergence theorem. The L^1 convergence in (k3) follows from (k1). For the a.s. convergence in (k3), let $Z_n := \sup\{|X_m - X| : m \geq n\}$. Then $Z_n \leq |X| + |Y|$, $|X| + |Y| - Z_n \uparrow |X| + |Y|$ a.s., so that by (k_2), $\mathbb{E}(|X| + |Y| - Z_n|\mathcal{G}) \uparrow \mathbb{E}(|X| + |Y||\mathcal{G})$ a.s. Hence $\mathbb{E}(Z_n|\mathcal{G}) \downarrow 0$ a.s., and by (e), $|\mathbb{E}(X_n|\mathcal{G}) - \mathbb{E}(X|\mathcal{G})| \leq \mathbb{E}(|X_n - X||\mathcal{G}) \leq \mathbb{E}(Z_n|\mathcal{G}) \to 0$ a.s.

If one takes $\mathcal{G} = \sigma(U)$, then (ℓ) follows by the Fubini–Tonelli theorem (if one uses the change of variables formula to do integrations on the product space $(S_1 \times S_2, \mathcal{S}_1 \otimes \mathcal{S}_2, Q_1 \times Q_2)$, where Q_1, Q_2 are the distributions of U and V, respectively). For the general case, first consider ψ of the form $\psi(u, v) = \sum_{i=1}^n f_i(u)g_i(v)$ with f_i and g_i bounded and measurable (on $(S_1, \mathcal{S}_1)$ and $(S_2, \mathcal{S}_2)$, respectively), $1 \leq i \leq n$. In this case, for every $G \in \mathcal{G}$, one has $h(U) = \sum_{i=1}^n f_i(U)\mathbb{E}g_i(V)$, and

$$
\begin{aligned}
\int_G \psi(U, V)dP &\equiv \mathbb{E}\left(\mathbf{1}_G \sum_{i=1}^n f_i(U)g_i(V)\right) \\
&= \sum_{i=1}^n \mathbb{E}(\mathbf{1}_G f_i(U) \cdot g_i(V)) = \sum_{i=1}^n \mathbb{E}(\mathbf{1}_G f_i(U)) \cdot \mathbb{E}g_i(V) \\
&= \mathbb{E}\left(\mathbf{1}_G \left\{\sum_{i=1}^n f_i(U) \cdot \mathbb{E}g_i(V)\right\}\right) = \mathbb{E}(\mathbf{1}_G h(U)) \\
&\equiv \int_G h(U)dP.
\end{aligned}
$$

The case of arbitrary $\psi(U, V) \in L^1(\Omega, \mathcal{F}, P)$ follows by the convergence result (i), noting that functions of the form $\sum_{i=1}^n f_i(u)g_i(v)$ are dense in $L^1(S_1 \times S_2, S_1 \otimes S_2, Q_1 \times Q_2)$ (Exercise 2).

For the proof of (m) observe that for bounded, measurable g, one has using the substitution property that $\mathbb{E}(Xg(Y, Z)) = \mathbb{E}(\mathbb{E}[Xg(Y, Z)|\sigma(Z)]) = \mathbb{E}\varphi(Z)$, where $\varphi(z) = \mathbb{E}(Xg(Y, z)) = \mathbb{E}(\mathbb{E}[Xg(Y, z)|\sigma(Y)]) = \mathbb{E}(g(Y, z)\mathbb{E}[X|\sigma(Y)])$. In particular, $\mathbb{E}(Xg(Y, Z)) = \mathbb{E}(\mathbb{E}[X|\sigma(Y)]g(Y, Z))$ completes the proof of (m). ∎

The following inequality illustrates a clever application of these properties, including conditional Jensen's inequality, for its proof.[3] First, let $X \in L^p$, $p > 1$, be a nonnegative random variable and define

$$v_p(X) = \mathbb{E}X^p - (\mathbb{E}X)^p.$$

In particular $v_2(X)$ is the machine formula for variance of X, and the Neveu–Chauvin inequality is the all-important additive equality for variance of a sum of independent random variables; i.e., Corollary 2.4.

Proposition 2.11 *(Neveu–Chauvin Inequality)* Let $X_1, X_2, \ldots X_n$ be nonnegative, independent random variables in L^p for some $p > 1$, and let $c_1, \ldots, c_n$ be nonnegative constants. Then, for $1 < p \leq 2$,

$$v_p(\sum_{j=1}^n c_j X_j) \leq \sum_{j=1}^n c_j^p v_p(X_j).$$

Proof By induction it is sufficient to establish $v_p(X + Y) \leq v_p(X) + v_p(Y)$ for nonnegative independent random variables X, Y. That is, one must show

$$\mathbb{E}(X + Y)^p - (\mathbb{E}X + \mathbb{E}Y)^p \leq \mathbb{E}X^p + \mathbb{E}Y^p - (\mathbb{E}X)^p - (\mathbb{E}Y)^p.$$

Noting the concavity of $x \to (x+y)^p - x^p$ on $[0, \infty)$ for fixed $y \geq 0$ and $1 < p \leq 2$, it follows from the substitution property and Jensen's inequality that

$$\mathbb{E}\{(X + Y)^p - X^p|\sigma(Y)\} \leq [\mathbb{E}(X) + Y]^p - [\mathbb{E}X]^p.$$

Thus, taking expected values, using independence and properties of conditional expectation,

$$\mathbb{E}(X + Y)^p - \mathbb{E}X^p \leq \mathbb{E}(\mathbb{E}X + Y)^p - (\mathbb{E}X)^p.$$

Applying this formula to Y and $\mathbb{E}X$ in place of X and Y, respectively, one has

$$\mathbb{E}(Y + \mathbb{E}X)^p - \mathbb{E}Y^p \leq (\mathbb{E}Y + \mathbb{E}X)^p - (\mathbb{E}Y)^p.$$

---

[3]This inequality appears in J. Neveu (1988): Multiplicative martingales for spatial branching processes, *Seminar on Stochastic Processes*, 223–242, with attribution to joint work with Brigitte Chauvin.

Thus

$$\mathbb{E}(X+Y)^p \leq \mathbb{E}(\mathbb{E}X+Y)^p + \mathbb{E}X^p - (\mathbb{E}X)^p$$
$$\leq \mathbb{E}Y^p + (\mathbb{E}Y + \mathbb{E}X)^p - (\mathbb{E}X)^p - (\mathbb{E}Y)^p + \mathbb{E}X^p$$
$$= (\mathbb{E}Y + \mathbb{E}X)^p + v_p(X) + v_p(Y).$$

As noted at the outset, this inequality is sufficient for the proof. ∎

Specializing the notion of conditional expectation to indicator functions $\mathbf{1}_A$ of sets A in $\mathcal{F}$, one defines the **conditional probability of A given** $\mathcal{G}$, denoted by $P(A|\mathcal{G})$, by

$$P(A|\mathcal{G}) := \mathbb{E}(\mathbf{1}_A|\mathcal{G}), \qquad A \in \mathcal{F}. \tag{2.18}$$

As before, $P(A|\mathcal{G})$ is a (unique) element of $L^1(\Omega, \mathcal{G}, P)$, and thus defined only up to "equivalence" by the (second) definition (2.16). That is, there are in general different versions of (2.18) differing from one another only on P-null sets in $\mathcal{G}$. In particular, the orthogonality condition may be expressed as follows:

$$P(A \cap G) = \int_G P(A|\mathcal{G})(\omega)P(d\omega), \quad \forall G \in \mathcal{G}. \tag{2.19}$$

It follows from properties (d), (e), (h) (linearity, order, and monotone convergence) in Theorem 2.10 that (outside $\mathcal{G}$-measurable P-null sets)

$$0 \leq P(A|\mathcal{G}) \leq 1, \quad P(\phi|\mathcal{G}) = 0, \quad P(\Omega|\mathcal{G}) = 1, \tag{2.20}$$

and that for every countable disjoint sequence $\{A_n\}_{n=1}^\infty$ in $\mathcal{F}$,

$$P(\cup_n A_n|\mathcal{G}) = \sum_n P(A_n|\mathcal{G}). \tag{2.21}$$

In other words, conditional probability, given $\mathcal{G}$, has properties like those of a probability measure. Indeed, under certain conditions one may choose for each $A \in \mathcal{F}$ a version of $P(A|\mathcal{G})$ such that $A \to P(A|\mathcal{G})(\omega)$ is a probability measure on $(\Omega, \mathcal{F})$ for every $\omega \in \Omega$. However, such a probability measure may not exist in the full generality in which conditional expectation is defined.[4] The technical difficulty in constructing the conditional probability measure (for each $\omega \in \Omega$) is that each one of the relations in (2.20) and (2.21) holds outside a P-null set, and individual P-null sets may pile up to a nonnull set. Such a probability measure, when it exists, is called a **regular conditional probability measure given** $\mathcal{G}$, and denoted by $P^{\mathcal{G}}(A)(\omega)$. It is more generally available as a probability measure (for each ω outside a P-null set) on appropriate sub-σ-fields of $\mathcal{F}$ (even if it is not a probability measure on all

---

[4]Counterexamples have been constructed, see for example, Halmos (1950), p. 210.

of $\mathcal{F}$). An important case occurs under the terminology of a **regular conditional distribution** of a random map Z (on $(\Omega, \mathcal{F}, P)$ into some measurable space $(S, \mathcal{S})$).

Definition 2.10 Let Y be a random map on $(\Omega, \mathcal{F}, P)$ into $(S, \mathcal{S})$. Let $\mathcal{G}$ be a sub-σ-field of $\mathcal{F}$. A **regular conditional distribution of Y given $\mathcal{G}$** is a function $(\omega, C) \to Q^{\mathcal{G}}(\omega, C) \equiv P^{\mathcal{G}}([Y \in C])(\omega)$ on $\Omega \times S$ such that

(i) $\forall\, C \in \mathcal{S}$, $Q^{\mathcal{G}}(\cdot, C) = P([Y \in C]|\mathcal{G})$ a.s. (and $Q^{\mathcal{G}}(\cdot, C)$ is $\mathcal{G}$-measurable),
(ii) $\forall\, \omega \in \Omega$, $C \to Q^{\mathcal{G}}(\omega, C)$ is a probability measure on $(S, \mathcal{S})$.

The following definition provides a topological framework in which one can be assured existence of a regular version of the conditional distribution of a random map.

Definition 2.11 A topological space S whose topology can be induced by a metric is said to be **metrizable**. If S is metrizable as a complete and separable metric space then S is referred to as a **Polish space**.

For our purposes a general existence theorem as the Doob–Blackwell theorem[5] stated in the footnote will be unnecessary for the present text since we will have an explicit expression of $Q^{\mathcal{G}}$ given directly when needed. Once $Q^{\mathcal{G}}$ is given, one can calculate $\mathbb{E}(f(Y)|\mathcal{G})$ (for arbitrary functions f on $(S, \mathcal{S})$ such that $f(Y) \in L^1$) as

$$\mathbb{E}(f(Y)|\mathcal{G}) = \int f(y) Q^{\mathcal{G}}(\cdot, dy). \qquad (2.22)$$

This formula holds for $f(y) = \mathbf{1}_C(y)\ \forall\, C \in \mathcal{S}$ by definition. The general result follows by approximation of f by simple functions, using linearity and convergence properties of conditional expectation (and of corresponding properties of integrals with respect to a probability measure $Q^{\mathcal{G}}(\omega, \cdot)$). Combining (2.22) with Theorem 2.10(b) yields the so-called **disintegration formula**

$$\mathbb{E}(f(Y)) = \int_{\Omega} \int f(y) Q^{\mathcal{G}}(\omega, dy) P(d\omega). \qquad (2.23)$$

The conditional Jensen inequality (i) of Theorem 2.10 follows from the existence of a regular conditional distribution of X, given $\mathcal{G}$. The following simple examples tie up the classical concepts of conditional probability with the more modern general framework presented above.

Example 6 Let $B \in \mathcal{F}$ be such that $P(B) > 0$, $P(B^c) > 0$. Let $\mathcal{G} = \sigma(B) \equiv \{\Omega, B, B^c, \emptyset\}$. Then for every $A \in \mathcal{F}$ one has

---

[5]The Doob–Blackwell theorem provides the existence of a regular conditional distribution of a random map Y, given a σ-field $\mathcal{G}$, taking values in a Polish space equipped with its Borel σ-field $\mathcal{B}(S)$. For a proof, see Breiman (1968), pp. 77–80.

$$P(A|\mathcal{G})(\omega) = \begin{cases} P(A|B) := \frac{P(A\cap B)}{P(B)}, & \text{if } \omega \in B \\ P(A|B^c) := \frac{P(A\cap B^c)}{P(B^c)}, & \text{if } \omega \in B^c. \end{cases}$$

More generally, let $\{B_n : n = 1, 2, \dots\}$ be a countable disjoint sequence in $\mathcal{F}$ such that $\cup_n B_n = \Omega$, called a **partition** of Ω. Let $\mathcal{G} = \sigma(\{B_n : n \geq 1\})$ ($\mathcal{G}$ is the class of all unions of sets in this countable collection). Then for every A in $\mathcal{F}$, assuming $P(B_n) > 0$, one has

$$P(A|\mathcal{G})(\omega) = \frac{P(A \cap B_n)}{P(B_n)} \quad \text{if } \omega \in B_n. \tag{2.24}$$

If $P(B_n) = 0$ then for $\omega \in B_n$, define $P(A|\mathcal{G})(\omega)$ to be some constant, say c, chosen arbitrarily (Exercise 2).

Remark 2.2 Let $Y \in L^1(\Omega, \mathcal{F}, P)$ and suppose X is a random map on $(\Omega, \mathcal{F}, P)$ with values in $(S, \mathcal{S})$. In view of Proposition 2.5, $\mathbb{E}(Y|\sigma(X))$ is a function of X, say $f(X)$, and thus constant on each event $[X = x]$, $x \in S$; i.e., $\mathbb{E}(Y|\sigma(X))(\omega) = f(X(\omega)) = f(x)$, $\omega \in [X = x] = \{\omega \in \Omega : X(\omega) = x\}$. In particular, the notation $\mathbb{E}(Y|X = x)$ may be made precise by defining $\mathbb{E}(Y|X = x) := f(x)$, $x \in S$.

Example 7 (*A Canonical Probability Space*) Let $(S_i, \mathcal{S}_i, \mu_i), i = 1, 2$ be two σ-finite measure spaces, that may serve as the image spaces of a pair of random maps. The *canonical probability model* is constructed on the image space as follows. Let $\Omega = S_1 \times S_2$, $\mathcal{F} = \mathcal{S}_1 \otimes \mathcal{S}_2$. Assume that the probability P is absolutely continuous with respect to $\mu = \mu_1 \times \mu_2$ with density f i.e., f is a nonnegative $\mathcal{F}$-measurable function such that $\int_\Omega f \, d\mu = 1$, and $P(A) = \int_A f \, du$, $A \in \mathcal{F}$. One may view P as the distribution of the joint coordinate maps (X, Y), where $X(\omega) = x$, $Y(\omega) = y$, for $\omega = (x, y) \in S_1 \times S_2$. The σ-field $\mathcal{G} = \{B \times S_2 : B \in \mathcal{S}_1\}$ is the σ-field generated by the first coordinate map X. A little thought leads naturally to a reasonable guess for a (regular) conditional distribution of Y given $\sigma(X)$. Namely, for every event $A = [Y \in C] = S_1 \times C$ ($C \in \mathcal{S}_2$), one has

$$P(A|\mathcal{G})(\omega) = \frac{\int_C f(x, y)\mu_2(dy)}{\int_{S_2} f(x, y')\mu_2(dy')} \quad \text{if } \omega = (x, v) (\in \Omega). \tag{2.25}$$

To check this, first note that by the Fubini–Tonelli theorem, the function D defined by the right-hand side of (2.25) is $\mathcal{G}$-measurable. Second, for every nonnegative bounded Borel measurable g on S_1, i.e., $\mathcal{G}$-measurable test random variables $Z = g(X)$, one has

$$
\begin{aligned}
\mathbb{E}(ZD) &= \int_{S_1 \times S_2} g(x) \frac{\int_C f(x,y)\mu_2(dy)}{\int_{S_2} f(x,y)\mu_2(dy)} f(x,y')\mu_1(dx)\mu_2(dy') \\
&= \int_{S_1} g(x) \left\{ \int_{S_2} \frac{\int_C f(x,y)\mu_2(dy)}{\int_{S_2} f(x,y')\mu_2(dy')} f(x,y')\mu_2(dy') \right\} \mu_1(dx) \\
&= \int_{S_1} g(x) \left\{ \frac{\int_C f(x,y)\mu_2(dy)}{\int_{S_2} f(x,y')\mu_2(dy')} \cdot \int_{S_2} f(x,y')\mu_2(dy') \right\} \mu_1(dx) \\
&= \int_{S_1} g(x) \left\{ \int_C f(x,y)\mu_2(dy) \right\} \mu_1(dx) = \mathbb{E}(\mathbf{1}_{S_1 \times C} g(X)) \\
&= \mathbb{E}(Z\mathbf{1}_{[Y \in C]}).
\end{aligned}
$$

In particlular, $P(A|\mathcal{G}) = \mathbb{E}(\mathbf{1}_A|\mathcal{G}) = D$. The function $f(x,y)/\int_{S_2} f(x,y)\mu_2(dy)$ is called the **conditional pdf of Y given $X = x$**, and denoted by $f(y|x)$; i.e., the conditional pdf is simply the joint density (section) $y \to f(x,y)$ normalized to a probability density by dividing by the (marginal) pdf $f_X(x) = \int_{S_2} f(x,y)\mu_2(dy)$ of X. Let $A \in \mathcal{F} = S_1 \otimes S_2$. By the same calculations using Fubini–Tonelli one more generally obtains

$$
P(A|\mathcal{G})(\omega) = \int_{A_x} f(y|x)\mu_2(dy) \equiv \frac{\int_{A_x} f(x,y)\mu_2(dy)}{\int_{S_2} f(x,y)\mu_2(dy)} \quad \text{if } \omega \equiv (x,y'), \quad (2.26)
$$

where $A_x = \{ y \in S_2 : (x,y) \in A \}$.

One may change the perspective here a little and let $(\Omega, \mathcal{F}, P)$ be any probability space on which are defined two maps X and Y with values in $(S_1, \mathcal{S}_1)$ and $(S_2, \mathcal{S}_2)$, respectively. If the (joint) distribution of (X, Y) on $(S_1 \times S_2, \mathcal{S}_1 \otimes \mathcal{S}_2)$ has a pdf f with respect to a product measure $\mu_1 \times \mu_2$, where μ_i is a σ-finite measure on $(S, \mathcal{S}_i)$, $i = 1, 2$, then for $\mathcal{G} = \sigma(X)$, after using the change of variable formula mapping $\Omega \to S_1 \times S_2$, precisely the same calculations show that the (regular) conditional distribution of Y given $\mathcal{G}$ (or "given X") is given by

$$
P([Y \in C]|\mathcal{G})(\omega) = \frac{\int_C f(x,y)\mu_2(dy)}{\int_{S_2} f(x,y)\mu_2(dy)} \quad \text{if } X(\omega) = x, \quad (2.27)
$$

i.e., if $\omega \in [X = x] \equiv X^{-1}\{x\}$, $x \in S_1$. Note that the conditional probability is constant on $[X = x]$ as required for $\sigma(X)$-measurability; cf Proposition 2.5.

Two particular frameworks in which conditional probability and conditional expectation are very prominent are those of (i) Markov processes and (ii) martingales. The former is most naturally expressed as the property that the conditional distribution of future states of a process, given the past and present states coincides with the conditional distribution given the present, and the latter is the property that the conditional expectation of a future state given past and present states, is simply the present. The Markov property is illustrated in the next example, and martingales are the topic of the next chapter.

Example 8 *(Markov Property for General Random Walks on* $\mathbb{R}^k$ *)* Let $\{Z_n : n \geq 1\}$ be a sequence of independent and identically distributed (i.i.d.) k-dimensional random vectors defined on a probability space $(\Omega, \mathcal{F}, P)$. Let μ denote the distribution of Z_1 (hence of each Z_n). For arbitrary $x \in \mathbb{R}^k$, **a random walk starting at x with step-size distribution** μ is defined by the sequence $S_n^x := x + Z_1 + \cdots + Z_n$ $(n \geq 1)$, $S_0^x = x$.

For notational simplicity we will restrict to the case of $k = 1$ dimensional random walks, however precisely the same calculations are easily seen to hold for arbitrary $k \geq 1$ (Exercise 2). Let Q_x denote the distribution of $\{S_n^x := n \geq 0\}$ on the product space $(\mathbb{R}^\infty, \mathcal{B}^\infty)$. Here $\mathcal{B}^\infty$ is the σ-field generated by cylinder sets of the form $C = B_m \times \mathbb{R}^\infty := \{\mathbf{y} = (y_0, y_1, \dots) \in \mathbb{R}^\infty; (y_0, y_1, \dots, y_m) \in B_m\}$ with B_m a Borel subset of $\mathbb{R}^{m+1}$ $(m = 0, 1, 2, \dots)$. Note that $Q_x(B_m \times \mathbb{R}^\infty) = P((S_0^x, S_1^x \dots, S_m^x) \in B_m)$, so that $Q_x(B_m \times \mathbb{R}^\infty)$ may be expressed in terms of the m-fold product measure $\mu \times \mu \times \cdots \times \mu$, which is the distribution of $(Z_1, Z_2, \dots, Z_m)$. For our illustration, let $\mathcal{G}_n = \sigma(\{S_j^x : 0 \leq j \leq n\}) = \sigma(\{Z_1, Z_2, \dots, Z_n\})$ $(n \geq 1)$. We would like to establish the following property: **The conditional distribution of the "after-n process"** $S_n^{x+} := \{S_{n+m}^x : m = 0, 1, 2, \dots\}$ on $(\mathbb{R}^\infty, \mathcal{B}^\infty)$ **given** $\mathcal{G}_n$ **is** $Q_y|_{y = S_n^x} \equiv Q_{S_n^x}$. In other words, for the random walk $\{S_n^x : n \geq 0\}$, the conditional distribution of the *future evolution* defined by S_n^{x+}, given the *past states* $S_0^x, \dots, S_{n-1}^x$ and *present state* S_n^x, depends solely on the present state S_n^x, namely, $Q_{S_n^x}$ i.e., it is given by the regular conditional distribution $Q^{\mathcal{G}_n}(\omega, \cdot) = Q_{S_n^x(\omega)}(\cdot)$.

Theorem 2.12 *(Markov Property)* For every $n \geq 1$, the conditional distribution of S_n^{x+} given $\sigma(S_0^x, \dots, S_n^x)$ is a function only of S_n^x.

Proof To prove the theorem choose a cylinder set $C \in \mathcal{B}^\infty$. That is, $C = B_m \times \mathbb{R}^\infty$ for some $m \geq 0$. We want to show that

$$P([S_n^{x+} \in C]|\mathcal{G}_n) \equiv \mathbb{E}(\mathbf{1}_{[S_n^{x+} \in C]}|\mathcal{G}_n) = Q_{S_n^x}(C). \qquad (2.28)$$

Now $[S_n^{x+} \in C] = [(S_n^x, S_n^x + Z_{n+1}, \dots, S_n^x + Z_{n+1} + \cdots + Z_{n+m}) \in B_m]$, so that one may write

$$\mathbb{E}\left(\mathbf{1}_{[S_n^{x+} \in C]}|\mathcal{G}_n\right) = \mathbb{E}(\psi(U, V)|\mathcal{G}_n),$$

where $U = S_n^x$, $V = (Z_{n+1}, Z_{n+2}, \dots, Z_{n+m})$ and, for $u \in \mathbb{R}$ and $v \in \mathbb{R}^m$, $\psi(u, v) = \mathbf{1}_{B_m}(u, u + v_1, u + v_1 + v_2, \dots, u + v_1 + \cdots + v_m)$. Since S_n^x is $\mathcal{G}_n$-measurable and V is independent of $\mathcal{G}_n$, it follows from property (ℓ) of Theorem 2.10 that $\mathbb{E}(\psi(U, V)|\mathcal{G}_n) = h(S_n^x)$, where $h(u) = \mathbb{E}\psi(u, V)$. But

$$\begin{aligned} \mathbb{E}\psi(u, V) &= P((u, u + Z_{n+1}, \dots, u + Z_{n+1} + \cdots + Z_{n+m}) \in B_m) \\ &= P((u, u + Z_1, u + Z_1 + Z_2, \dots, u + Z_1 + \cdots + Z_m) \in B_m) \\ &= P((S_0^u, S_1^u, \dots, S_m^u) \in B_m) = Q_u(C). \end{aligned}$$

Therefore, $P([S_n^{x+} \in C]|\mathcal{G}_n) = (Q_u(C))_{u = S_n^x} = Q_{S_n^x}(C)$. We have now shown that the class $\mathcal{L}$ of sets $C \in \mathcal{B}^\infty$ for which "$P([S_n^{x+} \in C]|\mathcal{G}_n) = Q_{S_n^x}(C)$ a.s."

holds contains the class $\mathcal{C}$ of all cylinder sets. Since this class is a λ-system (see the convergence property (k) of Theorem 2.10) containing the π-system of cylinder sets that generate $\mathcal{B}^\infty$, it follows by the $\pi - \lambda$ theorem that $\mathcal{L} = \mathcal{B}^\infty$. ∎

Example 9 *(Recurrence of 1-d Simple Symmetric Random Walk)* Let X_0 be an integer-valued random variable, and $X_1, X_2, \ldots$ an i.i.d. sequence of symmetric Bernoulli ± 1-valued random variables, independent of X_0, defined on a probability space $(\Omega, \mathcal{F}, P)$. For $x \in \mathbb{Z}$, the sequence of random variables $S^x = \{S_0^x, S_1^x, \ldots\}$ defined by $S_0^x = x$, $S_n^x = x + X_1 + \cdots + X_n$, $n = 1, 2, \ldots$, is referred to as the *one-dimensional simple symmetric random walk on* $\mathbb{Z}$ *started at* x. Let $Q^x = P \circ (S^x)^{-1}$ denote the distribution of S^x. Observe that the conditional distribution of the after-one process $S_{+1}^x = (S_1^x, S_2^x, \ldots)$ given $\sigma(X_1)$ is the composition $\omega \to Q^{S_1(\omega)}$. Thus, letting $R = \cup_{n=1}^\infty \{(x_0, x_1, x_2, \ldots) \in \mathbb{Z}^\infty : x_n = x_0\}$, $[S^x \in R]$ denotes the event of eventual return to x, one has

$$
\begin{aligned}
Q^x(R) = P(S^x \in R) &= \mathbb{E}P(S^x \in R | \sigma(X_1)) \\
&= \mathbb{E}Q^{S_1^x}(R) = Q^{x+1}(R)P(S_1^x = x + 1) + Q^{x-1}(R)P(S_1^x = x - 1) \\
&= Q^{x+1}(R)\frac{1}{2} + Q^{x-1}(R)\frac{1}{2}.
\end{aligned}
\tag{2.29}
$$

Since $Q^x(R) \geq P(X_1 = 1, X_2 = -1) = 1/4 > 0$, it follows that the only solution to (2.29) is $Q^x(R) = 1$. Note that this does not depend on the particular choice of $x \in \mathbb{Z}$, and is therefore true for all $x \in \mathbb{Z}$. Thus the simple symmetric random walk started at any $x \in \mathbb{Z}$ is certain to eventually return to x.

Independence and conditional probability underly most theories of stochastic processes in fundamental ways. From the point of view of ideas developed thus far, the general framework is as follows. A **stochastic process** $\{X_t : t \in \Lambda\}$ on a probability space $(\Omega, \mathcal{F}, P)$ is a family of random maps $X_t : \Omega \to S_t, t \in \Lambda$, for measurable spaces $(S_t, \mathcal{S}_t), t \in \Lambda$. The index set Λ is most often of one of the following types: (i) $\Lambda = \{0, 1, 2, \ldots\}$. Then $\{X_t : t = 0, 1, 2, \ldots\}$ is referred to as a **discrete-parameter stochastic process**, usually with $S = \mathbb{Z}^k$ or $\mathbb{R}^k$. (ii) $\Lambda = [0, \infty)$. Then $\{X_t : t \geq 0\}$ is called a **continuous-parameter stochastic process**, usually with $S = \mathbb{Z}^k$ or $\mathbb{R}^k$.

Given an arbitrary collection of sets $S_t, t \in \Lambda$, the **product space**, denoted by $S = \prod_{t \in \Lambda} S_t \equiv \times_{t \in \Lambda} S_t$, is defined as the space of functions $\mathbf{x} = (x_t, t \in \Lambda)$ mapping Λ to $\cup_{t \in \Lambda} S_t$ such that $x_t \in S_t$ for each $t \in \Lambda$. This general definition applies to cases in which Λ is finite, countably infinite, or a continuum. In the case that each $S_t, t \in \Lambda$, is also a measurable space with respective σ-fields $\mathcal{S}_t$, the product σ-field, denoted by $\otimes_{t \in \Lambda} \mathcal{S}_t$, is defined as the σ-field generated by the collection $\mathcal{R}$ of **finite-dimensional rectangles** of the form $R = \{\mathbf{x} \in \prod_{t \in \Lambda} S_t : (x_{t_1}, \ldots, x_{t_k}) \in B_1 \times \cdots \times B_k\}$, for $k \geq 1$, $B_i \in \mathcal{S}_{t_i}$, $1 \leq i \leq k$. Alternatively, the product σ-field is the smallest σ-field of subsets of $\prod_{t \in \Lambda} S_t$ which makes each of the **coordinate projections**, $\pi_s(\mathbf{x}) = x_s, \mathbf{x} \in \prod_{t \in \Lambda} S_t, s \in \Lambda$, a measurable map. In this case the pair$(S = \prod_{t \in \Lambda} S_t, \otimes_{t \in \Lambda} \mathcal{S}_t)$ is the (measure-theoretic) **product space.**

As previously noted, it is natural to ask whether, as already known for finite products, given any family of probability measures Q_t (on $(S_t, \mathcal{S}_t)$), $t \in \Lambda$, can one construct a probability space $(\Omega, \mathcal{F}, P)$ on which are defined random maps X_t $(t \in \Lambda)$ such that (i) X_t has distribution Q_t $(t \in \Lambda)$ and (ii) $\{X_t : t \in \Lambda\}$ is a family of independent maps ? Indeed, on the product space $(S \equiv \times_{t \in \Lambda} S_t, \mathcal{S} \equiv \otimes_{t \in \Lambda} \mathcal{S}_t)$ it will be shown in Chapter VIII that there exists such a product probability measure $Q = \prod_{t \in \Lambda} Q_t$; and one may take $\Omega = S$, $\mathcal{F} = \mathcal{S}$, $P = Q$, $X_t(\omega) = x_t$ for $\omega = (x_t, t \in \Lambda) \in S$. The practical utility of such a construction for infinite Λ is somewhat limited to the case when Λ is denumerable. However such a product probability space for a countable sequence of random maps $X_1, X_2, \ldots$ is remarkably useful.

The important special case of a sequence $X_1, X_2, \ldots$ of **independent** and **identically distributed** random maps is referred to as an **i.i.d.** sequence. An example of the construction of an i.i.d. (coin tossing) sequence $\{X_n\}_{n=1}^{\infty}$ of Bernoulli-valued random variables with values in $\{0, 1\}$ and defined on a probability space $(\Omega, \mathcal{F}, P)$ with prescribed distribution $P(X_1 = 1) = p = 1 - P(X_1 = 0)$, for given $p \in [0, 1]$, is given in Exercise 37. As remarked above, the general existence of infinite product measures will be proved in Chapter VIII. This is a special case of the **Kolmogorov extension theorem** proved in Chapter VIII in the case that $(S, \mathcal{S})$ has some extra topological structure; see Exercise 37 for a simple special case illustrating how one may exploit topological considerations. Existence of an infinite product probability measure will also be seen to follow in full measure-theoretic generality, i.e., without topological requirements on the image spaces, from the **Tulcea extension theorem** discussed in Chapter VIII.

Exercise Set II

1. Suppose that $X_1, \ldots, X_n$ are independent random maps defined on a probability space $(\Omega, \mathcal{F}, P)$. Show that the product measure $Q = P \circ (X_1, \ldots, X_n)^{-1}$ is given by $Q_1 \times \cdots \times Q_n$, where $Q_i = P \circ X_i^{-1}$. Also show that any subset of $\{X_1, \ldots, X_n\}$ comprises independent random maps.

2. Suppose $\mathbf{X}_1, \mathbf{X}_2, \ldots \mathbf{X}_n$ are independent k-dimensional random vectors having distributions $Q_1, Q_2, \ldots, Q_n$, respectively. Prove that the distribution of $\mathbf{X}_1 + \mathbf{X}_2 + \cdots + \mathbf{X}_n$, $n \geq 2$, is given by the n-fold convolution $Q^{*n} = Q_1 * Q_2 * \cdots * Q_n$ inductively defined by $Q^{*n}(B) = \int_{\mathbb{R}^k} Q^{*(n-1)}(B - x) Q_n(dx)$, where $B - x := \{y - x : y \in B\}$ for Borel sets $B \subset \mathbb{R}^k$, $Q^{*(1)} = Q_1$.

3. Let $X_1, \ldots, X_n$ be i.i.d. random variables with finite variance $\sigma^2 = \mathbb{E}(X_1 - \mathbb{E}X_1)^2$, and finite central fourth moment $\mu_4 = \mathbb{E}(X_1 - \mathbb{E}X_1)^4$. Let $S_n = \sum_{j=1}^{n} X_j$. (a) Show $\mathbb{E}(S_n - n\mathbb{E}X_1)^4 = n\mu_4 + 3n(n-1)\sigma^4$. (b) For the case in which $X_1, \ldots, X_n$ is the i.i.d. Bernoulli $0 - 1$ valued sequence given in Example 3 show that both the variance and the fourth central moment of $n\hat{p}_n \equiv S_n$ are maximized at $p = 1/2$, and determine the respective maximum values.

4. Let X_1, X_2 be random maps with values in σ-finite measure spaces $(S_1, \mathcal{S}_1, \mu_1)$ and $(S_2, \mathcal{S}_2, \mu_2)$, respectively. Assume that the distribution of (X_1, X_2) has a pdf f with respect to product measure $\mu_1 \times \mu_2$, i.e., f is a nonnegative measurable function such that

$$P((X_1, X_2) \in B) = \int_B f(x_1, x_2)\mu_1 \times \mu_2(dx_1 \times dx_2), \ B \in \mathcal{S}_1 \otimes \mathcal{S}_2.$$

Show that X_1 and X_2 are independent if and only if $f(x_1, x_2) = f_1(x_1)f_2(x_2) \mu_1 \times \mu_2 - a.e.$ for some $0 \leq f_i$, $\int_{S_i} f_i d\mu_i = 1$, $(i = 1, 2)$.

5. Suppose that $X_1, X_2, \ldots$ is a sequence of independent random variables on $(\Omega, \mathcal{F}, P)$. Show that the two families $\{X_1, X_3, X_5, \ldots\}$ and $\{X_2, X_4, X_6, \ldots\}$ are independent. [*Hint*: Express $\sigma(X_1, X_3, \ldots) = \sigma(\mathcal{C}_1)$, for the π-system $\mathcal{C}_1 = \{[(X_1, X_3, \ldots, X_{2m-1}) \in A_1 \times \cdots \times A_m] : A_j \in \mathcal{B}, 1 \leq j \leq m, m \geq 1\}$, and similarly for the even indices.]

6. (a) Show that the coordinate variables $(U_{n,1}, \ldots, U_{n,n})$ in Exercise 31 of Chapter I are independent, and each is uniformly distributed over $[0, 2]$. (b) Show that if $\frac{X_n}{n} \to c$ in probability as $n \to \infty$ for a constant $c < 0$, then $X_n \to -\infty$ in probability in the sense that for any $\lambda < 0$, $P(X_n > \lambda) \to 0$ as $n \to \infty$. (c) Taking logarithms and using (a) and (b) together with a Chebychev inequality, show that $U_{n,1} \cdots U_{n,n} \to 0$ in probability as $n \to \infty$. (d) Use Jensen's inequality to extend (c) to any independent, positive, mean one random variables $(U_{n,1}, \ldots, U_{n,n})$ having finite second moment.

7. Show that if (U_1, U_2) is uniformly distributed on the disc $D = \{(x, y) : x^2 + y^2 \leq 1\}$, i.e., distributed as a multiple $(\frac{1}{\pi})$ of Lebesgue measure on D, then X and Y are not independent. Compute $\text{Cov}(X, Y)$.

8. Let $X_1, X_2, \ldots, X_n$ be i.i.d. random variables defined on $(\Omega, \mathcal{F}, P)$ and having (common) distribution Q.

   (i) Suppose $Q(dx) = \lambda e^{-\lambda x} \mathbf{1}_{[0,\infty)}(x)dx$, for some $\lambda > 0$, referred to as the *exponential distribution with parameter* λ. Show that $X_1 + \cdots + X_n$ has distribution $Q^{*n}(dx) = \lambda^n \frac{x^{n-1}}{(n-1)!} e^{-\lambda x} \mathbf{1}_{[0,\infty)}(x)dx$. This latter distribution is referred to as a *gamma distribution* with parameters n, λ.

   (ii) Suppose $Q(dx) = \frac{1}{\sqrt{2\pi\sigma^2}} e^{-\frac{(x-\mu)^2}{\sigma^2}} dx$; referred to as the *Gaussian* or *normal* distribution with parameters $\mu \in \mathbb{R}, \sigma^2 > 0$. Show that $X_1 + \cdots + X_n$ has a normal distribution with parameters $n\mu$ and $n\sigma^2$.

   (iii) Let X be a standard normal $N(0, 1)$ random variable. Find the distribution Q of X^2, and compute Q^{*2}. [*Hint*: By the $\pi - \lambda$ theorem it is sufficient to compute a pdf for $Q(-\infty, x], x \in \mathbb{R}$ to determine $Q(A)$ for all Borel sets A. Also, $\int_0^1 u^{-\frac{1}{2}}(1-u)^{-\frac{1}{2}}du = \pi$.]

9. Let X_1, X_2 be random maps on $(\Omega, \mathcal{F}, P)$ taking values in the measurable spaces $(S_1, \mathcal{S}_1), (S_2, \mathcal{S}_2)$, respectively. Show that the joint distribution of (X_1, X_2) on $(S_1 \times S_2, \mathcal{S}_1 \otimes \mathcal{S}_2)$ is product measure if and only if $\sigma(X_1)$ and $\sigma(X_2)$ are independent σ-fields.

10. (i) Let V_1 take values ± 1 with probability $1/4$ each, and 0 with probability $1/2$. Let $V_2 = V_1^2$. Show that $\text{Cov}(V_1, V_2) = 0$, though V_1, V_2 are not independent.

(ii) Show that random maps V_1, V_2 are independent if and only if $f(V_1)$ and $g(V_2)$ are uncorrelated for all pairs of real-valued Borel-measurable functions f, g such that $f(V_1), g(V_2) \in L^2$.

11. Suppose that $X_1, X_2, \ldots$ is a sequence of random variables on $(\Omega, \mathcal{F}, P)$ each having the same distribution $Q = P \circ X_n^{-1}$. (i) Show that if $\mathbb{E}|X_1| < \infty$ then $P(|X_n| > n \ i.o.) = 0$. [*Hint*: First use (1.10) to get $\mathbb{E}|X_1| = \int_0^\infty P(|X_1| > x) dx$, and then apply Borel–Cantelli.] (ii) Assume that $X_1, X_2, \ldots$ are also independent with $\mathbb{E}|X_n| = \infty$. Show that $P(|X_n| > n \ i.o.) = 1$.

12. Suppose that $Y_1, Y_2, \ldots$ is an i.i.d. sequence of nonconstant random variables. Show that $\liminf_{n \to \infty} Y_n < \limsup_{n \to \infty} Y_n$ a.s. In particular $P(\lim_{n \to \infty} Y_n$ exists$) = 0$. [*Hint*: Use there must be numbers $a < b$ such that $P(Y_1 < a) > 0$ and $P(Y_1 > b) > 0$. Use Borel–Cantelli II.]

13. [*Percolation*] Formulate and extend Example 2 by determining the critical probability for percolation on the rooted b-ary tree defined by $T = \cup_{j=0}^\infty \{1, 2, \ldots, b\}^j$ for a natural number $b \geq 3$.

14. Suppose that $Y_1, Y_2, \ldots$ is a sequence of i.i.d. positive random variables with $P(Y_1 = 1) < 1$ and $\mathbb{E}Y_1 = 1$. (a) Show that $X_n = Y_1 Y_2 \ldots Y_n \to 0$ a.s. as $n \to \infty$; [*Hint*: Consider $\mathbb{E}X_n^t$ for fixed $0 < t < 1$, and apply Chebychev inequality, Jensen inequality and Borel–Cantelli I.] (b) Is the sequence $X_1, X_2, \ldots$ uniformly integrable ?

15. (i) Consider three independent tosses of a balanced coin and let A_i denote the event that the outcomes of the ith and $(i+1)$st tosses match, for $i = 1, 2$. Let A_3 be the event that the outcomes of the third and first match. Show that A_1, A_2, A_3 are pairwise independent but not independent. (ii) Show that $A_1, \ldots, A_n, A_j \in \mathcal{S}_j, 1 \leq j \leq n$, are independent events if and only if the 2^n equations $P(C_1 \cap \cdots \cap C_n) = \prod_{j=1}^n P(C_j)$ hold for all choices of $C_j = A_j$ or $C_j = A_j^c, 1 \leq j \leq n$. (iii) Show that $A_1, \ldots, A_n, A_j \in \mathcal{S}_j, 1 \leq i \leq n$, are independent events if and only if the $2^n - n - 1$ equations $P(A_{j_1} \cap \cdots \cap A_{j_m}) = \prod_{i=1}^m P(A_{j_i}), m = 2, \ldots, n$, hold.

16. Suppose that $A_1, A_2, \ldots$ is a sequence of independent events, each having the same probability $p = P(A_n) > 0$ for each $n = 1, 2, \ldots$. Show that the event $[A_n \ e.o.] := \cup_{n=1}^\infty A_n$ has probability one, where *e.o.* denotes *eventually occurs*.

17. Let $(\Omega, \mathcal{F}, P)$ be an arbitrary probability space and suppose $A_1, A_2, \ldots$ is a sequence of *independent* events in $\mathcal{F}$ with $P(A_n) < 1, \forall n$. Suppose $P(\cup_{n=1}^\infty A_n) = 1$. (i) Show that $P(A_n \ i.o.) = 1$. (ii) Give an example of independent events for which $P(\cup_{n=1}^\infty A_n) = 1$ but $P(A_n \ i.o.) < 1$.

18. Let $(\Omega, \mathcal{F}, P)$ be an arbitrary probability space and suppose $\{A_n\}_{n=1}^\infty$ is a sequence of *independent* events in $\mathcal{F}$ such that $\sum_{n=1}^\infty P(A_n) \geq 2$. Let E denote the event that none of the A_n's occur for $n \geq 1$. (i) Show that $E \in \mathcal{F}$. (ii) Show that $P(E) \leq \frac{1}{e^2}$. [*Hint*: $1 - x \leq e^{-x}, x \geq 0$.]

19. Suppose that $X_1, X_2, \ldots$ is an i.i.d. sequence of Bernoulli 0 or 1-valued random variables with $P(X_n = 1) = p, P(X_n = 0) = q = 1 - p$. Fix $r \geq 1$ and let $R_n := [X_n = 1, X_{n+1} = 1, \ldots, X_{n+r-1} = 1]$ be the event of a run of 1's of length at least r starting from n.

(i) Show that $P(R_n \text{ i.o.}) = 1$ if $0 < p \leq 1$.

(ii) Suppose r is allowed to grow with n, say $r_n = [\theta \log_2 n]$ in defining the event R_n; here $[x]$ denotes the largest integer not exceeding x. In the case of a balanced coin ($p = 1/2$), show that if $\theta > 1$ then $P(R_n \text{ i.o.}) = 0$, [*Hint*: Borel–Cantelli Lemma 1], and if $0 < \theta \leq 1$ then $P(R_n \text{ i.o.}) = 1$. [*Hint*: Consider a subsequence $R_{n_k} = [X_{n_k} = 1, \ldots, X_{n_k + r_{n_k} - 1} = 1]$ with n_1 sufficiently large that $\theta \log_2 n_1 > 1$, and $n_{k+1} = n_k + r_{n_k}, k \geq 1$. Compare $\sum_{k=1}^{\infty} n_k^{-\theta} \equiv \sum_{k=1}^{\infty} \frac{n_k^{-\theta}}{n_{k+1} - n_k}(n_{k+1} - n_k)$ to an integral $\int_{n_1}^{\infty} f(x) dx$ for an appropriately selected function f.]

20. Show that in the case $\mathbb{E} \log^+ \varepsilon_1 = \infty$, the formal power series of Example 5 is almost surely divergent for any $x \neq 0$. [*Hint*: Consider $0 < |x| < 1, |x| \geq 1$, separately.]

21. Let $\mathcal{C}$ denote the collection of functions of the form $\sum_{i=1}^{n} f_i(u) g_i(v)$, $(u, v) \in S_1 \times S_2$, where $f_i, g_i, 1 \leq i \leq n$, are bounded Borel-measurable functions on the probability spaces $(S_1, \mathcal{S}_1, Q_1)$ and $(S_2, \mathcal{S}_2, Q_2)$, respectively. Show that $\mathcal{C}$ is dense in $L^1(S_1 \times S_2, \mathcal{S}_1 \otimes \mathcal{S}_2, Q_1 \times Q_2)$. [*Hint*: Use the method of approximation by simple functions.]

22. Suppose that X, Y are independent random variables on $(\Omega, \mathcal{F}, P)$. Assume that there is a number $a < 1$ such that $P(X \leq a) = 1$. Also assume that Y is exponentially distributed with mean one. Calculate $\mathbb{E}[e^{XY} | \sigma(X)]$. [*Hint*: Use the substitution property.]

23. Suppose that (X, Y) is uniformly distributed on the unit disk $D = \{(x, y) : x^2 + y^2 \leq 1\}$, i.e., has constant pdf on D. (i) Calculate the (marginal) distribution of X. (ii) Calculate the conditional distribution of Y given $\sigma(X)$. (iii) Calculate $\mathbb{E}(Y^2 | \sigma(X))$.

24. (i) Give a proof of (2.13) using the second, and therefore the first, definition of conditional expectation. [*Hint*: The only measurable random variables with respect to $\{\Omega, \emptyset, B, B^c\}$ are those of the form $c\mathbf{1}_B + d\mathbf{1}_{B^c}$, for $c, d \in \mathbb{R}$.] (ii) Prove (2.24), (2.26).

25. Suppose that X, N are independent random variables with standard normal distribution. Let $Y = X + bN$; i.e., X with an independent additive noise term bN. Calculate $\mathbb{E}(X | \sigma(Y))$. [*Hint*: Compute the conditional pdf of the regular conditional distribution of X given $\sigma(Y)$ by first computing the joint pdf of (X, Y). For this, either view (X, Y) as a function (linear transformation) of independent pair (X, N), or notice that the conditional pdf of Y given $\sigma(X)$ follows immediately from the substitution property. Also, the marginal of X is the convolution of two normal distribution and hence, normal. It is sufficient to compute the mean and variance of X. From here obtain the joint density as the product of the conditional density of Y given $\sigma(X)$ and the marginal of X.]

26. Suppose that U is uniformly distributed on $[0, 1]$ and $V = U(1 - U)$. Determine $\mathbb{E}(U | \sigma(V))$. Does the answer depend on the symmetry of the uniform distribution on $[0, 1]$?

27. Suppose X is a real-valued random variable with symmetric distribution Q about zero; i.e., X and $-X$ have the same distribution. (i) Compute $P(X > 0 | \sigma(|X|))$. (ii) Determine the (regular) conditional distribution of X given $\sigma(|X|)$. [*Hint*:

Keep in mind that $\int_{\mathbb{R}} f(x)\delta_{\{a\}}(dx) = f(a)$ for any $a \in \mathbb{R}$, and consider $\mathbb{E}f(X)g(|X|)$ for $\mathbb{E}|f(X)| < \infty$ and bounded, Borel measurable g. Partition the integral by $1 = \mathbf{1}_{(0,\infty)}(|X|) + \mathbf{1}_{\{0\}}(|X|)$.]

28. Suppose that $X_1, X_2, \ldots$ is an i.i.d. sequence of square-integrable random variables with mean μ and variance $\sigma^2 > 0$, and N is a nonnegative integer-valued random variable, independent of $X_1, X_2, \ldots$. Let $S = \sum_{j=1}^{N} X_j$, with the convention $S = 0$ on the event $[N = 0]$. Compute the mean and variance of S. [*Hint*: Condition on N.]

29. (a) Let $X_1, \ldots, X_n$ be an i.i.d. sequence of random variables on $(\Omega, \mathcal{F}, P)$ and let $S_n = X_1 + \cdots + X_n$. Assume $\mathbb{E}|X_1| < \infty$. Show that $\mathbb{E}(X_j | \sigma(S_n)) = \mathbb{E}(X_1 | \sigma(S_n))$. [*Hint*: Use Fubini–Tonelli.] Calculate $\mathbb{E}(X_j | \sigma(S_n))$. [*Hint*: Add up and use properties of conditional expectation.] (b) Generalize (a) to the case in which the distribution of $(X_1, \ldots, X_n)$ is invariant under permutations of the indices, i.e., the distribution of $(X_{\pi_1}, \ldots, X_{\pi_n})$ is the same for all permutations π of $(1, 2, \ldots, n)$.

30. Suppose that (X, Y) is distributed on $[0, 1] \times [0, 1]$ according to the pdf $f(x, y) = 4xy, 0 \le x, y \le 1$. Determine $\mathbb{E}[X | \sigma(X + Y)]$.

31. Suppose that $Y_1, \ldots, Y_n$ are i.i.d. exponentially distributed with mean one. Let $S_n = \sum_{j=1}^{n} Y_j$.

   (i) Calculate $\mathbb{E}(Y_1^2 | S_n)$. [*Hint*: In view of part (iii) of this problem, calculate the joint pdf of $(Y_1, Y_2 + \cdots + Y_n)$ and then that of (Y_1, S_n) by a change of variable under the linear transformation $(y, s) \mapsto (y, y+s)$, for an arbitrary distribution of Y_1 having pdf $g(y)$. Namely, $g(y)g^{*(n-1)}(s - y)/g^{*n}(s)$.]

   (ii) Calculate $\mathbb{E}(Y_1 Y_2 | S_n)$. [*Hint*: Consider $S_n^2 = \mathbb{E}(S_n^2 | S_n)$ along with the previous exercise.]

   (iii) Make the above calculations in the case that $Y_1, Y_2, \ldots Y_n$ are i.i.d. with standard normal distributions.

32. [*Conditional Chebyshev-type*] For $X \in L^p$, $p \ge 1$, prove for $\lambda > 0$, $P(|X| > \lambda | \mathcal{G}) \le \mathbb{E}(|X|^p | \mathcal{G})/\lambda^p$ a.s.

33. [*Conditional Cauchy–Schwarz*] For $X, Y \in L^2$ show that $|\mathbb{E}(XY | \mathcal{G})|^2 \le \mathbb{E}(X^2 | \mathcal{G}) \mathbb{E}(Y^2 | \mathcal{G})$.

34. Let Y be an exponentially distributed random variable on $(\Omega, \mathcal{F}, P)$. Fix $a > 0$.

   (i) Calculate $\mathbb{E}(Y | \sigma(Y \wedge a))$, where $Y \wedge a := \min\{Y, a\}$. [Hint: $[Y < a] = [Y \wedge a < a]$. Let g be a bounded Borel-measurable function and either make and verify an intuitive guess for $\mathbb{E}(Y | \sigma(Y \wedge a))$ (based on "lack of memory" of the exponential distribution), or calculate $\mathbb{E}(Yg(Y \wedge a))$ by integration by parts.]

   (ii) Determine the regular conditional distribution of Y given $\sigma(Y \wedge a)$.

   (iii) Repeat this problem for Y having arbitrary distribution Q on $[0, \infty)$, and a such that $0 < P(Y \le a) < 1$.

35. Let U, V be independent random maps with values in measurable spaces $(S_1, \mathcal{S}_1)$ and $(S_2, \mathcal{S}_2)$, respectively. Let $\varphi(u, v)$ be a measurable map on $(S_1 \times S_2, \mathcal{S}_1 \otimes \mathcal{S}_2)$ into a measurable space $(S, \mathcal{S})$. Show that a regular conditional distribution of

$\varphi(U, V)$, given $\sigma(V)$, is given by Q_V, where Q_v is the distribution of $\varphi(U, v)$. [*Hint*: Use the Fubini–Tonelli theorem or Theorem 2.10(I).]

36. Prove the Markov property for k-dimensional random walks with $k \geq 2$.

37. Let $\Omega = \{0, 1\}^\infty$ be the space of infinite binary 0–1 sequences, and let $\mathcal{F}_0$ denote the field of finite unions of sets of the form $A_n(\varepsilon_1, \ldots, \varepsilon_n) = \{\omega = (\omega_1, \omega_2, \ldots) \in \Omega : \omega_1 = \varepsilon_1, \ldots, \omega_n = \varepsilon_n\}$ for arbitrary $\varepsilon_i \in \{0, 1\}$, $1 \leq i \leq n, n \geq 1$. Fix $p \in [0, 1]$ and define $P_p(A_n(\varepsilon_1, \ldots, \varepsilon_n)) = p^{\sum_{i=1}^n \varepsilon_i}(1-p)^{n-\sum_{i=1}^n \varepsilon_i}$. (i) Show that the natural finitely additive extension of P_p to $\mathcal{F}_0$ defines a measure on the field $\mathcal{F}_0$. [*Hint*: By Tychonoff's theorem from topology, the set Ω is compact for the product topology, see Appendix B. Check that sets $C \in \mathcal{F}_0$ are both open and closed for the product topology, so that by compactness, any countable disjoint union belonging to $\mathcal{F}_0$ must be a finite union.] (ii) Show that P_p has a unique extension to $\sigma(\mathcal{F}_0)$. This probability P_p defines the infinite product probability, also denoted by $(p\delta_{\{1\}} + (1-p)\delta_{\{0\}})^\infty$. [*Hint*: Apply the Carathéodory extension theorem.] (iii) Show that the coordinate projections $X_n(\omega) = \omega_n, \omega = (\omega_1, \omega_2, \ldots) \in \Omega, n \geq 1$, define an i.i.d. sequence of (coin tossing) Bernoulli 0 or 1-valued random variables.

38. Prove that the set $\tilde{B}$ in the proof of Proposition 2.5 belongs to $\mathcal{B}(\mathbb{R}^k)$, and the function g, there, is Borel-measurable.

39. (i) Prove that the set of all simple function on $(\Omega, \mathcal{F}, P)$ is dense in $L^p, \forall p \geq 1$.
    (ii) Prove that L^p is dense in L^r (in L^r − norm) for $p > r \geq 1$ (in L^r-norm)

Chapter III
Martingales and Stopping Times

The notion of **martingale** has proven to be among the most powerful ideas to emerge in probability in the past century. This chapter provides a foundation for this theory together with some illuminating examples and applications. For a prototypical illustration of the martingale property, let $Z_1, Z_2, \ldots$ be a sequence of independent integrable random variables and let $X_n = Z_1 + \cdots + Z_n$, $n \geq 1$. If $\mathbb{E}Z_j = 0$, $j \geq 1$, then one clearly has

$$\mathbb{E}(X_{n+1}|\mathcal{F}_n) = X_n, \quad n \geq 1,$$

where $\mathcal{F}_n := \sigma(X_1, \ldots, X_n)$.

Definition 3.1 *(First Definition of Martingale)* A sequence of integrable random variables $\{X_n : n \geq 1\}$ on a probability space $(\Omega, \mathcal{F}, P)$ is said to be a **martingale** if, writing $\mathcal{F}_n := \sigma(X_1, X_2, \ldots, X_n)$,

$$\mathbb{E}(X_{n+1}|\mathcal{F}_n) = X_n \text{ a.s. } (n \geq 1). \tag{3.1}$$

This definition extends to any (finite or infinite) family of integrable random variables $\{X_t : t \in T\}$, where T is a linearly ordered set: Let $\mathcal{F}_t = \sigma(X_s : s \leq t)$. Then $\{X_t : t \in T\}$ is a **martingale** if

$$\mathbb{E}(X_t|\mathcal{F}_s) = X_s \text{ a.s } \forall s < t \ (s, t \in T). \tag{3.2}$$

In the previous case of a sequence $\{X_n : n \geq 1\}$, as one can see by taking successive conditional expectations $\mathbb{E}(X_n|\mathcal{F}_m) = \mathbb{E}[\mathbb{E}(X_n|\mathcal{F}_{n+1})|\mathcal{F}_m] = \mathbb{E}(X_{n+1}|\mathcal{F}_m) = \cdots = \mathbb{E}(X_{m+1}|\mathcal{F}_m) = X_m$, (3.1) is equivalent to

$$\mathbb{E}(X_n|\mathcal{F}_m) = X_m \quad \text{a.s.} \quad \forall \, m < n. \tag{3.3}$$

© Springer International Publishing AG 2016
R. Bhattacharya and E.C. Waymire, *A Basic Course in Probability Theory*,
Universitext, DOI 10.1007/978-3-319-47974-3_III

Thus, (3.1) is a special case of (3.2). Most commonly, $T = \mathbb{N}$ or $\mathbb{Z}^+$, or $T = [0, \infty)$. Note that if $\{X_t : t \in T\}$ is a martingale, one has the *constant expectations property:* $\mathbb{E}X_t = \mathbb{E}X_s \, \forall \, s, t \in T$.

Remark 3.1 Let $\{X_n : n \geq 1\}$ be a martingale sequence. Define its associated **martingale difference sequence** by $Z_1 := X_1, Z_{n+1} := X_{n+1} - X_n \ (n \geq 1)$. Note that for $X_n \in L^2(\Omega, \mathcal{F}, P), n \geq 1$, the martingale differences are uncorrelated. In fact, for $X_n \in L^1(\Omega, \mathcal{F}, P), n \geq 1$, one has

$$
\begin{aligned}
\mathbb{E}Z_{n+1} f(X_1, X_2, \ldots, X_n) &= \mathbb{E}[\mathbb{E}(Z_{n+1} f(X_1, \ldots, X_n)|\mathcal{F}_n)] \\
&= \mathbb{E}[f(X_1, \ldots, X_n)\mathbb{E}(Z_{n+1}|\mathcal{F}_n)] = 0 \qquad (3.4)
\end{aligned}
$$

for all bounded $\mathcal{F}_n$ measurable functions $f(X_1, \ldots, X_n)$. If $X_n \in L^2(\Omega, \mathcal{F}, P) \, \forall \, n \geq 1$, then (3.1) implies, and is equivalent to, the fact that $Z_{n+1} \equiv X_{n+1} - X_n$ is orthogonal to $L^2(\Omega, \mathcal{F}_n, P)$. It is interesting to compare this orthogonality to that of independence of Z_{n+1} and $\{Z_m : m \leq n\}$. Recall that Z_{n+1} is independent of $\{Z_m : 1 \leq m \leq n\}$ or, equivalently, of $\mathcal{F}_n = \sigma(X_1, \ldots, X_n)$ if and only if $\overline{g}(Z_{n+1})$ is orthogonal to $L^2(\Omega, \mathcal{F}_n, P)$ for all bounded measurable $\overline{g}$ such that $\mathbb{E}\overline{g}(Z_{n+1}) = 0$. Thus independence translates as $0 = \mathbb{E}\{[g(Z_{n+1}) - \mathbb{E}g(Z_{n+1})] \cdot f(X_1, \ldots, X_n)\} = \mathbb{E}\{g(Z_{n+1}) \cdot f(X_1, \ldots, X_n)\} - \mathbb{E}g(Z_{n+1}) \cdot \mathbb{E}f(X_1, \ldots, X_n)$, for all bounded measurable g on $\mathbb{R}$ and for all bounded measurable f on $\mathbb{R}^n$.

Example 1 *(Independent Increment Process)* Let $\{Z_n : n \geq 1\}$ be an independent sequence having *zero means,* and X_0 an integrable random variable independent of $\{Z_n : n \geq 1\}$. Then

$$
X_0, \ X_n := X_0 + Z_1 + \cdots + Z_n \equiv X_{n-1} + Z_n \ (n \geq 1) \qquad (3.5)
$$

is a martingale sequence.

Definition 3.2 If one has inequality in place of (3.1), namely,

$$
\mathbb{E}(X_{n+1}|\mathcal{F}_n) \geq X_n \text{ a.s.} \quad \forall n \geq 1, \qquad (3.6)
$$

then $\{X_n : n \geq 1\}$ is said to be a submartingale. More generally, if the index set T is as in (3.2), then $\{X_t : t \in T\}$ is a **submartingale** if

$$
\mathbb{E}(X_t|\mathcal{F}_s) \geq X_s \, \forall \, s < t \ (s, t \in T). \qquad (3.7)
$$

If instead of $\geq$, one has $\leq$ in (3.7) (3.8), the process $\{X_n : n \geq 1\}$ ($\{X_t : t \in T\}$) is said to be a **supermartingale**.

In Example 1, if $\mathbb{E}Z_k \geq 0 \, \forall \, k$, then the sequence $\{X_n : n \geq 1\}$ of partial sums of independent random variables is a submartingale. If $\mathbb{E}Z_k \leq 0$ for all k, then $\{X_n : n \geq 1\}$ is a supermartingale. In Example 3, it follows from the triangle inequality for conditional expectations that the sequence $\{Y_n := |X_n| : n \geq 1\}$ is a submartingale. The following proposition provides an important generalization of this latter example.

Proposition 3.1 (a) If $\{X_n : n \geq 1\}$ is a martingale and $\varphi(X_n)$ is a convex and integrable function of X_n, then $\{\varphi(X_n) : n \geq 1\}$ is a submartingale. (b) If $\{X_n\}$ is a submartingale, and $\varphi(X_n)$ is a convex and nondecreasing integrable function of X_n, then $\{\varphi(X_n) : n \geq 1\}$ is a submartingale.

Proof The proof is obtained by an application of the conditional Jensen's inequality given in Theorem 2.10. In particular, for (a) one has

$$\mathbb{E}(\varphi(X_{n+1})|\mathcal{F}_n) \geq \varphi(\mathbb{E}(X_{n+1}|\mathcal{F}_n)) = \varphi(X_n). \tag{3.8}$$

Now take the conditional expectation of both sides with respect to $\mathcal{G}_n \equiv \sigma(\varphi(X_1)$, $\ldots, \varphi(X_n)) \subset \mathcal{F}_n$, to get the martingale property of $\{\varphi(X_n) : n \geq 1\}$. Similarly, for (b), for convex and nondecreasing φ one has in the case of a submartingale that

$$\mathbb{E}(\varphi(X_{n+1})|\mathcal{F}_n) \geq \varphi(\mathbb{E}(X_{n+1}|\mathcal{F}_n)) \geq \varphi(X_n), \tag{3.9}$$

and taking conditional expectation in (3.9), the desired submartingale property follows. ∎

Proposition 3.1 immediately extends to martingales and submartingales indexed by an arbitrary linearly ordered set T.

Example 2 (a) If $\{X_t : t \in T\}$ is a martingale, $\mathbb{E}|X_t|^p < \infty$ $(t \in T)$ for some $p \geq 1$, then $\{|X_t|^p : t \in T\}$ is a submartingale. (b) If $\{X_t : t \in T\}$ is a submartingale, then for every real c, $\{Y_t := \max(X_t, c)\}$ is a submartingale. In particular, $\{X_t^+ := \max(X_t, 0)\}$ is a submartingale.

Remark 3.2 It may be noted that in (3.8), (3.9), the σ-field $\mathcal{F}_n$ is $\sigma(X_1, \ldots, X_n)$, and not $\sigma(\varphi(X_1), \ldots, \varphi(X_n))$, as seems to be required by the first definitions in (3.1), (3.6). It is, however, more convenient to give the definition of a **martingale (or a submartingale) with respect to a filtration** $\{\mathcal{F}_n\}$ for which (3.1) holds (or respectively, (3.6) holds) assuming at the outset that X_n is $\mathcal{F}_n$-measurable $(n \geq 1)$ (or, as one often says, $\{X_n\}$ is $\{F_n\}$-**adapted**). One refers to this sequence as an $\{\mathcal{F}_n\}$-**martingale** (respectively $\{\mathcal{F}_n\}$-**submartingale**). An important example of an $\mathcal{F}_n$ larger than $\sigma(X_1, \ldots, X_n)$ is given by "adding independent information" via $\mathcal{F}_n = \sigma(X_1, \ldots, X_n) \vee \mathcal{G}$, where $\mathcal{G}$ is a σ-field independent of $\sigma(X_1, X_2, \ldots)$, and $\mathcal{G}_1 \vee \mathcal{G}_2$ denotes the smallest σ-field containing $\mathcal{G}_1 \cup \mathcal{G}_2$. We formalize this with the following definition; also see Exercise 13.

Definition 3.3 *(Second General Definition)* Let T be an arbitrary linearly ordered set and suppose $\{X_t : t \in T\}$ is a stochastic process with (integrable) values in $\mathbb{R}$ and defined on a probability space $(\Omega, \mathcal{F}, P)$. Let $\{\mathcal{F}_t : t \in T\}$ be a nondecreasing collection of sub-σ-fields of $\mathcal{F}$, referred to as a **filtration**, i.e., $\mathcal{F}_s \subset \mathcal{F}_t$ if $s \leq t$. Assume that for each $t \in T$, X_t is **adapted** to $\mathcal{F}_t$ in the sense that X_t is $\mathcal{F}_t$-measurable. We say that $\{X_t : t \in T\}$ is a **martingale**, respectively **submartingale, supermartingale**, with respect to the filtration $\{\mathcal{F}_t\}$ if $\mathbb{E}[X_t|\mathcal{F}_s] = X_s$, $\forall s, t \in T, s \leq t$, respectively $\geq X_s, \forall s, t \in T, s \leq t$, or $\leq X_s \ \forall s, t \in T, s \leq t$.

Example 3 Let X be an integrable random variable on $(\Omega, \mathcal{F}, P)$ and let $\{\mathcal{F}_n : n \geq 1\}$ be a filtration of $\mathcal{F}$. One may check that the stochastic process defined by

$$X_n := \mathbb{E}(X|\mathcal{F}_n) \ (n \geq 1) \tag{3.10}$$

is an $\{\mathcal{F}_n\}$-martingale.

Note that for submartingales the expected values are nondecreasing, while those of supermartingales are nonincreasing. Of course, martingales continue to have constant expected values under this more general definition.

Theorem 3.2 (*Doob's Maximal Inequality*) Let $\{X_1, X_2, \ldots, X_n\}$ be an $\{\mathcal{F}_k : 1 \leq k \leq n\}$-martingale, or a nonnegative submartingale, and $\mathbb{E}|X_n|^p < \infty$ for some $p \geq 1$. Then, for all $\lambda > 0$, $M_n := \max\{|X_1|, \ldots, |X_n|\}$ satisfies

$$P(M_n \geq \lambda) \leq \frac{1}{\lambda^p} \int_{[M_n > \lambda]} |X_n|^p dP \leq \frac{1}{\lambda^p} \mathbb{E}|X_n|^p. \tag{3.11}$$

Proof Let $A_1 = [|X_1| \geq \lambda]$, $A_k = [|X_1| < \lambda, \ldots, |X_{k-1}| < \lambda, |X_k| \geq \lambda]$ $(2 \leq k \leq n)$. Then $A_k \in \mathcal{F}_k$ and $[A_k : 1 \leq k \leq n]$ is a (disjoint) partition of $[M_n \geq \lambda]$. Therefore,

$$P(M_n \geq \lambda) = \sum_{k=1}^{n} P(A_k) \leq \sum_{k=1}^{n} \frac{1}{\lambda^p} \mathbb{E}(\mathbf{1}_{A_k}|X_k|^p) \leq \sum_{k=1}^{n} \frac{1}{\lambda^p} E(\mathbf{1}_{A_k}|X_n|^p)$$

$$= \frac{1}{\lambda^p} \int_{[M_n \geq \lambda]} |X_n|^p dP \leq \frac{\mathbb{E}|X_n|^p}{\lambda^p}.$$

∎

Remark 3.3 The classical **Kolmogorov maximal inequality** for sums of i.i.d. mean zero, square-integrable random variables is a special case of **Doob's maximal inequality** obtained by taking $p = 2$ for the martingales of Example 1 having square-integrable increments.

Corollary 3.3 Let $\{X_1, X_2, \ldots, X_n\}$ be an $\{\mathcal{F}_k : 1 \leq k \leq n\}$-martingale such that $\mathbb{E}|X_n|^p < \infty$ for some $p \geq 2$, and $M_n = \max\{|X_1|, \ldots, |X_n|\}$. Then $\mathbb{E}M_n^p \leq p^q \mathbb{E}|X_n|^p$.

Proof A standard application of the Fubini–Tonelli theorem (see (1.10)) provides the second moment formula

$$\mathbb{E}M_n^p = p \int_0^\infty x^{p-1} P(M_n > x) dx.$$

Noting that $p - 1 \geq 1$ to first apply the Doob maximal inequality (3.11), one then makes another application of the Fubini–Tonelli theorem, and finally the Hölder inequality, noting $pq - q = p$ for the conjugacy $\frac{1}{p} + \frac{1}{q} = 1$, to obtain

$$\mathbb{E}M_n^p \leq p \int_0^\infty \mathbb{E}\left(|X_n|^{p-1}\mathbf{1}_{[M_n \geq x]}\right) dx = p\mathbb{E}\left(|X_n|^{p-1}M_n\right)$$

$$\leq p(\mathbb{E}|X_n|^{(p-1)q})^{\frac{1}{q}}(\mathbb{E}M_n^p)^{\frac{1}{p}}.$$

Divide both sides by $(\mathbb{E}|M_n|^p)^{\frac{1}{p}}$ and use monotonicity of $x \to x^{\frac{1}{q}}, x \geq 0$, to complete the proof. ∎

Doob also obtained a bound of this type for $p > 1$ with a smaller constant $q^p \leq p^q$ when $p \geq 2$, but it also requires a bit more clever estimation than in the above proof. Doob's statement and proof are as follows.

Theorem 3.4 (*Doob's Maximal Inequality for Moments*) Suppose that $\{X_1, X_2, \dots, X_n\}$ is an $\{\mathcal{F}_k : 1 \leq k \leq n\}$-martingale, or a nonnegative submartingale, and let $M_n = \max\{|X_1|, \dots, |X_n|\}$. Then

1. $\mathbb{E}M_n \leq \frac{e}{e-1}\left(1 + \mathbb{E}|X_n| \log^+ |X_n|\right)$.
2. If $\mathbb{E}|X_n|^p < \infty$ for some $p > 1$, then $\mathbb{E}M_n^p \leq q^p \mathbb{E}|X_n|^p$, where q is the conjugate exponent defined by $\frac{1}{q} + \frac{1}{p} = 1$, i.e., $q = \frac{p}{p-1}$.

Proof For any nondecreasing function F_1 on $[0, \infty)$ with $F_1(0) = 0$, one may define a corresponding Lebesgue–Stieltjes measure $\mu_1(dy)$. Use the integration by parts Proposition 1.4, to get

$$\mathbb{E}F_1(M_n) = \int_{[0,\infty)} P(M_n \geq y) F_1(dy)$$

$$\leq \int_{[0,\infty)} \left[\frac{1}{y} \int_{[M_n \geq y]} |X_n| dP\right] F_1(dy)$$

$$= \int_\Omega |X_n| \left(\int_{[0,M_n]} \frac{1}{y} F_1(dy)\right) dP, \tag{3.12}$$

where the integrability follows from Theorem 3.2 (with $p = 1$). For the first part, consider the function $F_1(y) = y\mathbf{1}_{[1,\infty)}(y)$. Then $y - 1 \leq F_1(y)$, and one gets

$$\mathbb{E}(M_n - 1) \leq \mathbb{E}F_1(M_n) \leq \int_\Omega |X_n| \int_{[1,\max\{1,M_n\}]} y\frac{1}{y} dy\right) dP$$

$$= \int_\Omega |X_n| \log(\max\{1, M_n\}) dP$$

$$= \int_{[M_n \geq 1]} |X_n| \log M_n dP. \tag{3.13}$$

Now use the inequality (proved in the remark below)

$$a \log b \leq a \log^+ a + \frac{b}{e}, \quad a, b \geq 0, \tag{3.14}$$

to further arrive at

$$\mathbb{E}M_n \leq 1 + \mathbb{E}|X_n| \log^+ |X_n| + \frac{\mathbb{E}M_n}{e}. \tag{3.15}$$

This establishes the inequality for the case $p = 1$. For $p > 1$ take $F_1(y) = y^p$. Then

$$\begin{aligned}
\mathbb{E}M_n^p &\leq \mathbb{E}\left(|X_n| \int_{[0,M_n]} py^{p-2}dy\right) \\
&= \mathbb{E}\left(|X_n| \frac{p}{p-1} M_n^{p-1}\right) \\
&\leq \frac{p}{p-1}(\mathbb{E}|X_n|^p)^{\frac{1}{p}} (\mathbb{E}M_n^{(p-1)q})^{\frac{1}{q}} \\
&= q(\mathbb{E}|X_n|^p)^{\frac{1}{p}} (\mathbb{E}M_n^p)^{\frac{1}{q}}. \tag{3.16}
\end{aligned}$$

The bound for $p > 1$ now follows by dividing by $(\mathbb{E}M_n^p)^{\frac{1}{q}}$ and a little algebra. ∎

Remark 3.4 To prove the inequality (3.14) it is sufficient to consider the case $1 < a < b$, since it obviously holds otherwise. In this case it may be expressed as

$$\log b \leq \log a + \frac{b}{ae},$$

or

$$\log \frac{b}{a} \leq \frac{b}{ae}.$$

But this follows from the fact that $f(x) = \frac{\log x}{x}$, $x > 1$, has a maximum value $\frac{1}{e}$.

Corollary 3.5 Let $\{X_t : t \in [0,T]\}$ be a right-continuous nonnegative $\{\mathcal{F}_t\}$-submartingale with $\mathbb{E}|X_T|^p < \infty$ for some $p \geq 1$. Then $M_T := \sup\{X_s : 0 \leq s \leq T\}$ is $\mathcal{F}_T$-measurable and, for all $\lambda > 0$,

$$P(M_T > \lambda) \leq \frac{1}{\lambda^p} \int_{[M_T > \lambda]} X_T^p dP \leq \frac{1}{\lambda^p} \mathbb{E}X_T^p. \tag{3.17}$$

Proof Consider the nonnegative submartingale $\{X_0, \ldots, X_{Ti2^{-n}}, \ldots, X_T\}$, for each $n = 1, 2, \ldots$, and let $M_n := \max\{X_{iT2^{-n}} : 0 \leq i \leq 2^n\}$. For $\lambda > 0$, $[M_n > \lambda] \uparrow [M_T > \lambda]$ as $n \uparrow \infty$. In particular, M_T is $\mathcal{F}_T$-measurable. By Theorem 3.2,

$$P(M_n > \lambda) \leq \frac{1}{\lambda^p} \int_{[M_n > \lambda]} X_T^p dP \leq \frac{1}{\lambda^p} \mathbb{E}X_T^p.$$

Letting $n \uparrow \infty$, (3.17) is obtained. ∎

We finally come to the notions of **stopping times**, and **optional times** which provide a powerful probabilistic tool to analyze processes by viewing them at appropriate random times.

Definition 3.4 Let $\{\mathcal{F}_t : t \in T\}$ be a filtration on a probability space $(\Omega, \mathcal{F}, P)$, with T a linearly ordered index set to which one may adjoin, if necessary, a point '∞' as the largest point of $T \cup \{\infty\}$. A random variable $\tau : \Omega \to T \cup \{\infty\}$ is an $\{\mathcal{F}_t\}$-**stopping time** if $[\tau \leq t] \in \mathcal{F}_t \, \forall \, t \in T$. If $[\tau < t] \in \mathcal{F}_t$ for all $t \in T$ then τ is called an **optional time**.

Most commonly, T in this definition is $\mathbb{N}$ or $\mathbb{Z}^+$, or $[0, \infty)$, and τ is related to an $\{\mathcal{F}_t\}$-adapted process $\{X_t : t \in T\}$.

The intuitive idea of τ as a stopping-time strategy is that to "stop by time t, or not," according to τ, is determined by the knowledge of the past up to time t, and does not require "a peek into the future."

Example 4 Let $\{X_t : t \in T\}$ be an $\{\mathcal{F}_t\}$-adapted process with values in a measurable space $(S, \mathcal{S})$, with a linearly ordered index set. (a) If $T = \mathbb{N}$ or $\mathbb{Z}^+$, then for every $B \in \mathcal{S}$,

$$\tau_B := \inf\{t \geq 0 : X_t \in B\} \tag{3.18}$$

is an $\{\mathcal{F}_t\}$-stopping time. (b) If $T = \mathbb{R}_+ \equiv [0, \infty)$, S is a metric space $\mathcal{S} = \mathcal{B}(S)$, and B is *closed*, $t \mapsto X_t$ is continuous, then τ_B is an $\{\mathcal{F}_t\}$-stopping time. (c) If $T = \mathbb{R}_+$, S is a topological space, $t \mapsto X_t$ is right-continuous, and B is *open*, then $[\tau_B < t] \in \mathcal{F}_t$ for all $t \geq 0$, and hence τ_B is an optional time; see Definition 3.4.

We leave the proofs of (a)–(c) as Exercise 11. Note that (b), (c) imply that under the hypothesis of (b), τ_B is an optional time if B is open or closed; recall Definition 3.4.

Definition 3.5 Let $\{\mathcal{F}_t : t \in T\}$ be a filtration on $(\Omega, \mathcal{F})$. Suppose that τ is a $\{\mathcal{F}_t\}$-stopping time. The **pre-τ σ-field** $\mathcal{F}_\tau$ comprises all $A \in \mathcal{F}$ such that $A \cap [\tau \leq t] \in \mathcal{F}_t$ for all $t \in T$.

Heuristically, $\mathcal{F}_\tau$ comprises events determined by information available only up to time τ. For example, if T is discrete with elements $t_1 < t_2 < \cdots$, and $\mathcal{F}_t = \sigma(X_s : 0 \leq s \leq t) \subset \mathcal{F}, \forall t$, where $\{X_t : t \in T\}$ is a process with values in some measurable space $(S, \mathcal{S})$, then $\mathcal{F}_\tau = \sigma(X_{\tau \wedge t} : t \geq 0)$; (Exercise 9). The stochastic process $\{X_{\tau \wedge t} : t \geq 0\}$ is referred to as the **stopped process**. The notation $\wedge$ is defined by $a \wedge b = \min\{a, b\}$. Similarly $\vee$ is defined by $a \vee b = \max\{a, b\}$.

If τ_1, τ_2 are two $\{\mathcal{F}_t\}$-stopping times and $\tau_1 \leq \tau_2$, then it is simple to check that

$$\mathcal{F}_{\tau_1} \subset \mathcal{F}_{\tau_2}. \tag{3.19}$$

Suppose $\{X_t\}$ is an $\{\mathcal{F}_t\}$-adapted process with values in a measurable space $(S, \mathcal{S})$, and τ is an $\{\mathcal{F}_t\}$-stopping time. For many purposes the following notion of adapted joint measurability of $(t, \omega) \mapsto X_t(\omega)$ is important.

Definition 3.6 Let $T = [0, \infty)$ or $T = [0, t_0]$ for some $t_0 < \infty$. A stochastic process $\{X_t : t \in T\}$ with values in a measurable space $(S, \mathcal{S})$ is **progressively measurable** with respect to $\{\mathcal{F}_t\}$ if for each $t \in T$, the map $(s, \omega) \mapsto X_s(\omega)$, from $[0, t] \times \Omega$ to

S is measurable with respect to the σ-fields $\mathcal{B}[0,t] \otimes \mathcal{F}_t$ (on $[0,t] \times \Omega$) and $\mathcal{S}$ (on S). Here $\mathcal{B}[0,t]$ is the Borel σ-field on $[0,t]$, and $\mathcal{B}[0,t] \otimes \mathcal{F}_t$ is the usual product σ-field.

Proposition 3.6 (a) Suppose $\{X_t : t \in T\}$ is progressively measurable, and τ is a stopping time. Then X_τ is $\mathcal{F}_\tau$-measurable, i.e., $[X_\tau \in B] \cap [\tau \le t] \in \mathcal{F}_t$ for each $B \in \mathcal{S}$ and each $t \in T$. (b) Suppose S is a metric space and $\mathcal{S}$ its Borel σ-field. If $\{X_t : t \in T\}$ is right-continuous, then it is progressively measurable.

Proof (a) Fix $t \in T$. On the set $\Omega_t := [\tau \le t]$, X_τ is the composition of the maps (i) $f(\omega) := (\tau(\omega), \omega)$, from $\omega \in \Omega_t$ into $[0,t] \times \Omega_t$, and (ii) $g(s, \omega) = X_s(\omega)$ on $[0,t] \times \Omega_t$ into S. Now f is $\tilde{\mathcal{F}}_t$-measurable on Ω_t, where $\tilde{\mathcal{F}}_t := \{A \cap \Omega_t : A \in \mathcal{F}_t\}$ is the **trace** σ-**field** on Ω_t, and $\mathcal{B}[0,t] \otimes \tilde{\mathcal{F}}_t$ is the σ-field on $[0,t] \times \Omega_t$. Next the map $g(s, \omega) = X_s(\omega)$ on $[0,t] \times \Omega$ into S is $\mathcal{B}[0,t] \otimes \mathcal{F}_t$-measurable. Therefore, the restriction of this map to the measurable subset $[0,t] \times \Omega_t$ is measurable on the trace σ-field $\{A \cap ([0,t] \times \Omega_t) : A \in \mathcal{B}[0,t] \otimes \mathcal{F}_t\}$. Therefore, the composition X_τ is $\tilde{\mathcal{F}}_t$-measurable on Ω_t, i.e., $[X_\tau \in B] \cap [\tau \le t] \in \tilde{\mathcal{F}}_t \subset \mathcal{F}_t$ and hence $[X_\tau \in B] \in \mathcal{F}_\tau$, for $B \in \mathcal{S}$.

(b) Fix $t \in T$. Define, for each positive integer n, the stochastic process $\{X_s^{(n)} : 0 \le s \le t\}$ by

$$X_s^{(n)} := X_{j2^{-n}t} \text{ for } (j-1)2^{-n}t \le s < j2^{-n}t \quad (1 \le j \le 2^n), \ X_t^{(n)} = X_t.$$

Since $\{(s, \omega) \in [0,t] \times \Omega : X_s^{(n)}(\omega) \in B\} = \cup_{j=1}^{2^n}([j-1)2^{-n}t, j2^{-n}t) \times \{\omega : X_{j2^{-n}t}(\omega) \in B\}) \cup (\{t\} \times \{\omega : X_t(\omega) \in B\}) \in \mathcal{B}[0,t] \otimes \mathcal{F}_t$, it follows that $\{X_s^{(n)}\}$ is progressively measurable. Now $X_t^{(n)}(\omega) \to X_t(\omega)$ for all (t, ω) as $n \to \infty$, in view of the right-continuity of $t \mapsto X_t(\omega)$. Hence $\{X_t : t \in T\}$ is progressively measurable. ∎

Remark 3.5 It is often important to relax the assumption of 'right-continuity' of $\{X_t : t \in T\}$ to "a.s. right-continuity." To ensure progressive measurability in this case, it is convenient to take $\mathcal{F}, \mathcal{F}_t$ to be P-**complete**, i.e., if $P(A) = 0$ and $B \subset A$ then $B \in \mathcal{F}$ and $B \in \mathcal{F}_t \forall t$. Then modify X_t to equal $X_0 \forall t$ on the P-null set $N = \{\omega : t \to X_t(\omega) \text{ is not right-continuous}\}$. This modified $\{X_t : t \in T\}$, together with $\{\mathcal{F}_t : t \in T\}$ satisfy the hypothesis of part (b) of Proposition 3.6.

The following proposition is distinguished as a characterization of the uniformly integrable martingales as conditional expectations.

Proposition 3.7 (a) Let Y be integrable and $\mathcal{F}_n (n = 1, 2, \ldots)$ a filtration. Then the martingale $Y_n = \mathbb{E}(Y|\mathcal{F}_n)$ is uniformly integrable. (b) Suppose Y_n is a $\mathcal{F}_n$-martingale $(n = 1, 2, \ldots)$ such that $Y_n \to Y$ in L^1. Then $Y_n = \mathbb{E}(Y|\mathcal{F}_n), n \ge 1$.

Proof (a) Note that $P(|Y_n| > \lambda) \le \frac{\mathbb{E}|Y_n|}{\lambda} \le \frac{\mathbb{E}|Y|}{\lambda}$ for $\lambda > 0$. Hence, given $\varepsilon > 0$, there exist $\lambda > 0$ such that $P(|Y_n| > \lambda) < \varepsilon$ for all n. Therefore, $\mathbb{E}(\mathbf{1}_{[|Y_n|>\lambda]}|Y_n|) \le \mathbb{E}(\mathbf{1}_{[|Y_n|>\lambda]}\mathbb{E}(|Y||\mathcal{F}_n)) = \mathbb{E}(\mathbf{1}_{[|Y_n|>\lambda]}|Y|) \to 0$ as $\lambda \to \infty$ uniformly in n (see Exercise 14). (b) Let $A \in \mathcal{F}_m$. Then $\mathbb{E}(\mathbf{1}_A Y_m) = \mathbb{E}(\mathbf{1}_A Y_n)$ for all $n \ge m$. Taking the limit

as $n \to \infty$, we have $\mathbb{E}(1_A Y_m) = \mathbb{E}(1_A Y)$ for all $A \in \mathcal{F}_m$. That is, $Y_m = \mathbb{E}(Y|\mathcal{F}_m)$.
∎

Remark 3.6   One can show that a $\mathcal{F}_n$-martingale $\{Z_n : n \geq 1\}$ has the representation $Z_n = \mathbb{E}(Z|\mathcal{F}_n)$ iff it is uniformly integrable, and then $Z_n \to Z$ a.s. and in L^1. Indeed, a uniformly integrable martingale converges a.s. and in L^1 by the martingale convergence theorem; see Theorems 1.10 and 3.12.

One of the important implications of the martingale property is that of *constant expected values*. Let us consider a substantially stronger property. Consider a discrete parameter martingale sequence $X_0, X_1, \ldots$, and stopping time τ with respect to some filtration $\mathcal{F}_n, n \geq 0$. Let m be an integer and suppose that $\tau \leq m$. For $G \in \mathcal{F}_\tau$, write $g = 1_G$. One has that $G \cap [\tau = k] = (G \cap [\tau \leq k])\backslash(G \cap [\tau \leq k-1]) \in \mathcal{F}_k$ from the definition of $\mathcal{F}_\tau$. It follows from the martingale property $\mathbb{E}(X_m|\mathcal{F}_k) = X_m$, one has

$$\mathbb{E}(gX_\tau) = \sum_{k=0}^m \mathbb{E}(g1_{[\tau=k]}X_k)$$

$$= \sum_{k=0}^m \mathbb{E}(g1_{[\tau=k]}\mathbb{E}(X_m|\mathcal{F}_k))$$

$$= \sum_{k=0}^m \mathbb{E}(g1_{[\tau=k]}X_m) = \mathbb{E}(gX_m). \tag{3.20}$$

Thus the constancy of expectations $\mathbb{E}X_n = \mathbb{E}X_0$ property of martingales extends to certain stopping times τ in place of n. However, as illustrated in Example 5, below, this requires some further conditions on τ than merely being a stopping time to extend to unbounded cases. The following theorem provides precisely such conditions.

Theorem 3.8 *(Optional Stopping)* Let $\{X_t : t \in T\}$ be a right-continuous $\{\mathcal{F}_t\}$-martingale, where $T = \mathbb{N}$ or $T = [0, \infty)$. (a) If $\tau_1 \leq \tau_2$ are bounded stopping times, then

$$\mathbb{E}(X_{\tau_2}|\mathcal{F}_{\tau_1}) = X_{\tau_1}. \tag{3.21}$$

(b) *(Optional Sampling)*. If τ is a stopping time (not necessarily finite), then $\{X_{\tau \wedge t} : t \in T\}$ is an $\{\mathcal{F}_{\tau \wedge t}\}_{t \in T}$-martingale.

(c) Suppose $\tau_1 \leq \tau_2$ are stopping times such that (i) $P(\tau_2 < \infty) = 1$, and (ii) $X_{\tau_2 \wedge t}(t \in T)$ is uniformly integrable. Then

$$\mathbb{E}(X_{\tau_2}|\mathcal{F}_{\tau_1}) = X_{\tau_1}. \tag{3.22}$$

In particular,

$$\mathbb{E}(X_{\tau_2}) = \mathbb{E}(X_{\tau_1}) = \mathbb{E}(X_0).$$

Proof First we consider the case $T = \mathbb{N}$. In the case that τ_2 is bounded by some positive integer m, it follows from the above calculation (3.20) that

$$\mathbb{E}(X_m|\mathcal{F}_{\tau_2}) = X_{\tau_2}.$$

Thus if $\tau_1 \leq \tau_2 \leq m$, then one also has

$$\mathbb{E}(X_m|\mathcal{F}_{\tau_1}) = X_{\tau_1}.$$

In other words, the two term sequence X_{τ_1}, X_{τ_2} is a martingale with respect to $\mathcal{F}_{\tau_1}, \mathcal{F}_{\tau_2}$. Hence it follows from the "smoothing property"of conditional expectation that $X_{\tau_1} = \mathbb{E}(X_m|\mathcal{F}_{\tau_1}) = \mathbb{E}(\mathbb{E}(X_m|\mathcal{F}_{\tau_2})|\mathcal{F}_{\tau_1}) = \mathbb{E}(X_{\tau_2}|\mathcal{F}_{\tau_1})$. This proves (a) in the discrete parameter case. Part (b) follows directly from (a) since $\tau_1 := \tau \wedge n \leq \tau_2 := n < \infty$ for any n, and both of these τ_1, τ_2, so-defined, are stopping times. For (c), let $G \in \mathcal{F}_{\tau_1}$. Then, $G \cap [\tau_1 \leq n] \in \mathcal{F}_n$. Also $G \cap [\tau_1 \leq n] \in \mathcal{F}_{\tau_1 \wedge n}$. To see this, for $m \geq n$, $G \cap [\tau_1 \leq n]) \cap [\tau_1 \wedge n \leq m] = G \cap [\tau_1 \leq n] \in \mathcal{F}_n \subset \mathcal{F}_m$, and if $m < n$ then $(G \cap [\tau_1 \leq n]) \cap [\tau_1 \wedge n \leq m] = G \cap [\tau_1 \leq m] \in \mathcal{F}_m$. So in either case $G \cap [\tau_1 \leq n] \in \mathcal{F}_{\tau_1 \wedge n}$. Also $\tau_1 \wedge n \leq \tau_2 \wedge n$. By part (a), $\mathbb{E}(X_{\tau_2 \wedge n}|\mathcal{F}_{\tau_1 \wedge n}) = X_{\tau_1 \wedge n}$. Thus

$$\mathbb{E}(g\mathbf{1}_{[\tau_1 \leq n]}X_{\tau_2 \wedge n}) = \mathbb{E}(g\mathbf{1}_{[\tau_1 \leq n]}X_{\tau_1 \wedge n}), \quad g = \mathbf{1}_G. \tag{3.23}$$

So, by the uniform integrability of $X_{\tau_2 \wedge n}$, and the fact that $X_{\tau_2 \wedge n} \to X_{\tau_2}$ a.s. as $n \to \infty$, one has $X_{\tau_2 \wedge n} \to X_{\tau_2}$ in L^1. Now observe that the uniform integrability of $X_{\tau_2 \wedge n}$ implies that of $X_{\tau_1 \wedge n}, n \geq 1$, as follows: Since $X_{\tau_2 \wedge n}, n \geq 1$, is uniformly integrable, it converges in L^1 (and a.s.) to X_{τ_2}, and $X_{\tau_2 \wedge n} = \mathbb{E}(X_{\tau_2}|\mathcal{F}_{\tau_2 \wedge n})$. Therefore $X_{\tau_1 \wedge n} = \mathbb{E}(X_{\tau_2 \wedge n}|\mathcal{F}_{\tau_1 \wedge n}) = \mathbb{E}[\mathbb{E}(X_{\tau_2}|\mathcal{F}_{\tau_2 \wedge n})|\mathcal{F}_{\tau_1 \wedge n}] = \mathbb{E}(X_{\tau_2}|\mathcal{F}_{\tau_1 \wedge n})$. Uniform integrability of $X_{\tau_1 \wedge n}, n \geq 1$, now follows from Proposition 3.7(a). Putting this uniform integrability together, it follows that the left side of (3.23) converges to $\mathbb{E}(gX_{\tau_2})$ and the right side to $\mathbb{E}(gX_{\tau_1})$. Since this is for any $g = \mathbf{1}_G, G \in \mathcal{F}_{\tau_1}$, the proof of (b) follows.

Next we consider the case $T = [0, \infty)$. Let $\tau_1 \leq \tau_2 \leq t_0$ a.s. The idea for the proof is, as above, to check that $\mathbb{E}[X_{t_0}|\mathcal{F}_{\tau_i}] = X_{\tau_i}$, for each of the stopping times $(i = 1, 2)$ simply by virtue of their being bounded. Once this is established, the result (a) follows by smoothing of conditional expectation, since $\mathcal{F}_{\tau_1} \subset \mathcal{F}_{\tau_2}$. That is, it will then follow that

$$\mathbb{E}[X_{\tau_2}|\mathcal{F}_{\tau_1}] = \mathbb{E}[\mathbb{E}(X_{t_0}|\mathcal{F}_{\tau_2})|\mathcal{F}_{\tau_1}] = \mathbb{E}[X_{t_0}|\mathcal{F}_{\tau_1}] = X_{\tau_1}.$$

So let τ denote either of τ_i, $i = 1, 2$, and consider $\mathbb{E}[X_{t_0}|\mathcal{F}_\tau]$. For each $n \geq 1$ consider the nth dyadic subdivision of $[0, t_0]$ and define $\tau^{(n)} = (k + 1)2^{-n}t_0$ if $\tau \in [k2^{-n}t_0, (k + 1)2^{-n}t_0)(k = 0, 1, \ldots, 2^n - 1)$, and $\tau^{(n)} = t_0$ if $\tau = t_0$. Then $\tau^{(n)}$ is a stopping time and $\mathcal{F}_\tau \subset \mathcal{F}_{\tau^{(n)}}$ (since $\tau \leq \tau^{(n)}$). For $G \in \mathcal{F}_\tau$, exploiting the martingale property $\mathbb{E}[X_{t_0}|\mathcal{F}_{(k+1)2^{-n}t_0}] = X_{t_{(k+1)2^{-n}t_0}}$, one has

$$\mathbb{E}(\mathbf{1}_G X_{t_0}) = \sum_{k=0}^{2^n-1} \mathbb{E}(\mathbf{1}_{G \cap [\tau^{(n)} = (k+1)2^{-n} t_0]} X_{t_0})$$

$$= \sum_{k=0}^{2^n-1} \mathbb{E}(\mathbf{1}_{G \cap [\tau^{(n)} = (k+1)2^{-n} t_0]} X_{(k+1)2^{-n} t_0})$$

$$= \sum_{k=0}^{2^n-1} \mathbb{E}(\mathbf{1}_{G \cap [\tau^{(n)} = (k+1)2^{-n} t_0]} X_{\tau^{(n)}})$$

$$= \mathbb{E}(\mathbf{1}_G X_{\tau^{(n)}}) \rightarrow \mathbb{E}(\mathbf{1}_G X_\tau). \tag{3.24}$$

The last convergence is due to the L^1-convergence criterion of Theorem 1.10 in view of the following checks: (1) X_t is right-continuous (and $\tau^{(n)} \downarrow \tau$), so that $X_{\tau^{(n)}} \rightarrow X_\tau$ a.s., and (2) $X_{\tau^{(n)}}$ is uniformly integrable, since by the submartingale property of $\{|X_t| : t \in T\}$,

$$\mathbb{E}(\mathbf{1}_{[|X_{\tau^{(n)}}| > \lambda]} |X_{\tau^{(n)}}|) = \sum_{k=0}^{2^n-1} \mathbb{E}(\mathbf{1}_{[\tau^{(n)} = (k+1)2^{-n} t_0] \cap [|X_{\tau^{(n)}}| > \lambda]} |X_{(k+1)2^{-n} t_0}|)$$

$$\leq \sum_{k=0}^{2^n-1} \mathbb{E}(\mathbf{1}_{[\tau^{(n)} = (k+1)2^{-n} t_0] \cap [|X_{\tau^{(n)}}| > \lambda]} |X_{t_0}|)$$

$$= \mathbb{E}(\mathbf{1}_{[|X_{\tau^{(n)}}| > \lambda]} |X_{t_0}|) \rightarrow \mathbb{E}(\mathbf{1}_{[|X_\tau| > \lambda]} |X_{t_0}|).$$

Since the left side of (3.24) does not depend on n, it follows that

$$\mathbb{E}(\mathbf{1}_G X_{t_0}) = \mathbb{E}(\mathbf{1}_G X_\tau) \quad \forall\, G \in \mathcal{F}_\tau,$$

i.e., $\mathbb{E}(X_{t_0} | \mathcal{F}_\tau) = X_\tau$ applies to both $\tau = \tau_1$ and $\tau = \tau_2$. The result (a) therefore follows by the smoothing property of conditional expectations noted at the start of the proof.

As in the discrete parameter case, (b) follows immediately from (a). For if $s < t$ are given, then $\tau \wedge s$ and $\tau \wedge t$ are both bounded by t, and $\tau \wedge s \leq \tau \wedge t$.

(c) Since $\tau < \infty$ a.s., $\tau \wedge t$ equals τ for sufficiently large t (depending on ω), outside a P-null set. Therefore, $X_{\tau \wedge t} \rightarrow X_\tau$ a.s. as $t \rightarrow \infty$. By assumption (ii), $X_{\tau \wedge t}$ $(t \geq 0)$ is uniformly integrable. Hence $X_{\tau \wedge t} \rightarrow X_\tau$ in L^1. In particular, $\mathbb{E}(X_{\tau \wedge t}) \rightarrow \mathbb{E}(X_\tau)$ as $t \rightarrow \infty$. But $\mathbb{E} X_{\tau \wedge t} = \mathbb{E} X_0 \; \forall\, t$, by (b). ∎

Remark 3.7 If $\{X_t : t \in T\}$ in Theorem 3.8 is taken to be a submartingale, then instead of the equality sign "=" in (3.21), (3.22), one gets "≤."

Remark 3.8 The *stopping time approximation technique* used in the proof of Theorem 3.8, to obtain a decreasing sequence $\tau^{(1)} \geq \tau^{(2)} \geq \cdots$ of discrete stopping times converging to τ, is adaptable to any number of situations involving the analysis of processes having right-continuous sample paths.

The following proposition and its corollary are often useful for verifying the hypothesis of Theorem 3.8 in examples.

Proposition 3.9 Let $\{Z_n : n \in \mathbb{N}\}$ be real-valued random variables such that for some $\varepsilon > 0, \delta > 0$, one has

$$P(Z_{n+1} > \varepsilon \mid \mathcal{G}_n) \geq \delta \text{ a.s.} \quad \forall n = 0, 1, 2, \ldots$$

or

$$P(Z_{n+1} < -\varepsilon \mid \mathcal{G}_n) \geq \delta \text{ a.s.} \quad \forall n = 0, 1, 2, \ldots, \tag{3.25}$$

where $\mathcal{G}_n = \sigma\{Z_1, \ldots, Z_n\}$ $(n \geq 1)$, $\mathcal{G}_0 = \{\emptyset, \Omega\}$. Let $S_n^x = x + Z_1 + \cdots + Z_n$ $(n \geq 1)$, $S_0^x = x$, and let $a < x < b$. Let τ be the first escape time of $\{S_n^x\}$ from (a, b), i.e., $\tau = \tau^x = \inf\{n \geq 1 : S_n^x \in (a, b)^c\}$. Then $\tau < \infty$ a.s. and

$$\sup_{\{x: a < x < b\}} \mathbb{E} e^{\tau z} < \infty \quad \text{for} \quad -\infty < z < \frac{1}{n_0}\left(\log \frac{1}{1 - \delta_0}\right), \tag{3.26}$$

writing $[y]$ for the integer part of y,

$$n_0 = \left[\frac{b - a}{\varepsilon}\right] + 1, \qquad \delta_0 = \delta^{n_0}. \tag{3.27}$$

Proof Suppose the first relation in (3.25) holds. Clearly, if $Z_j > \varepsilon \; \forall j = 1, 2, \ldots, n_0$, then $S_{n_0}^x > b$, so that $\tau \leq n_0$. Therefore, $P(\tau \leq n_0) \geq P(Z_1 > \varepsilon, \ldots, Z_{n_0} > \varepsilon) \geq \delta^{n_0}$, by taking successive conditional expectations (given $\mathcal{G}_{n_0-1}, \mathcal{G}_{n_0-2}, \ldots, \mathcal{G}_0$, in that order). Hence $P(\tau > n_0) \leq 1 - \delta^{n_0} = 1 - \delta_0$. For every integer $k \geq 2$, $P(\tau > kn_0) = P(\tau > (k-1)n_0, \tau > kn_0) = \mathbb{E}[\mathbf{1}_{[\tau > (k-1)n_0]}P(\tau > kn_0 | \mathcal{G}_{(k-1)n_0})] \leq (1 - \delta_0)P(\tau > (k-1)n_0)$, since, on the set $[\tau > (k-1)n_0]$, $P(\tau \leq kn_0 | \mathcal{G}_{(k-1)n_0}) \geq P(Z_{(k-1)n_0+1} > \varepsilon, \ldots, Z_{kn_0} > \varepsilon | \mathcal{G}_{(k-1)n_0}) \geq \delta^{n_0} = \delta_0$. Hence, by induction, $P(\tau > kn_0) \leq (1 - \delta_0)^k$. Hence $P(\tau = \infty) = 0$, and for all $z > 0$,

$$\mathbb{E} e^{z\tau} = \sum_{r=1}^{\infty} e^{zr} P(\tau = r) \leq \sum_{k=1}^{\infty} e^{zkn_0} \sum_{r=(k-1)n_0+1}^{kn_0} P(\tau = r)$$

$$\leq \sum_{k=1}^{\infty} e^{zkn_0} P(\tau > (k-1)n_0) \leq \sum_{k=1}^{\infty} e^{zkn_0} (1 - \delta_0)^{k-1}$$

$$= e^{zn_0}(1 - (1 - \delta_0)e^{zn_0})^{-1} \quad \text{if} \quad e^{zn_0}(1 - \delta_0) < 1.$$

An entirely analogous argument holds if the second relation in (3.25) holds. ∎

The following corollary immediately follows from Proposition 3.9.

Corollary 3.10 Let $\{Z_n : n = 1, 2, \ldots\}$ be an i.i.d. sequence such that $P(Z_1 = 0) < 1$. Let $S_n^n = x + Z_1 + \cdots + Z_n$ $(n \geq 1)$, $S_0^x = x$, and $a < x < b$. Then the first escape time τ of the random walk from the interval (a, b) has a finite moment generating function in a neighborhood of 0.

Example 5 Let $Z_n (n \geq 1)$ be i.i.d. symmetric Bernoulli, $P(Z_i = +1) = P(Z_i = -1) = \frac{1}{2}$, and let $S_n^x = x + Z_1 + \cdots + Z_n (n \geq 1)$, $S_0^x = x$, be the simple symmetric random walk on the state space $\mathbb{Z}$, starting at x. Let $a \leq x \leq b$ be integers, $\tau_y := \inf\{n \geq 0 : S_n^x = y\}$, $\tau = \tau_a \wedge \tau_b = \inf\{n \geq 0 : S_n^x \in \{a, b\}\}$. Then $\{S_n^x : n \geq 0\}$ is a martingale and τ satisfies the hypothesis of Theorem 3.8(c) (Exercise 4). Hence

$$x \equiv \mathbb{E} S_0^x = \mathbb{E} S_\tau^x = a P(\tau_a < \tau_b) + b P(\tau_b < \tau_a) = a + (b - a) P(\tau_b < \tau_a),$$

so that

$$P(\tau_b < \tau_a) = \frac{x - a}{b - a}, \quad P(\tau_a < \tau_b) = \frac{b - x}{b - a}, \quad a \leq x \leq b. \tag{3.28}$$

Letting $a \downarrow -\infty$ in the first relation, and letting $b \uparrow \infty$ in the second, one arrives at the conclusion that the simple symmetric random walk reaches every state with probability one, no mater where it starts. This property is referred to as *recurrence*; also see Example 9, of Chapter II. To illustrate the importance of the hypothesis imposed on τ in Theorem 3.8(c), one may naively try to apply (3.22) to τ_b (see Exercise 4) and arrive at the silly conclusion $x = b$!

Example 6 One may apply Theorem 3.8(c) to a simple asymmetric random walk with $P(Z_i = 1) = p$, $P(Z_i = -1) = q \equiv 1 - p(0 < p < 1, p \neq 1/2)$, so that $X_n^x := S_n^x - (2p - 1)n \ (n \geq 1)$, $X_0^x \equiv x$, is a martingale. Then with $\tau_a, \tau_b, \tau = \tau_a \wedge \tau_b$ as above, one gets

$$x \equiv \mathbb{E} X_0^x = \mathbb{E} X_\tau^x = \mathbb{E} S_\tau^x - (2p - 1)\mathbb{E}\tau = a + (b - a) P(\tau_b < \tau_a) - (2p - 1)\mathbb{E}\tau. \tag{3.29}$$

Since we do not know $\mathbb{E}\tau$ yet, we can not quite solve (3.29). We therefore use a second martingale $(q/p)^{S_n^x} (n \geq 0)$. Note that $\mathbb{E}[(q/p)^{S_{n+1}^x} | \sigma\{Z_1, \ldots, Z_n\}] = (q/p)^{S_n^x} \cdot \mathbb{E}[(q/p)^{Z_{n+1}}] = (q/p)^{S_n^x}[(q/p)p + (q/p)^{-1}q] = (q/p)^{S_n^x} \cdot 1 = (q/p)^{S_n^x}$, proving the martingale property of the "exponential process" $Y_n := (q/p)^{S_n^x} = \exp(c S_n^x)$, $c = \ln(q/p), n \geq 0$. Note that $(q/p)^{S_{\tau \wedge n}^x} \leq \max\{(q/p)^y : a \leq y \leq b\}$, which is a finite number. Hence the hypothesis of uniform integrability holds. Applying (3.22) we get

$$(q/p)^x = (q/p)^a \cdot P(\tau_a < \tau_b) + (q/p)^b P(\tau_b < \tau_a),$$

or

$$P(\tau_b < \tau_a) = \frac{(q/p)^x - (q/p)^a}{(q/p)^b - (q/p)^a} \equiv \varphi(x) \quad (a \leq x \leq b). \tag{3.30}$$

Using this in (3.29) we get

$$\mathbb{E}\tau \equiv \mathbb{E}\tau_a \wedge \tau_b = \frac{x - a - (b - a)\varphi(x)}{1 - 2p}, \quad a \leq x \leq b. \tag{3.31}$$

Suppose now that $p < q$, i.e., $p < \frac{1}{2}$. Letting $a \downarrow -\infty$ in (3.30), one sees that the probability of ever reaching b starting from $x < b$ is $(\frac{p}{q})^{b-x} < 1$. Similarly if $p > \frac{1}{2}$, i.e., $q < p$, then the probability of ever reaching a starting from $x > a$ is $(\frac{q}{p})^{x-a} < 1$ (Exercise 6).

Very loosely speaking the submartingale and and supermartingale properties convey a sense of "monotonicity"in predictions of successive terms based on the past. This is so much so that the expected values comprise a monotone sequence of numbers. Recall from calculus that every sequence of real numbers bounded above (or below) must have a limit. Perhaps some form of "boundedness"at least seems worthy of consideration in the context of martingale convergence ? Indeed, as we now see, the implications are striking!

Let $\{Z_n : n = 1, 2, \dots\}$ be a $\{\mathcal{F}_n\}_{n=1}^{\infty}$-submartingale, and $a < b$ arbitrary real numbers. Recursively define successive *crossing times* of (a, b) by $\eta_1 = 1, \eta_2 = \inf\{n \geq 1 : Z_n \geq b\}, \eta_{2k-1} = \inf\{n \geq \eta_{2k-2} : Z_n \leq a\}, \eta_{2k} = \inf\{n \geq \eta_{2k-1} : Z_n \geq b\}$. In particular η_{2k} is the time of the k-th upcrossing of the interval (a, b) by the sequence $\{Z_n : n = 1, 2, \dots\}$. η_{2k} is also the time of the k-th upcrossing of $(0, b - a)$ by the sequence $X_n = \max(Z_n - a, 0) = (Z_n - a)^+, n \geq 1$. Note that these crossing times are in fact stopping times. Also, X_n is *nonnegative* and $X_n = 0$ if $Z_n \leq a$, and $X_n \geq b - a$ if $Z_n \geq b$.

For a positive integer N, consider their truncations $\tau_k = \eta_k \wedge N$, which are also $\{\mathcal{F}_n\}$-stopping times, in fact, *bounded* stopping times. Let $U_N = \max\{k : \eta_{2k} \leq N\}$ denote the number of upcrossings of (a, b) by $\{Z_n : n = 1, 2 \dots\}$ by the time N. Then U_N may also be viewed as the number of upcrossings of the interval $(0, b - a)$ by the submartingale $X_n (n = 1, 2, \dots)$.

Theorem 3.11 *(Doob's Upcrossing Inequality)* Let $\{Z_n : n \geq 1\}$ be a $\{\mathcal{F}_n\}$-submartingale, and $a < b$ arbitrary real numbers. Then the number U_N of upcrossings of (a, b) by time N satisfies

$$\mathbb{E}U_N \leq \frac{\mathbb{E}(Z_N - a)^+ - \mathbb{E}(Z_1 - a)^+}{b - a} \leq \frac{\mathbb{E}|Z_N| + |a|}{b - a}.$$

Proof For $k > U_N + 1$, $\eta_{2k} > N$, and $\eta_{2k-1} > N$ so that $\tau_{2k} = N$ and $\tau_{2k-1} = N$. Hence, $X_{\tau_{2k}} = X_N = X_{\tau_{2k-1}}$. If $k \leq U_N$ then $\eta_{2k} \leq N$, and $\eta_{2k-1} \leq N$ so that $X_{\tau_{2k}} \geq b - a, 0 = X_{\tau_{2k-1}} = X_{\eta_{2k-1}}$. Now suppose $k = U_N + 1$. Then $\eta_{2k} > N$ and $X_{\tau_{2k}} = X_N$. Also, either $\eta_{2k-1} \geq N$ so that $\tau_{2k-1} = N$, and $X_{\tau_{2k-1}} = X_N, X_{\tau_{2k}} - X_{\tau_{2k-1}} = 0$, or $\eta_{2k-1} < N$, in which case $\eta_{2k-1} = \tau_{2k-1}$ and $X_{\tau_{2k-1}} = 0$, so that $X_{\tau_{2k}} - X_{\tau_{2k-1}} = X_N \geq 0$. Thus, in any case, if $k = U_N + 1$, $X_{\tau_{2k}} - X_{\tau_{2k-1}} \geq 0$. Now choose a (nonrandom) integer $m > \frac{N}{2} + 2$. Then $m > U_N + 1$ and one has

$$X_N - X_1 = X_{\tau_{2m}} - X_1 = \sum_{k=1}^{m}(X_{\tau_{2k}} - X_{\tau_{2k-1}}) + \sum_{k=2}^{m}(X_{\tau_{2k-1}} - X_{\tau_{2k-2}})$$

$$= \sum_{k=1}^{U_N+1}(X_{\tau_{2k}} - X_{\tau_{2k-1}}) + \sum_{k=2}^{m}(X_{\tau_{2k-1}} - X_{\tau_{2k-2}})$$

$$\geq (b-a)U_N + \sum_{k=2}^{m}(X_{\tau_{2k-1}} - X_{\tau_{2k-2}}). \tag{3.32}$$

Taking expected values and using the fact from the optional sampling theorem that $\{X_{\tau_k} : k \geq 1\}$ is a submartingale, one has

$$\mathbb{E}X_N - \mathbb{E}X_1 \geq (b-a)\mathbb{E}U_N.$$

∎

Remark 3.9 Observe that the relations (3.32) do not require the submartingale assumption on $\{Z_n : n \geq 1\}$. It is merely a relationship among a sequence of numbers.

One of the most significant consequences of the uncrossing inequality is the following.

Theorem 3.12 *(Submartingale Convergence Theorem)* Let $\{Z_n : n \geq 1\}$ be a submartingale such that $\mathbb{E}(Z_n^+)$ is a bounded sequence. Then $\{Z_n : n \geq 1\}$ converges a.s. to a limit Z_∞. If $M := \sup_n \mathbb{E}|Z_n| < \infty$, then Z_∞ is a.s. finite and $\mathbb{E}|Z_\infty| \leq M$.

Proof Let $U(a, b)$ denote the total number of upcrossings of (a, b) by $\{Z_n : n \geq 1\}$. Then $U_N(a, b) \uparrow U(a, b)$ as $N \uparrow \infty$. Therefore, by the monotone convergence theorem

$$\mathbb{E}U(a, b) = \lim_{N\uparrow\infty} \mathbb{E}U_N(a, b) \leq \sup_N \frac{\mathbb{E}Z_N^+ + |a|}{b - a} < \infty. \tag{3.33}$$

In particular $U(a, b) < \infty$ almost surely, so that

$$P\left(\liminf Z_n < a < b < \limsup Z_n\right) = 0. \tag{3.34}$$

Since this holds for every pair $a, b = a + \frac{1}{m}$ with $a \varepsilon Q$ and m a positive integer, and the set of all such pairs is countable, one must have $\liminf Z_n = \limsup Z_n$ almost surely. Let Z_∞ denote the a.s. limit. By Fatou's lemma, $\mathbb{E}|Z_\infty| \leq \lim \mathbb{E}|Z_n|$. ∎

An immediate consequence of Theorem 3.12 is

Corollary 3.13 A nonnegative martingale $\{Z_n : n \geq 1\}$ converges almost surely to a finite limit Z_∞. Also, $\mathbb{E}Z_\infty \leq \mathbb{E}Z_1$.

Proof For a nonnegative martingale $\{Z_n : n \geq 1\}$, $|Z_n| = Z_n$ and therefore, $\sup \mathbb{E}|Z_n| = \sup \mathbb{E}Z_n = \mathbb{E}Z_1 < \infty$. Hence the Corollary follows from Theorem 3.12. ∎

The following Corollary provides an illustrative application of this theory.

Corollary 3.14 Suppose $X_1, X_2, \ldots$ is a sequence of independent, nonnegative random variables such that $\sum_{n=1}^{\infty} \mathbb{E}X_n < \infty$. Then $\sum_{n=1}^{\infty} X_n$ converges almost surely.

Proof Since $Z_n = \sum_{j=1}^{n}(X_j - \mathbb{E}X_j), n \geq 1$, is a martingale with $\mathbb{E}Z_n^+ \leq 2\sum_{j=1}^{\infty} \mathbb{E}X_j < \infty$ for all n, one has that $Z_{\infty} = \lim_{n \to \infty} Z_n$ exists. Thus $\sum_{j=1}^{n} X_j = Z_n + \sum_{j=1}^{n} \mathbb{E}X_j$ has an a.s. limit as $n \to \infty$. ∎

Doob's upcrossing inequality (Theorem 3.11) also applies to the so-called *reverse martingales, submartingales* defined as follows.

Definition 3.7 Let $\mathcal{F}_n, n \geq 1$, be a decreasing sequence of sub-sigmafields of $\mathcal{F}$, i.e., $\mathcal{F} \supset \mathcal{F}_n \supset \mathcal{F}_{n+1}, n = 1, 2, \ldots$. A sequence $\{X_n : n \geq 1\}$ of integrable random variables on $(\Omega, \mathcal{F}, P)$ is said to be a ***reverse submartingale*** with respect to $\mathcal{F}_n, n \geq 1$, if X_n is $\mathcal{F}_n$-measurable and $\mathbb{E}(X_n|\mathcal{F}_{n+1}) \geq X_{n+1}, \forall n$. If one has equality for each n then the sequence is called a ***reverse martingale*** with respect to $\mathcal{F}_n, n \geq 1$.

Theorem 3.15 *(Reverse submartingale convergence theorem)* Let $\{X_n : n \geq 1\}$ be a reverse submartingale with respect to a decreasing sequence $\mathcal{F}_n, n \geq 1$. Then X_n converges almost surely to an integrable random variable Z as $n \to \infty$.

Proof For each $N > 1$, $\{X_N, X_{N-1}, \ldots, X_1\}$ is a submartingale with respect to the filtration $\{\mathcal{F}_N, \mathcal{F}_{N-1}, \ldots, \mathcal{F}_1\}$. Thus, with U_N denoting the number of up crossings of (a, b) by $\{X_1, \ldots, X_N\}$, Doob's inequality yields $\mathbb{E}U_N \leq \frac{\mathbb{E}|X_1| + |a|}{b-a}$. Arguing as in the proof of of the submartingale convergence theorem (Theorem 3.12), the desired result follows. ∎

Remark 3.10 The martingale proof of the strong law of large numbers provides a beautiful illustration of Theorem 3.15 that will be given in Chapter V. Viewed this way, it will follow easily from the reverse martingale convergence theorem that the limit of the sample averages of an i.i.d. sequence of integrable random variables exists. However something more is needed to identify the limit (as the expected value).

Example 7 Let X be an integrable random variable on $(\Omega, \mathcal{F}, P)$ and $\mathcal{F}_n, (n \geq 1)$, a decreasing sequence of sigmafields $\mathcal{F}_n \subset \mathcal{F}, n \geq 1$. Then $X_n = \mathbb{E}(X|\mathcal{F}_n), n \geq 1$, is a reverse martingale. Thus $X_n \to Z$ a.s. as $n \to \infty$, for some integrable random variable Z. Note that $\{X_n : n \geq 1\}$ is uniformly integrable since $\int_{[|X_n| > \lambda]} |X_n| dP \leq \int_{[|X_n| > \lambda]} \mathbb{E}(|X||\mathcal{F}_n) dP = \int_{[|X_n| > \lambda]} \mathbb{E}(|X||\mathcal{F}_1) dP$. Hence $X_n \to Z$ in L^1 as well.

Example 8 It follows from the Corollary that the martingales $\{Z_n := \prod_{j=1}^{n} X_j\}$ converge almost surely to an integrable random variable Z_{∞}, if $\{X_n\}_{n=1}^{\infty}$ is an independent nonnegative sequence with $\mathbb{E}X_n = 1$ for all n. In the case $\{X_n\}_{n=1}^{\infty}$ is *i.i.d.* and $P(X_1 = 1) < 1$, it is an interesting fact that the limit of $\{Z_n : n \geq 1\}$ is 0 a.s., as shown by the following proposition.

Proposition 3.16 Let $\{X_n : n \geq 1\}$ be an i.i.d. sequence of nonnegative random variables with $\mathbb{E}X_1 = 1$. Then $\{Y_n := \prod_{j=1}^n X_j\}$ converges almost surely to 0, provided $P(X_1 = 1) < 1$.

Proof First assume $P(X_1 = 0) > 0$. Then $P(X_n = 0 \text{ for some } n) = 1 - P(X_n > 0$ for all $n) = 0$, since $P(X_j > 0 \text{ for } 1 \leq j \leq n) = (P(X_1 > 0))^n$. But if $X_m = 0$ then $Z_n = 0$ for all $n \geq m$. Therefore, $P(Z_n = 0 \text{ for all sufficiently large } n) = 1$.

Assume now $P(X_1 > 0) = 1$. Consider the *i.i.d.* sequence $\{\log X_n\}_{n=1}^\infty$. Since $x \to \log x$ is concave one has, by Jensen's inequality, $\mathbb{E} \log X_1 \leq \log \mathbb{E}X_1 = 0$. Since $P(X_1 = 1) < 1$, for any $0 < h < 1$, X_1^h is not degenerate (i.e., not almost surely the constant 1.). Hence the Jensen inequality is *strict*. Therefore, $\mathbb{E}X_1^h < 1$. Thus, using Fatou's lemma,

$$0 \leq \mathbb{E}Z_\infty^h \leq \liminf_{n \to \infty} \mathbb{E}Z_n^h = \liminf_{n \to \infty} (\mathbb{E}X_1^h)^n = 0.$$

It follows that $Z_\infty^h = 0$ a.s. ∎

Example 9 (*Binary Multiplicative Cascade Measure*) Suppose that one is given a countable collection $\{X_v : v \in \cup_{n=1}^\infty \{0, 1\}^n\}$ of positive, mean one random variables indexed by the set of vertices $\partial T = \cup_{n=1}^\infty \{0, 1\}^n$ of a binary tree. For $v = (v_1, \ldots, v_n) \in \{0, 1\}^n$ we write $|v| = n$. For a given "generation"$n \geq 1$, one may consider a corresponding partition of the unit interval $[0, 1)$ into 2^n subintervals $[\frac{k-1}{2^n}, \frac{k}{2^n})$, $k = 1, \ldots 2^n$, and assign mass (area) $(\prod_{j=1}^n X_{v|j})2^{-n}$ to the interval indexed by v of length 2^{-n}, where $v|j = (v_1, \ldots, v_j)$, $v = (v_1, \ldots, v_n) \in \{0, 1\}^n$, to create a *random bar graph*. The total area in the graph is then given by

$$Z_n = \sum_{|v|=n} \prod_{j=1}^n X_{v|j} 2^{-n}, n = 1, 2, \ldots. \tag{3.35}$$

One may check that $\{Z_n : n \geq 1\}$ is a positive martingale. Thus $\lim_{n \to \infty} Z_n = Z_\infty$ exists almost surely. Moreover, Z_∞ satisfies the recursion

$$Z_\infty = X_0 Z_\infty(0)\frac{1}{2} + X_0 Z_\infty(1)\frac{1}{2}, \tag{3.36}$$

where $Z_\infty(0)$, $Z_\infty(1)$ are mutually independent, and independent of X_0, X_1, and have the same distribution as Z_∞. Let $0 < h \leq 1$. Then by sub-linearity of $z \to z^h$, $z \geq 0$, one has, for a generic random variable X distributed as an X_v,

$$\mathbb{E}Z_\infty^h \geq 2^{1-h}\mathbb{E}X^h\mathbb{E}Z_\infty^h.$$

Thus, if $2^{1-h}\mathbb{E}X^h > 1$ for some $0 < h \leq 1$ then $Z_\infty = 0$ a.s; for otherwise $\mathbb{E}Z_\infty^h > 0$ and one gets the reverse inequality $2^{1-h}\mathbb{E}X^h \leq 1$, or $\chi(h) := \log \mathbb{E}X^h - (h-1)\log 2 \leq 0$. Since $h \to \chi(h), 0 < h \leq 1$ is convex, this is equivalent to $\chi'(1^-) = \mathbb{E}X \log X \leq \log 2$. Thus if $\mathbb{E}X \log X > \log 2$ then $Z_\infty = 0$ a.s. Of course if $X = 1$ a.s. then $Z_\infty = 1$ a.s. as well. In some contexts, the quantity $\mathbb{E}X \log X$ is referred to as a *disorder parameter* and $\log 2$ is a *branching rate*. The heuristic condition for a nonzero limit is that the branching rate be sufficiently large relative to the disorder. This will be confirmed in Chapter V. Let us consider the case in which X is uniform on $[0, 2]$. In this case, as will be verified in Chapter V, one can solve the recursion (3.36), to obtain that Z_∞ has a Gamma distribution with density $ze^{-z}, z \geq 0$. As an alternative for now, we will apply the Chebyshev method from (Chapter I, Example 5) to derive lower bound estimates on $P(Z_\infty \leq z), z \geq 0$. One may check that for X uniformly distributed on $[0, 2]$, $\mathbb{E}X^k = \frac{2^k}{k+1}, k = 1, 2, \ldots$. Moreover, using induction (on k) one sees that the unique positive solution to the equation of moments corresponding to (3.36), namely

$$
\mathbb{E}Z_\infty^k = 2^{-k} \sum_{j=0}^{k} \binom{k}{j} \frac{2^j}{j+1} \mathbb{E}Z_\infty^j \frac{2^{k-j}}{k-j+1} \mathbb{E}Z_\infty^{k-j}
$$

$$
= \sum_{j=0}^{k} \binom{k}{j} \frac{\mathbb{E}Z_\infty^j}{1+j} \frac{\mathbb{E}Z_\infty^{k-j}}{1+k-j}, \tag{3.37}
$$

is given by $\mathbb{E}Z_\infty^k = (k+1)!$. Thus the Chebyshev method yields

$$
P(Z_\infty \leq x) \geq \begin{cases} 0 & \text{if } x \leq 2, \\ 1 - \frac{6}{x^2} & \text{if } 2 < x \leq 3, \\ \cdots & \\ 1 - \frac{(k+1)!}{x^k} & \text{if } k+1 < x \leq k+2, k = 2, 3, \ldots. \end{cases} \tag{3.38}
$$

Example 10 *(Ruin Probability in Insurance Risk)* The *Cramér–Lundberg*, and more generally *Sparre Andersen*, models of insurance markets involve insurance claims of strictly positive random amounts $X_1, X_2, \ldots$ arriving at random time times $T_1, T_2, \ldots$, together with a constant premium rate $c > 0$ per unit time. The two sequences $\{X_n : n \geq 1\}$ of claim sizes and arrival times $\{T_n : n \geq 1\}$ are assumed to be independent. Moreover, the inter-arrival times $A_n = T_n - T_{n-1}, n \geq 1, T_0 = 0$, are assumed to be i.i.d. positive random variables with $\mathbb{E}A_1 = \lambda < \infty$. For a company with initial capital reserves $u > 0$, the probability of ruin is defined by

$$
\psi(u) = P(\cup_{n=1}^{\infty}[\sum_{j=1}^{n} X_j > u + c \sum_{j=1}^{n} A_j]) = P(\cup_{n=1}^{\infty}[\sum_{j=1}^{n} Y_j > u]), \tag{3.39}
$$

where $Y_j := X_j - cA_j$. The common distribution of the i.i.d. sequence $\{Y_j : j \geq 1\}$ is assumed to satisfy the so-called *Net Profit Condition* (NPC)

$$
\mathbb{E}Y_1 = \mathbb{E}X_1 - c\mathbb{A}_1 > 0. \tag{3.40}
$$

Observe that if $\mathbb{E}Y_1$ is nonnegative and finite then, by the strong law of large numbers (SLLN), one has

$$\psi(u) \equiv 1 \quad \forall u. \tag{3.41}$$

To avoid the trivial case in which $\psi(u) = 0 \,\forall\, u > 0$, one may assume

$$P(Y_1 > 0) > 0. \tag{3.42}$$

The Cramér–Lundberg model refers to the case in which the $A_n, n \geq 1$, are i.i.d. *exponentially* distributed, while the more general model described above is referred to as the Sparre Andersen model. For the present let us assume that the claim size distribution is *light tailed* in the sense that

$$\mathbb{E}e^{qX_1} < \infty \quad \text{for some } q > 0. \tag{3.43}$$

With this one obtains the following bound on the ruin probability as a function of the initial capital.

Proposition 3.17 *(Lundberg Inequality)* In the non-degenerate case (3.42), the Sparre Andersen model satisfying the NPC (3.40), and the light-tailed claim size distribution condition (3.43), there is a unique parameter $q = R > 0$ such that $\mathbb{E}e^{qY_1} = 1$. Moreover

$$\psi(u) \leq \exp(-Ru), \quad \forall u > 0. \tag{3.44}$$

Proof Observe that the light-tailed condition (3.43) implies that there is an $h, 0 < h \leq \infty$ such that

$$0 < m(q) := \mathbb{E}e^{qY_1} < \infty, \quad \text{for } 0 \leq q < h, \quad \lim_{q \downarrow h} m(q) = \infty.$$

Also $m(0) = 1, m'(0) = \mathbb{E}Y_1 < 0$ (or $m'(0^+) < 0$ if $m(q) = \infty \forall q < 0$), and $m''(q) = \mathbb{E}Y_1^2 \exp(qY_1) > 0, \forall q > 0$, with $m(q) \to \infty$ as $q \uparrow h$. Thus $m(q)$ decreases from $m(0) = 1$ to a minimum in $(0, 1)$ at some $\tilde{q}$ before increasing without bound as $q \uparrow h$. It follows that there is a unique $q = R > 0$ such that $m(q) = 1$. To prove the asserted Lundberg bound, let $\tau = \inf\{n \geq 1 : S_n > u\}$, where $S_n = Y_1 + \cdots + Y_n, n \geq 1, S_0 = 0$. Then τ is a stopping time with respect to the filtration $\mathcal{F}_n = \sigma(Y_1, \ldots, Y_n), n \geq 1, \mathcal{F}_0 = \{\Omega, \emptyset\}$. Then

$$\psi(u) = P(\tau < \infty).$$

Next write $W_n = u - S_n, n \geq 1, W_0 = u$. Then $M_n = \exp\{-RW_n\}, n \geq 0$, is an $\mathcal{F}_n$-martingale since,

$$\mathbb{E}(M_{n+1}|\mathcal{F}_n) = \mathbb{E}(e^{RY_{n+1}}M_n|\mathcal{F}_n) = M_n\mathbb{E}e^{RY_{n+1}} = M_n m(R) = M_n.$$

By the optional sampling theorem one then has

$$e^{-Ru} = \mathbb{E}M_0 = \mathbb{E}M_{\tau \wedge n} \geq \mathbb{E}M_{\tau \wedge n}\mathbf{1}_{[\tau \leq n]} = \mathbb{E}M_\tau\mathbf{1}_{[\tau \leq n]}, \forall n. \qquad (3.45)$$

Noting that $M_\tau > 1$ on $[\tau < \infty]$, it follows from (3.45) that $e^{-Ru} \geq \mathbb{E}\mathbf{1}_{[\tau \leq n]} = P(\tau \leq n)$ for all n. Let $n \uparrow \infty$ to obtain the asserted Lundberg bound. ∎

Remark 3.11 The parameter R is generally referred to as the *Lundberg coefficient*, or *adjustment coefficient*. It can be shown that the exponential decay rate provided by the Lundberg inequality cannot be improved under the conditions of the theorem. In the Cramér–Lundberg model the true asymptotic rate is given by $\psi(u) \sim ce^{-Ru}$, as $u \to \infty$, for a constant $c < 1$; here $\sim$ demotes asymptotic equality in the sense that the ratio of the two sides converges to one as $u \to \infty$.[1]

Exercise Set III

1. (i) If τ_1 and τ_2 are $\{\mathcal{F}_t\}$-stopping times, then show that so are $\tau_1 \wedge \tau_2$ and $\tau_1 \vee \tau_2$.
    (ii) Show that $\tau + c$ is an $\{\mathcal{F}_t\}$-stopping time if τ is an $\{\mathcal{F}_t\}$-stopping time, $c > 0$, and $\tau + c \in T \cup \{\infty\}$. (iii) Show that (ii) is false if $c < 0$.
2. If τ is a discrete random variable with values $t_1 < t_2 < \cdots$ in a finite or countable set T in $\mathbb{R}^+$, then (i) τ is an $\{\mathcal{F}_t\}_{t \in T}$-stopping time if and only if $[\tau = t] \in \mathcal{F}_t \; \forall \, t \in T$; (ii) τ is an $\{\mathcal{F}_t\}$-stopping time if and only if it is an $\{\mathcal{F}_t\}$-optional time.
3. (*Wald's Identity*) Let $\{Y_j : j \geq 1\}$ be an i.i.d. sequence with finite mean μ, and take $Y_0 = 0, a.s.$ Let τ be an $\{\mathcal{F}_n\}$-stopping time, where $\mathcal{F}_n = \sigma(Y_j : j \leq n)$. Write $S_n = \sum_{j=0}^n Y_j$. If $\mathbb{E}\tau < \infty$ and $\mathbb{E}|S_\tau - S_{\tau \wedge m}| \to 0$ as $m \to \infty$, prove that $\mathbb{E}S_\tau = \mu\mathbb{E}\tau$. [*Hint*: $\{S_n - n\mu : n \geq 0\}$ is a martingale.]
4. In Example 5 for $\tau = \tau_a \wedge \tau_b$, show that (i) $\mathbb{E}\tau < \infty \; \forall \, a \leq x \leq b$, and $|S_{(\tau) \wedge n}| \leq \max\{|a|, |b|\} \; \forall \, n \geq 0$, is uniformly integrable, (ii) $P(\tau_a < \infty) = 1 \; \forall \, x, a$, but $\{S_{\tau_a \wedge n} : n \geq 0\}$ is not uniformly integrable. (iii) For Example 5 also show that $Y_n := S_n^2 - n, n \geq 0$, is a martingale and $\{Y_{\tau \wedge n} : n \geq 0\}$ is uniformly integrable. Use this to calculate $\mathbb{E}\tau$. [*Hint*: Use triangle inequality estimates on $|Y_{\tau \wedge n}| \leq |S_{\tau \wedge n}|^2 + \tau \wedge n|.$]
5. (*A cautionary example*) Let $\Omega = \{1, 2\}^3$, and assume all outcomes equally likely. For $\omega = (\omega_1, \omega_2, \omega_3) \in \Omega$, let $Y_i(\omega_1, \omega_2, \omega_3) = \delta_{\omega_i, \omega_{i+1}}, (i = 1, 2)$, and $X(\omega_1, \omega_2, \omega_3) = \delta_{\omega_3, \omega_1}$. Define $J(\omega) = 2$ if $Y_1(\omega) = 1$, and $J(\omega) = 1$ if $Y_1(\omega) = 0, \omega \in \Omega$. Then show that Y_1 and X are independent, as are Y_2 and X, and J and X. However Y_J is not independent of X.
6. (*Transience of asymmetric simple random walk*) Let $\theta(c|x)$ denote the probability that the simple random walk starting at x ever reaches c. Use (3.30) to prove (i) $\theta(b|x) = (\frac{p}{q})^{b-x}$ for $x < b$ if $p < 1/2$, and (ii) $\theta(a|x) = (\frac{q}{p})^{x-a}$ for $x > a$ if $p > 1/2$.
7. Let $Z_1, Z_2, \ldots$ be i.i.d. ± 1-valued Bernoulli random variables with $P(Z_n = 1) = p, P(Z_n = -1) = 1 - p, n \geq 1$, where $0 < p < 1/2$. Let $S_n = Z_1 + \cdots + Z_n, n \geq 1, S_0 = 0$.

---

[1]See Theorem 5.12 of S. Ramasubramanian (2009) for the asymptotic equality in the case of the Cramér-Lundberg model.

(i) Show that $P(\sup_{n\geq 0} S_n > y) \leq (\frac{p}{q})^y, y \geq 0$. [*Hint:* Apply a maximal inequality to $X_n = (q/p)^{S_n}$.]

(ii) Show for $p < 1/2$ that $\mathbb{E} \sup_{n\geq 0} S_n \leq \frac{1}{q-p}$. [*Hint:* Use (1.10), noting that the distribution function is a step function. Also see Exercise 30 of Chapter I.]

8. Suppose that $Z_1, Z_2, \ldots$ is a sequence of independent random variables with $\mathbb{E} Z_n = 0$ such that $\sum_n \mathbb{E} Z_n^2 < \infty$. Show that $\sum_{n=1}^{\infty} Z_n := \lim_N \sum_{n=1}^{N} Z_n$ exists a.s.[2] [*Hint:* Let $S_j = \sum_{k=1}^{j} Z_k$ and show that $\{S_j\}$ is a.s. a Cauchy sequence. For this note that $Y_n := \max_{k,j\geq n} |S_k - S_j|$ is a.s. a decreasing sequence and hence has a limit a.s. Apply Kolmogorov's maximal inequality to $\max_{n\leq j\leq N} |S_j - S_n|$ to show that the limit in probability is zero, and hence a.s. zero; also see Chapter I, Exercise 34.]

(i) For what values of θ will $\sum_{n=1}^{\infty} Z_n$ converge a.s. if $P(Z_n = n^{-\theta}) = P(Z_n = -n^{-\theta}) = 1/2$?

(ii) (Random Signs[3]) Suppose each X_n is symmetric Bernoulli ± 1-valued. Show that the series $\sum_{n=1}^{\infty} X_n a_n$ converges with probability one if $\{a_n\}$ is any square-summable sequence of real numbers.

(iii) Show that $\sum_{n=1}^{\infty} X_n \sin(n\pi t)/n$ converges a.s. for each t if the X_n's are i.i.d. standard normal.

9. Let $\{X_t : t \in T\}$ be a stochastic process on $(\Omega, \mathcal{F})$ with values in some measurable space $(S, \mathcal{S})$, T a discrete set with elements $t_1 < t_2 < \cdots$. Define $\mathcal{F}_t = \sigma(X_s : 0 \leq s \leq t) \subset \mathcal{F}, t \in T$. Assume that τ is an $\{\mathcal{F}_t\}$-stopping time and show that $\mathcal{F}_\tau = \sigma(X_{\tau \wedge t} : t \in T)$; i.e., $\mathcal{F}_\tau$ is the σ-field generated by the stopped process $\{X_{\tau \wedge t} : t \in T\}$.

10. Prove (3.19). Also prove that an $\{\mathcal{F}_t\}$-stopping time is an $\{\mathcal{F}_t\}$-optional time; recall Definition 3.4.

11. (i) Prove that τ_B defined by (3.18) is an $\{\mathcal{F}_t\}$-stopping time if B is closed and $t \mapsto X_t$ is continuous with values in a metric space (S, ρ). [*Hint:* For $t > 0$, B closed, $[\tau_B \leq t] = \cap_{n\in\mathbb{N}} \cup_{r\in Q\cap[0,t]} [\rho(X_r, B) \leq \frac{1}{n}]$, where Q is the set of rationals.] (ii) Prove that if $t \mapsto X_t$ is right-continuous, τ_B is an optional time for B open. [*Hint:* For B open, $t > 0$, $[\tau_B < t] = \cup_{r\in Q\cap(0,t)} [X_r \in B]$.] (iii) If $T = \mathbb{N}$ or $\mathbb{Z}^+$, prove that τ_B is a stopping time for all $B \in \mathcal{S}$.

12. Prove that if τ is an optional time with respect to a filtration $\{\mathcal{F}_t : 0 \leq t < \infty\}$, then τ is an optional time with respect to $\{\mathcal{F}_{t+} : 0 \leq t < \infty\}$, where $\mathcal{F}_{t+} := \cap_{\varepsilon>0}\mathcal{F}_{t+\varepsilon}$. Deduce that under the hypothesis of Example 4(b), if B is open or closed, then τ_B is a stopping time with respect to $\{\mathcal{F}_{t+} : 0 \leq t < \infty\}$.

13. Let $\{\mathcal{F}_t : t \in T\}$ and $\{\mathcal{G}_t : t \in T\}$ be two filtrations of $(\Omega, \mathcal{F})$, each adapted to $\{X_t : t \in T\}$, and assume $\mathcal{F}_t \subset \mathcal{G}_t, \forall t \in T$. Show that if $\{X_t : t \in T\}$ is a

---

[2] A more comprehensive treatment of this class of problems is given in Chapter VIII.

[3] Historically this is the problem that lead Hugo Steinhaus to develop an axiomatic theory of repeated coin tossing based on his reading of Lebesgue's newly developed integral and measure on the real number line. The problem is revisited in Chapter VIII.

$\{\mathcal{G}_t\}$-martingale (or sub or super) then it is an $\{\mathcal{F}_t\}$-martingale (or respectively sub or super).

14. (*Uniform absolute continuity*) Let Y be an integrable random variable. Prove that, given $\varepsilon > 0$, there is a $\delta > 0$ such that $\int_A |Y| dP < \varepsilon$ for every A with $P(A) < \delta$. [*Hint*: Prove by contradiction: There cannot exist a sequence A_n, $P(A_n) < \frac{1}{n}$, and $\int_{A_n} |Y| dP > \varepsilon (n = 1, 2, \dots).$]

Chapter IV
Classical Central Limit Theorems

In view of the great importance of the *central limit theorem* (CLT), we shall give a general but self-contained version due to Lindeberg.[1] This version is applicable to nonidentically distributed summands and provides the foundation to the following **CLT paradigm**, which permeates the sciences: "The sum of a large number of 'small' independent random terms is approximately normally distributed."

Let us first define a notion of **convergence in distribution** or **weak convergence** for sequences of probabilities defined on the finite-dimensional Euclidean space $S = \mathbb{R}^k$.

Definition IV.1 A sequence $\{Q_n\}_{n=1}^{\infty}$ of probabilities on the Borel σ-field of $\mathbb{R}^k$ is said to converge **weakly** or, equivalently, **in distribution** to a probability Q on $\mathcal{B}^k$, denoted $Q_n \Rightarrow Q$ as $n \to \infty$, if

$$\lim_{n \to \infty} \int_{\mathbb{R}^k} g(x) Q_n(dx) = \int_{\mathbb{R}^k} g(x) Q(dx),$$

for all bounded continuous functions $g : \mathbb{R}^k \to \mathbb{R}$. A sequence $X_n, n = 1, 2, \ldots$ of k-dimensional random vectors with respective distributions $Q_n, n = 1, 2, \ldots$, is said to converge in distribution to a k-dimensional random vector X with distribution Q, if the sequence $Q_n, n \geq 1$, converges weakly to Q.

The following theorem provides some alternative useful conditions for weak convergence. Additional useful methods are developed in Exercise 9 of Chapter VII.

---

[1]This approach has received recent attention of Terence Tao (2015, SPA Conference, Oxford) as "the most effective way to deal with local universality for non-Hermitian random matrices." In this context, it is referred to as *Lindeberg's exchange strategy*, e.g., see Tao (2012). The "exchange" refers to a substitution with a normal random variable and should not be confused another technical use of related terminology for permutation invariance.

© Springer International Publishing AG 2016
R. Bhattacharya and E.C. Waymire, *A Basic Course in Probability Theory*,
Universitext, DOI 10.1007/978-3-319-47974-3_IV

Theorem 4.1 (*Finite-Dimensional Weak Convergence*) Let $\{Q_n\}_{n=1}^{\infty}$, Q be probabilities on the Borel σ-field of $\mathbb{R}^k$. The following are equivalent statements:

(a) $Q_n \Rightarrow Q$.
(b) $\int_{\mathbb{R}^k} f \, dQ_n \rightarrow \int_{\mathbb{R}^k} f \, dQ$ for all (bounded) continuous f vanishing outside a compact set.
(c) $\int_{\mathbb{R}^k} f \, dQ_n \rightarrow \int_{\mathbb{R}^k} f \, dQ$ for all infinitely differentiable functions f vanishing outside a compact set.
(d) Let $F_n(x) := Q_n((-\infty, x_1] \times \cdots \times (-\infty, x_k])$, and $F(x) := Q((-\infty, x_1] \times \cdots \times (-\infty, x_k])$, $x \in \mathbb{R}^k$, $n = 1, 2, \ldots$. Then $F_n(x) \rightarrow F(x)$ as $n \rightarrow \infty$, for every point of continuity x of F.

Proof We give the proof for the one-dimensional case $k = 1$. The case $k \geq 2$ requires little difference in proof for (a)–(c) and is left as Exercise 1 for these parts. The equivalence of (a) and (d) for the case $k \geq 2$ will be derived in Chapter VII. First let us check that (b) is sufficient. It is obviously necessary by definition of weak convergence. Assume (b) and let f be an arbitrary bounded continuous function, $|f(x)| \leq c$ for all x. The idea is to construct a continuous approximation to f having compact support. For notational convenience write $\{x \in \mathbb{R} : |x| \geq N\} = \{|x| \geq N\}$, etc. Given $\varepsilon > 0$ there exists N such that $Q(\{|x| \geq N\}) < \varepsilon/4c$. Define θ_N by $\theta_N(x) = 1, |x| \leq N$, $\theta_N(x) = 0, |x| \geq N + 1$, and linearly interpolate for $N \leq |x| \leq N + 1$. Then,

$$\underline{\lim}_{n \to \infty} Q_n(\{|x| \leq N + 1\}) \geq \underline{\lim}_{n \to \infty} \int \theta_N(x) dQ_n(x) = \int \theta_N(x) dQ(x)$$

$$\geq Q(\{|x| \leq N\}) > 1 - \frac{\varepsilon}{4c},$$

so that

$$\overline{\lim}_{n \to \infty} Q_n(\{|x| > N + 1\}) \equiv 1 - \underline{\lim}_{n \to \infty} Q_n(\{|x| \leq N + 1\}) < \frac{\varepsilon}{4c}. \quad (4.1)$$

Now define $f_N := f\theta_{N+1}$. Noting that $f = f_N$ on $\{|x| \leq N + 1\}$ and that on $\{|x| > N+1\}$ one has $|f(x)| \leq c$, upon first writing $f = f\mathbf{1}_{\{|x| \leq N+1\}} + f\mathbf{1}_{\{|x| > N+1\}}$, and then further writing $\int_{\mathbb{R}} f_N \mathbf{1}_{\{|x| \leq N+1\}} dQ_n = \int_{\mathbb{R}} f_N dQ_n - \int_{\{N+1 < |x| \leq N+2\}} f_N dQ_n$ (and similarly for the integral with respect to Q), one has from the triangle inequality and the bound on f and f_N that

$$\overline{\lim}_{n \to \infty} \left| \int_{\mathbb{R}} f \, dQ_n - \int_{\mathbb{R}} f \, dQ \right| \leq \overline{\lim}_{n \to \infty} \left| \int_{\mathbb{R}} f_N \, dQ_n - \int_{\mathbb{R}} f_N \, dQ \right|$$

$$+ \overline{\lim}_{n \to \infty} (2c Q_n(\{|x| > N + 1\})$$

$$+ 2c Q(\{|x| > N + 1\}))$$

$$< 2c \frac{\varepsilon}{4c} + 2c \frac{\varepsilon}{4c} = \varepsilon.$$

Since $\varepsilon > 0$ is arbitrary, $\int_{\mathbb{R}} f \, dQ_n \to \int_{\mathbb{R}} f \, dQ$. So (a) and (b) are equivalent. Let us now show that (b) and (c) are equivalent. It is enough to prove (c) is sufficient for (b). For this we construct an approximation to f that is C^∞ and has compact support. For each $\varepsilon > 0$ define the function

$$\rho_\varepsilon(x) = d(\varepsilon) \exp\left\{-\frac{1}{1 - x^2/\varepsilon^2}\right\} \mathbf{1}_{[-\varepsilon, \varepsilon]}(x), \qquad (4.2)$$

where $d(\varepsilon)$ is so chosen as to make $\int \rho_\varepsilon(x) dx = 1$. One may check that $\rho_\varepsilon(x)$ is infinitely differentiable in x. Now let f be a continuous function that vanishes outside a finite interval. Then f is uniformly continuous, and therefore, $\delta(\varepsilon) = \sup\{|f(x) - f(y)| : |x - y| \le \varepsilon\} \to 0$ as $\varepsilon \downarrow 0$. Define

$$f^\varepsilon(x) = \int_{\mathbb{R}} f(y) \rho_\varepsilon(x - y) dy = f * \rho_\varepsilon(x) := \int_{-\varepsilon}^{\varepsilon} f(x - y) \rho_\varepsilon(y) dy, \qquad (4.3)$$

and note that since $f^\varepsilon(x)$ is infinitely differentiable, vanishes outside a compact set, and is an average over values of f within the interval $(x - \varepsilon, x + \varepsilon)$, $|f^\varepsilon(x) - f(x)| \le \delta(\varepsilon)$ for all ε. Hence,

$$\left|\int_{\mathbb{R}} f \, dQ_n - \int_{\mathbb{R}} f^\varepsilon \, dQ_n\right| \le \delta(\varepsilon) \quad \text{for all } n, \qquad \left|\int_{\mathbb{R}} f \, dQ - \int_{\mathbb{R}} f^\varepsilon \, dQ\right| \le \delta(\varepsilon),$$

$$\left|\int_{\mathbb{R}} f \, dQ_n - \int_{\mathbb{R}} f \, dQ\right| \le \left|\int_{\mathbb{R}} f \, dQ_n - \int_{\mathbb{R}} f^\varepsilon \, dQ_n\right| + \left|\int_{\mathbb{R}} f^\varepsilon \, dQ_n - \int_{\mathbb{R}} f^\varepsilon \, dQ\right|$$

$$+ \left|\int_{\mathbb{R}} f^\varepsilon \, dQ - \int_{\mathbb{R}} f \, dQ\right|$$

$$\le 2\delta(\varepsilon) + \left|\int_{\mathbb{R}} f^\varepsilon \, dQ_n - \int_{\mathbb{R}} f^\varepsilon \, dQ\right| \to 2\delta(\varepsilon) \text{ as } n \to \infty.$$

Since $\delta(\varepsilon) \to 0$ as $\varepsilon \to 0$ it follows that $\int_{\mathbb{R}} f \, dQ_n \to \int_{\mathbb{R}} f \, dQ$, as claimed. Next let F_n, F be the distribution functions of Q_n, Q, respectively ($n = 1, 2, \ldots$), and suppose that (a) holds. Then we want to show that $F_n(x) \to F(x)$ at all points x of continuity of the limit distribution function F. Fix such a point of continuity x_0 of F. Given $\varepsilon > 0$ there is an $\eta = \eta(\varepsilon) > 0$, such that $|F(x) - F(x_0)| < \varepsilon$ for $|x - x_0| < \eta$. Let $\psi_\varepsilon^+(x) = 1, x \le x_0$ and $\psi_\varepsilon^+(x) = 0, x > x_0 + \eta$. Extend ψ_ε^+ to $(x_0, x_0 + \eta)$ by linear interpolation. Similarly, define $\psi_\varepsilon^-(x) = 0, x \ge x_0$ and $\psi_\varepsilon^-(x) = 1, x \le x_0 - \eta$. Extend ψ_ε^- to $(x_0 - \eta, x_0)$. Then, using the definition of weak convergence of Q_n to Q,

$$\limsup_{n \to \infty} F_n(x_0) \le \limsup_{n \to \infty} \int_{\mathbb{R}} \psi_\varepsilon^+(x) Q_n(dx) = \int_{\mathbb{R}} \psi_\varepsilon^+(x) Q(dx) \le F(x_0 + \eta)$$

$$< F(x_0) + \varepsilon,$$

and

$$\liminf_{n\to\infty} F_n(x_0) \le \liminf_{n\to\infty} \int_{\mathbb{R}} \psi_\varepsilon^-(x) Q_n(dx) = \int_{\mathbb{R}} \psi_\varepsilon^-(x)(x) Q(dx) \ge F(x_0 - \eta)$$
$$> F(x_0) - \varepsilon.$$

Since $\varepsilon > 0$ is arbitrary, one has

$$\limsup_{n\to\infty} F_n(x_0) \le F(x_0) \le \liminf_{n\to\infty} F_n(x_0).$$

In particular, $F_n(x_0) \to F(x_0)$ as $n \to \infty$. To show that the converse (d) implies (a) also holds, suppose $F_n(x) \to F(x)$ at all points of continuity of a distribution function F. Note that since F is nondecreasing and bounded between 0 and 1, it can have at most countably many discontinuities, i.e., only finitely many jumps of size larger than $1/n$ for any $n = 1, 2, \ldots$. Consider a continuous function f that vanishes outside the compact set K contained in an interval $[a, b]$ where a, b are selected as points of continuity of F. The idea is to construct an approximation to f by a step function with jumps at points of continuity of F. Given any $\varepsilon > 0$ one may partition $[a, b]$ into a finite number of subintervals whose endpoints are all points of continuity of F, and obtain a uniform approximation of f to within $\varepsilon > 0$ by a step function f_ε having constant values of f at the endpoint over each respective subinterval. Then, $\int_{\mathbb{R}} f_\varepsilon dQ_n \to \int_{\mathbb{R}} f_\varepsilon dQ$ as $n \to \infty$. Thus $|\int_{\mathbb{R}} f dQ_n - \int_{\mathbb{R}} f dQ| \le \int_{\mathbb{R}} |f - f_\varepsilon| dQ_n + |\int_{\mathbb{R}} f_\varepsilon dQ_n - \int_{\mathbb{R}} f_\varepsilon dQ| + \int_{\mathbb{R}} |f_\varepsilon - f| dQ \le 2\varepsilon + |\int_{\mathbb{R}} f_\varepsilon dQ_n - \int_{\mathbb{R}} f_\varepsilon dQ|$. Since $\varepsilon > 0$ is arbitrary, one readily obtains (b) and hence (a). ∎

In preparation for the proof of the central limit theorem, the following simple lemma is easily checked by an integration by parts left as Exercise 3.

Lemma 1 *(A Second-Order Taylor Expansion)* Let f be a real-valued function of $\mathbb{R}$ such that f, f', f'', f''' are bounded. Then for $x, h \in \mathbb{R}$,

$$f(x + h) = f(x) + hf'(x) + \frac{h^2}{2!} f''(x) + h^2 \int_0^1 (1 - \theta)\{f''(x + \theta h) - f''(x)\} d\theta.$$

Theorem 4.2 *(Lindeberg's CLT)* For each n, let $X_{n,1}, \ldots, X_{n,k_n}$ be an independent array of random variables satisfying

$$\mathbb{E}X_{n,j} = 0, \quad \sigma_{n,j} := (\mathbb{E}X_{n,j}^2)^{1/2} < \infty, \quad \sum_{j=1}^{k_n} \sigma_{n,j}^2 = 1, \qquad (4.4)$$

and, for each $\varepsilon > 0$,

$$\text{(Lindeberg condition)} \quad \lim_{n\to\infty} \sum_{j=1}^{k_n} \mathbb{E}(X_{n,j}^2 \mathbf{1}_{[|X_{n,j}|>\varepsilon]}) = 0. \qquad (4.5)$$

Then, $\sum_{j=1}^{k_n} X_{n,j}$ converges in distribution to the standard normal law $N(0, 1)$.

Proof Let $\{Z_j : j \geq 1\}$ be a sequence of i.i.d. $N(0, 1)$ random variables, independent of $\{X_{n,j} : 1 \leq j \leq k_n\}$. Write

$$Z_{n,j} := \sigma_{n,j} Z_j \qquad (1 \leq j \leq k_n), \tag{4.6}$$

so that $\mathbb{E} Z_{n,j} = 0 = \mathbb{E} X_{n,j}$, $\mathbb{E} Z_{n,j}^2 = \sigma_{n,j}^2 = \mathbb{E} X_{n,j}^2$. Define

$$U_{n,m} := \sum_{j=1}^{m} X_{n,j} + \sum_{j=m+1}^{k_n} Z_{n,j} \qquad (1 \leq m \leq k_n - 1),$$

$$U_{n,0} := \sum_{j=1}^{k_n} Z_{n,j}, \qquad U_{n,k_n} := \sum_{j=1}^{k_n} X_{n,j}, \tag{4.7}$$

$$V_{n,m} := U_{n,m} - X_{n,m} \qquad (1 \leq m \leq k_n).$$

Let f be a real-valued function of $\mathbb{R}$ such that f, f', f'', f''' are bounded. Taking $x = V_{n,m}$, $h = X_{n,m}$ in the Taylor expansion Lemma 1, one has

$$\mathbb{E} f(U_{n,m}) \equiv \mathbb{E} f(V_{n,m} + X_{n,m}) = \mathbb{E} f(V_{n,m}) + \mathbb{E}(X_{n,m} f'(V_{n,m}))$$
$$+ \tfrac{1}{2} \mathbb{E}(X_{n,m}^2 f''(V_{n,m})) + \mathbb{E}(R_{n,m}), \tag{4.8}$$

where

$$R_{n,m} := X_{n,m}^2 \int_0^1 (1 - \theta)\{f''(V_{n,m} + \theta X_{n,m}) - f''(V_{n,m})\} \, d\theta. \tag{4.9}$$

Since $X_{n,m}$ and $V_{n,m}$ are independent, and $\mathbb{E} X_{n,m} = 0$, $\mathbb{E} X_{n,m}^2 = \sigma_{n,m}^2$, (4.8) reduces to

$$\mathbb{E} f(U_{n,m}) = \mathbb{E} f(V_{n,m}) + \frac{\sigma_{n,m}^2}{2} \mathbb{E} f''(V_{n,m}) + \mathbb{E}(R_{n,m}). \tag{4.10}$$

Also $U_{n,m-1} = V_{n,m} + Z_{n,m}$, and $V_{n,m}$ and $Z_{n,m}$ are independent. Therefore, exactly as above one gets, using $\mathbb{E} Z_{n,m} = 0$, $\mathbb{E} Z_{n,m}^2 = \sigma_{n,m}^2$,

$$\mathbb{E} f(U_{n,m-1}) = \mathbb{E} f(V_{n,m}) + \frac{\sigma_{n,m}^2}{2} \mathbb{E} f''(V_{n,m}) + \mathbb{E} R'_{n,m}), \tag{4.11}$$

where

$$R'_{n,m} := Z_{n,m}^2 \int_0^1 (1 - \theta)\{f''(V_{n,m} + \theta Z_{n,m}) - f''(V_{n,m})\} \, d\theta. \tag{4.12}$$

Hence,

$$|\mathbb{E}f(U_{n,m}) - \mathbb{E}f(U_{n,m-1})| \leq \mathbb{E}|R_{n,m}| + \mathbb{E}|R'_{n,m}| \qquad (1 \leq m \leq k_n). \qquad (4.13)$$

Now, given an arbitrary $\varepsilon > 0$,

$$
\begin{aligned}
\mathbb{E}|R_{n,m}| &= \mathbb{E}(|R_{n,m}|\mathbf{1}_{[|X_{n,m}|>\varepsilon]}) + \mathbb{E}(|R_{n,m}|\mathbf{1}_{[|X_{n,m}|\leq\varepsilon]}) \\
&\leq \mathbb{E}\left[X_{n,m}^2\mathbf{1}_{[|X_{n,m}|>\varepsilon]}\int_0^1 (1-\theta)2\|f''\|_\infty\,d\theta\right] \\
&\quad + \mathbb{E}\left[X_{n,m}^2\mathbf{1}_{[|X_{n,m}|\leq\varepsilon]}\int_0^1 (1-\theta)|X_{n,m}|\|f'''\|_\infty\,d\theta\right] \\
&\leq \|f''\|_\infty\mathbb{E}(X_{n,m}^2\mathbf{1}_{[|X_{n,m}|>\varepsilon]}) + \tfrac{1}{2}\varepsilon\sigma_{n,m}^2\|f'''\|_\infty. \qquad (4.14)
\end{aligned}
$$

We have used the notation $\|g\|_\infty := \sup\{|g(x)| : x \in \mathbb{R}\}$. By (4.4), (4.5), and (4.14),

$$\lim \sum_{m=1}^{k_n} \mathbb{E}|R_{n,m}| \leq \tfrac{1}{2}\varepsilon\|f'''\|_\infty.$$

Since $\varepsilon > 0$ is arbitrary,

$$\lim \sum_{m=1}^{k_n} \mathbb{E}|R_{n,m}| = 0. \qquad (4.15)$$

Also,

$$
\begin{aligned}
\mathbb{E}|R'_{n,m}| &\leq \mathbb{E}\left[Z_{n,m}^2\int_0^1 (1-\theta)\|f'''\|_\infty|Z_{n,m}|\,d\theta\right] = \frac{1}{2}\|f'''\|_\infty\mathbb{E}|Z_{n,m}|^3 \\
&= \tfrac{1}{2}\|f'''\|_\infty\sigma_{n,m}^3\mathbb{E}|Z_1|^3 \leq c\sigma_{m,n}^3 \leq c\left(\max_{1\leq m\leq k_n}\sigma_{m,n}\right)\sigma_{n,m}^2, \qquad (4.16)
\end{aligned}
$$

where $c = \tfrac{1}{2}\|f'''\|_\infty\mathbb{E}|Z_1|^3$. Now, for each $\delta > 0$,

$$\sigma_{n,m}^2 = \mathbb{E}(X_{n,m}^2\mathbf{1}_{[|X_{n,m}|>\delta]}) + \mathbb{E}(X_{n,m}^2\mathbf{1}_{[X_{n,m}|\leq\delta]}) \leq \mathbb{E}(X_{n,m}^2\mathbf{1}_{[|X_{n,m}|>\delta]}) + \delta^2,$$

which implies that

$$\max_{1\leq m\leq k_n}\sigma_{n,m}^2 \leq \sum_{m=1}^{k_n}\mathbb{E}(X_{n,m}^2\mathbf{1}_{[|X_{n,m}|>\delta]}) + \delta^2.$$

Therefore, by (4.5),

$$\max_{1\leq m\leq k_n}\sigma_{n,m} \to 0 \qquad \text{as } n \to \infty. \qquad (4.17)$$

From (4.16) and (4.17), one gets

$$\sum_{m=1}^{k_n} \mathbb{E}|R'_{n,m}| \le c \left(\max_{1 \le m \le k_n} \sigma_{n,m} \right) \to 0 \qquad \text{as } n \to \infty. \qquad (4.18)$$

Combining (4.15) and (4.18), one finally gets, on telescoping the difference between (4.10) and (4.11),

$$|\mathbb{E}f(U_{n,k_n}) - \mathbb{E}f(U_{n,0})| \le \sum_{m=1}^{k_n} (\mathbb{E}|R_{n,m}| + \mathbb{E}|R'_{n,m}|) \to 0 \qquad \text{as } n \to \infty. \qquad (4.19)$$

But $U_{n,0}$ is a standard normal random variable. Hence,

$$\mathbb{E}f\left(\sum_{m=1}^{k_n} X_{n,j} \right) - \int_{\mathbb{R}} f(y)(2\pi)^{-1/2} \exp\{-\tfrac{1}{2}y^2\}\,dy \to 0 \qquad \text{as } n \to \infty.$$

By Theorem 4.1, the proof is complete.   ∎

It has been shown by Feller[2] that in the presence of the **uniform asymptotic negligibility** (u.a.n.) condition (4.17), the Lindeberg condition is also *necessary* for the central limit theorem to hold.

Corollary 4.3 *(The Classical CLT)* Let $\{X_j : j \ge 1\}$ be i.i.d. $\mathbb{E}X_j = \mu$, $0 < \sigma^2 :=$ Var $X_j < \infty$. Then $\sum_{j=1}^{n}(X_j - \mu)/(\sigma\sqrt{n})$ converges in distribution to $N(0, 1)$.

Proof Let $X_{n,j} = (X_j - \mu)/(\sigma\sqrt{n})$, $k_n = n$, and apply Theorem 4.2.   ∎

Remark 4.1 Note that the case $k_n = n$ corresponds to an exact **triangular array** of random variables. The general framework of the Lindeberg CLT is referred to as a triangular array as well.

Corollary 4.4 *(Lyapounov's CLT)* For each n let $X_{1,n}, X_{2,n}, \ldots, X_{n,k_n}$ be k_n independent random variables such that

$$\sum_{j=1}^{k_n} \mathbb{E}X_{n,j} = \mu, \qquad \sum_{j=1}^{k_n} \text{Var } X_{n,j} = \sigma^2 > 0,$$

(Lyapounov condition) $$\lim_{n \to \infty} \sum_{j=1}^{k_n} \mathbb{E}|X_{n,j} - \mathbb{E}X_{n,j}|^{2+\delta} = 0 \qquad (4.20)$$

for some $\delta > 0$. Then $\sum_{j=1}^{k_n} X_{n,j}$ converges in distribution to the Gaussian law with mean μ and variance σ^2.

---

[2] Billingsley (1986), p. 373.

Proof By normalizing one may assume, without loss of generality, that

$$\mathbb{E}X_{n,j} = 0, \qquad \sum_{j=1}^{k_n} \mathbb{E}X_{n,j}^2 = 1.$$

It then remains to show that the hypothesis of the corollary implies the Lindeberg condition (4.5). This is true, since for every $\varepsilon > 0$,

$$\sum_{j=1}^{k_n} \mathbb{E}(X_{n,j}^2 \mathbf{1}[X_{n,j} > \varepsilon]) \le \sum_{j=1}^{k_n} \mathbb{E}\frac{|X_{n,j}|^{2+\delta}}{\varepsilon^\delta} \to 0 \tag{4.21}$$

as $n \to \infty$, by (4.20). ∎

Based on such results, there can be no question of the importance of the Gaussian distribution to probability theory. The expression of the indefinite integral $\int e^{-\frac{z^2}{2}}\,dz$ in terms of elementary functions is known to be impossible, however, the following is an often useful property for estimating probabilities for this distribution.

Lemma 2 *(Feller's Tail Probability Estimates)* For a standard normal random variable Z,

$$(z^{-1} - z^{-3})\sqrt{\frac{2}{\pi}}\exp\{-z^2/2\} \le P(|Z| \ge z) \le \sqrt{\frac{2}{\pi z^2}}\exp\{-z^2/2\}, z > 1.$$

In particular,

$$\lim_{z \to \infty} \frac{P(|Z| > z)}{\sqrt{\frac{2}{\pi z^2}}\exp\{-z^2/2\}} = 1.$$

Proof One may obtain simple upper and lower bounds on the integrand by perfect derivatives as follows: $\int_z^\infty e^{-\frac{x^2}{2}}\,dx \le \int_z^\infty \frac{x}{z}e^{-\frac{x^2}{2}}\,dx$, and for the other direction $-\frac{d}{dx}\{(\frac{1}{x} - \frac{1}{x^3})e^{-\frac{x^2}{2}}\} = (1 - \frac{3}{x^4})e^{-\frac{x^2}{2}} \le e^{-\frac{x^2}{2}}$. ∎

Example 1 *(Sampling Design Revisited)* Recall from Example 3 in Chapter II, the computed sample size was $n = 2,154$ using the 4th moment Chebyshev inequality when estimating the sample proportion required to obtain an estimate of the Bernoulli parameter p of a Binomial distribution to within ± 0.03 with probability at least 0.95. However, using the central limit theorem one obtains that the interval $[\hat{p}_n - \xi_{.005}\sqrt{p(1-p)/n}, \hat{p}_n + \xi_{.005}\sqrt{p(1-p)/n}]$ contains p with probability at least 0.99. Thus, the central limit theorem requires n to be sufficiently large that $\xi_{0.005}\sqrt{p(1-p)/n} \le 0.05$. Since $p(1-p) \le 1/4$, it is sufficient that $n \ge 1,071$.[3]

---

[3]In general, the error of approximation in the CLT is $O(n^{-\frac{1}{2}})$; see the Berry-Esseen Theorem 4.6. For the binomial case the approximation is best near $p = 1/2$. However, even for p near the tail such

Theorem 4.5 *(Polya's Uniformity)* Suppose that F_n is a sequence of distribution functions converging pointwise to a continuous distribution function F on $\mathbb{R}$. Then the convergence is uniform, i.e.,

$$\lim_{n\to\infty} \sup_x |F_n(x) - F(x)| = 0.$$

Proof Partition the range of F as $0, \frac{1}{N}, \frac{2}{N}, \ldots \frac{N}{N} = 1$, and use continuity and monotonicity to get $-\infty \leq x_0 < x_1 < \cdots < x_N \leq \infty$ such that $F(x_j) = \frac{j}{N}$. On $[x_{j-1}, x_j], 1 \leq j \leq N$, one has

$$F_n(x) - F(x) \leq F_n(x_j) - F(x_{j-1}) = F_n(x_j) - F(x_j) + \frac{1}{N},$$

and

$$F_n(x) - F(x) \geq F_n(x_{j-1}) - F(x_j) = F_n(x_{j-1}) - F(x_{j-1}) - \frac{1}{N}.$$

In particular

$$\sup_{x\in\mathbb{R}} |F_n(x) - F(x)| \leq \max_{0\leq j\leq N} |F_n(x_j) - F(x_j)| + \frac{1}{N}.$$

Observe that in the case $x_0 = -\infty$, one has $F_n(x_0) - F(x_0) = 0$, and if $x_N = \infty$, then $F_n(x_N) - F(x_N) = 1 - 1 = 0$. Let $\varepsilon > 0$ and fix $N > \frac{1}{\varepsilon}$. Then $\sup_{x\in\mathbb{R}} |F_n(x) - F(x)| \leq \max_{1\leq j\leq N} |F_n(x_j) - F(x_j)| + \varepsilon$. Now let $n \to \infty$ and use pointwise convergence at finitely many points to get

$$\limsup_{n\to\infty} \sup_{x\in\mathbb{R}} |F_n(x) - F(x)| \leq \varepsilon.$$

Since $\varepsilon > 0$ is arbitrary, the proof is complete. ∎

The following (uniform) rate of convergence will be proven in Chapter VI using Fourier transform methods.

Theorem 4.6 *(Berry-Esseen Convergence Rate)* Let $X_1, X_2, \ldots$ be an i.i.d. sequence of random variables having finite third moments $\mathbb{E}|X_1|^3 < \infty$, with mean μ and variance σ^2. Then, for $S_n = X_1 + \cdots + X_n, n \geq 1$, one has

$$\sup_{x\in\mathbb{R}} |P(\frac{S_n - n\mu}{\sigma\sqrt{n}} \leq x) - \Phi(x)| \leq c\frac{\mathbb{E}|X_1|^3}{\sigma^3\sqrt{n}}, c = 0.5600.$$

(Footnote 3 continued)
as $p = 0.9$ or, $p = 0.1$, the approximation is quite good with $n \geq 500$; e.g., see the calculations in R. Bhattacharya, L. Lin and M. Majumdar (2013): Problems of ruin and survival in economics: applications of limit theorems in probability, *Sankhya*, **75B** 145–180.

Observe that the most crucial property of the normal distribution used in the proof of Theorem 4.2 is that the sum of independent normal random variables is normal. In fact, the normal distribution $N(0, 1)$ may be realized as the distribution of the sum of independent normal random variables having zero means and variances σ_i^2 for any arbitrarily specified set of nonnegative numbers σ_i^2 adding up to 1; a form of *infinite divisibility*[4] of the normal distribution.

Definition 4.2 A probability Q on $(\mathbb{R}^k, \mathcal{B}^k)$ is said to be **infinitely divisible** if for each integer $n \geq 1$ there is a probability Q_n such that $Q = Q_n^{*n}$.

Other well-known distributions possessing the infinite divisibility property are the **Poisson distribution**, as well as the class of **stable laws** introduced in Chapter VI in the context of the Holtzmark Example 3.

The following multidimensional version of Corollary 4.3 will also be obtained by the method of characteristic functions in Chapter VI.

Theorem 4.7 *(Multivariate Classical CLT)* Let $\{\mathbf{X}_n : n = 1, 2, \ldots\}$ be a sequence of i.i.d. random vectors with values in $\mathbb{R}^k$. Let $\mathbb{E}\mathbf{X}_1 = \mu \in \mathbb{R}^k$ (defined componentwise) and assume that the dispersion matrix (i.e., variance–covariance matrix) $\mathbf{D}$ of $\mathbf{X}_1$ exists. Then as $n \to \infty$, $n^{-1/2}(\mathbf{X}_1 + \cdots + \mathbf{X}_n - n\mu)$ converges in distribution to the Gaussian probability measure with mean zero and dispersion matrix $\mathbf{D}$.

Exercise Set IV

1. Prove the equivalence of (a)–(c) of Theorem 4.1 in the case $k \geq 2$.

2. Suppose that $\{(X_n, Y_n)\}_{n=1}^\infty$ is a sequence of pairs of real-valued random variables that converge in distribution to (X, Y). Show that $X_n + Y_n$ converges in distribution to $X + Y$. [*Hint*: The map $h : \mathbb{R} \times \mathbb{R} \to \mathbb{R}$ given by $h(x, y) = x + y$ is continuous.]

3. Give a proof of Lemma 1. [*Hint*: Use integration by parts.]

4. Define a one-dimensional normal distribution with mean μ and variance $\sigma^2 = 0$ to be δ_μ, the Dirac measure concentrated at μ. For dimensions $k > 1$, given $\mu \in \mathbb{R}^k$, and a nonnegative-definite (possibly singular) $k \times k$ matrix D, a **multivariate normal distribution** $\Phi_{D,\mu}$ is defined to be the distribution of $\mu + \sqrt{D}\mathbf{Z}$, where $\mathbf{Z}$ is k-dimensional standard normal, and $\sqrt{D}$ denotes a nonnegative-definite symmetric matrix such that $\sqrt{D}\sqrt{D} = D$. Extend the classical CLT (Corollary 4.3) to a version with such possible limits.

5. Suppose that $\{X_j\}_{j=1}^\infty, \ldots$ is a sequence of independent random variables with X_j distributed uniformly on $[-j, j]$, $j \geq 1$. Show that for a suitable choice of scaling constants c_n, the rescaled sum $c_n^{-1}(X_1 + \cdots + X_n)$ is asymptotically normal with mean 0 and variance 1 as $n \to \infty$.

---

[4]Infinitely divisible distributions are naturally associated with stochastic processes having independent increments. This connection is thoroughly developed in a companion text on stochastic processes.

6. $\{X_j\}_{j=1}^{\infty}$ is a sequence of independent random variables uniformly bounded by $M > 0$. Assume $\sigma_n^2 = \sum_{k=1}^{n} \text{Var}(X_k) \to \infty$ as $n \to \infty$. Show that the central limit theorem holds under suitable centering and scaling.

7. Suppose that $U_1, U_2, \ldots$ is an i.i.d. sequence of uniformly distributed random variables on $(0, 1)$. Calculate the limit distribution of $T_n = (eG_n)^{\sqrt{n}}$, where $G_n = (\prod_{j=1}^{n} U_j)^{\frac{1}{n}}$ is the (sample) geometric mean [*Hint*: Consider $\ln T_n$.]

8. Suppose that $\{X_m\}_{m=1}^{\infty}$ is a sequence of independent random variables with X_m distributed as $P(X_m = 1) = P(X_m = -1) = p_m$, $P(X_m = 0) = 1 - 2p_m$, $m \geq 1$, where $\sum_{m=1}^{\infty} p_m = \infty$. Use each of the methods of (a) Lindeberg, (b) Lyapounov to give a proof that, for a suitable choice of scaling constants c_n; the rescaled sum $c_n^{-1}(X_1 + \cdots + X_n)$ is asymptotically normal with mean 0 and variance 1 as $n \to \infty$.

Chapter V
Classical Zero–One Laws, Laws of Large Numbers and Large Deviations

The term **law** has various meanings within probability. It is sometimes used synonymously with *distribution* of a random variable. However, it also may refer to an event or phenomenon that occurs in some predictable sense, as in a "law of averages." The latter is the context of the present section. For example, if $X_0, X_1, \ldots$ is a sequence of independent random variables and $B_n \in \mathcal{B}, n \geq 0$, then, in view of the Borel–Cantelli lemmas, one may conclude that the event $[X_n \in B_n \ i.o.]$ will occur with probability one, or its complement is certain to occur. Before taking up the laws of large numbers, we consider two standard zero–one laws of this type. In particular, observe that the event $A = [X_n \in B_n \ i.o.]$ is special in that it does not depend on any finite number of values of the sequence $X_0, X_1, X_2, \ldots$. Such an event is referred to as a tail event. That is, an event $E \in \sigma(X_0, X_1, X_2, \ldots)$ is said to be a **tail event** if $E \in \sigma(X_n, X_{n+1}, \ldots)$ for every $n \geq 0$. The collection of all tail events is given by the **tail** $\sigma-$**field** $\mathcal{T} := \cap_{n=0}^{\infty}\sigma(X_n, X_{n+1}, \ldots)$.

Theorem 5.1 (*Kolmogorov Zero–One Law*) A tail event for a sequence of independent random variables has probability either zero or one.

Proof To see this first check that $\sigma(X_0, X_1, \ldots) = \sigma(\mathcal{F}_0)$, where $\mathcal{F}_0 := \cup_{k=0}^{\infty}\sigma(X_0, \ldots, X_k)$ is a field and, in particular, a π-system. For $E \in \mathcal{F}_0$, one has $E = [(X_0, \ldots, X_k) \in C]$ for some $k \geq 0, C \in \mathcal{B}^{k+1}$. Thus if A is a tail event then $A \in \sigma(X_{k+1}, \ldots)$ and hence A is independent of E; i.e., A is independent of $\mathcal{F}_0$. Also, the class $\mathcal{L}$ of events independent of A is a λ-system (containing $\mathcal{F}_0$). Therefore $\mathcal{L}$ contains $\sigma(\mathcal{F}_0)$. In particular, A is independent of itself and hence $P(A) = P(A \cap A) = P(A)P(A)$. The only solutions to the equation $x^2 = x$ are 0 and 1. ∎

Not all tail events for the sums of random variables need be tail events for the terms of the series. Let $S_n = X_1 + \cdots + X_n, n \geq 1$, be a random walk on the k-dimensional integer lattice $\mathbb{Z}^k$, for example. An event of the form $[S_n = 0 \ i.o.]$ is not covered by Kolmogorov's zero–one law, since it is not a tail event for the terms $X_1, X_2, \ldots$, and

© Springer International Publishing AG 2016
R. Bhattacharya and E.C. Waymire, *A Basic Course in Probability Theory*,
Universitext, DOI 10.1007/978-3-319-47974-3_V

the sums $S_1, S_2, \ldots$ are not independent. However, there is a special way in which such tail events for the sums depend on the sequence $X_1, X_2, \ldots$ of i.i.d. summands captured by the following zero-one law.

Let $\mathcal{B}^\infty$ denote the (Borel) σ-field of subsets of $\mathbb{R}^\infty = \{(x_1, x_2, \ldots) : x_i \in \mathbb{R}^1\}$ generated by events depending on finitely many coordinates. The following is a more general result than Theorem 5.1.

Theorem 5.2 *(Hewitt–Savage Zero–One Law)* Let $X_1, X_2, \ldots$ be an i.i.d. sequence of random variables. If an event $A = [(X_1, X_2, \ldots) \in B]$, where $B \in \mathcal{B}^\infty$, is invariant under finite permutations $(X_{i_1}, X_{i_2}, \ldots)$ of terms of the sequence $(X_1, X_2, \ldots)$, that is, $A = [(X_{i_1}, X_{i_2}, \ldots) \in B]$ for any finite permutation $(i_1, i_2, \ldots)$ of $(1, 2, \ldots)$, then $P(A) = 1$ or 0.

As noted above, the symmetric dependence with respect to $\{X_n\}_{n=1}^\infty$ applies, for example, to tail events expressed in terms of the partial sums $\{S_n\}_{n=1}^\infty$.

Proof To prove the Hewitt–Savage 0–1 law selects finite-dimensional approximants to A of the form $A_n = [(X_1, \ldots, X_n) \in B_n]$, $B_n \in \mathcal{B}^n$, such that $P(A \Delta A_n) \to 0$ as $n \to \infty$, where $E \Delta F := (E^c \cap F) \cup (E \cap F^c)$ is the *symmetric difference* of sets E, F; this approximation may be achieved from the Carathéodory extension formula (see Exercise 17). For each fixed n, let $(i_1, i_2, \ldots)$ be the permutation $(2n, 2n - 1, \ldots, 1, 2n + 1, \ldots)$ and define $\tilde{A}_n = [(X_{i_1}, \ldots, X_{i_n}) \in B_n]$. Then $\tilde{A}_n$, and A_n are independent with $P(A_n \cap \tilde{A}_n) = P(A_n)P(\tilde{A}_n) = (P(A_n))^2 \to (P(A))^2$ as $n \to \infty$. On the other hand, $P(A \Delta \tilde{A}_n) = P(A \Delta A_n) \to 0$, noting that $A \Delta \tilde{A}_n$ is obtained by a permutation from $A \Delta A_n$. Hence $P(A_n \Delta A) + P(\tilde{A}_n \Delta A) \to 0$ and, in particular, therefore $P(A_n \cap \tilde{A}_n) \to P(A)$ as $n \to \infty$. Thus $x = P(A)$ satisfies $x = x^2$. ∎

The classical strong law of large numbers (SLLN) refers to the almost sure limit of averages of a "large number" of i.i.d. random variables having finite first moment. While the zero–one laws are not required in the following proof, they do imply that the indicated limit of the averages is either sure to exist or sure not to exist.

A good warm-up exercise is to work out a proof using the Borel–Cantelli lemma I based on Chebyshev inequality estimates, assuming finite fourth moments (Exercise 1). The proof we present in this section is due to Etemadi.[1] It is based on Part 1 of the Borel–Cantelli lemmas. As remarked in Chapter III it is possible to obtain a proof from the reverse martingale convergence theorem. This is aided by the Hewitt–Savage zero–one law in identifying the limit as the expected value (constant). Before presenting Etemadi's proof, let's pause to consider that proof.

Theorem 5.3 *(Strong Law of Large Numbers: Reverse martingale proof)* Let $\{X_n : n \geq 1\}$ be an i.i.d. sequence of integrable random variables defined on a probability space $(\Omega, \mathcal{F}, P)$. Then with probability 1,

$$\lim_{n \to \infty} \frac{X_1 + \cdots + X_n}{n} = \mathbb{E}X_1. \tag{5.1}$$

[1]Etemadi, N. (1983): "On the Laws of Large Numbers for Nonnegative Random Variables," *J. Multivariate Analysis*, **13**, pp. 187–193.

Proof Let $\mathcal{G}_n$ denote the σ-field of symmetric events based on $\{X_1, \ldots, X_n\}$. One may simply check that

$$\mathbb{E}(X_1 | \mathcal{G}_n) = \frac{X_1 + \cdots + X_n}{n}, \quad n = 1, 2, \ldots.$$

Thus, using the reverse martingale convergence theorem, it follows that $\lim_{n \to \infty} \frac{X_1 + \cdots + X_n}{n}$ exists a.s. and in L^1; recall Example 7 of Chapter III. In particular, therefore, the limit has mean $\mathbb{E}X_1$. In fact, by the Hewitt–Savage zero–one law it follows that the limit is constant, and hence $\mathbb{E}X_1$. ∎

Theorem 5.4 (*Strong Law of Large Numbers: Etemadi proof*) Let $\{X_n : n \geq 1\}$ be a sequence of pairwise independent and identically distributed random variables defined on a probability space $(\Omega, \mathcal{F}, P)$. If $\mathbb{E}|X_1| < \infty$ then with probability 1,

$$\lim_{n \to \infty} \frac{X_1 + \cdots + X_n}{n} = \mathbb{E}X_1. \tag{5.2}$$

Proof Without loss of generality we may assume for the proof of the SLLN that the random variables X_n are nonnegative, since otherwise we can write $X_n = X_n^+ - X_n^-$, where $X_n^+ = \max(X_n, 0)$ and $X_n^- = -\min(X_n, 0)$ are both nonnegative random variables, and then the result in the nonnegative case yields that

$$\frac{S_n}{n} = \frac{1}{n} \sum_{k=1}^{n} X_k^+ - \frac{1}{n} \sum_{k=1}^{n} X_k^-$$

converges to $\mathbb{E}X_1^+ - \mathbb{E}X_1^- = \mathbb{E}X_1$ with probability 1.

Truncate the variables X_n by $Y_n = X_n \mathbf{1}_{[X_n \leq n]}$. Then $Y_1, Y_2, \ldots$ is a sequence of independent random variables, though *not* identically distributed. Moreover $\mathbb{E}Y_n^2 < \infty$ for all n; in fact Y_n has moments of all orders. Let $T_n = \sum_{k=1}^{n} Y_k$. By the dominated convergence theorem one has $\mathbb{E}Y_k = \mathbb{E}X_1 \mathbf{1}_{[X_1 \leq k]} \to \mathbb{E}X_1$ as $k \to \infty$. It follows simply that the same holds for the Caesaro mean

$$\lim_{n \to \infty} \frac{1}{n} \mathbb{E}T_n = \mathbb{E}X_1. \tag{5.3}$$

Next let us consider the sequence $\{T_n\}_{n=1}^{\infty}$ on the "fast" time scale $\tau_n = [\alpha^n]$, for a fixed $\alpha > 1$, where brackets [ ] denote the integer part. Let $\epsilon > 0$. Then by Chebyshev's inequality and pairwise independence,

$$P\left(\left|\frac{T_{\tau_n} - \mathbb{E}T_{\tau_n}}{\tau_n}\right| > \epsilon\right) \leq \frac{\mathrm{Var}(T_{\tau_n})}{\epsilon^2\tau_n^2} = \frac{1}{\epsilon^2\tau_n^2}\sum_{k=1}^{\tau_n}\mathrm{Var}\,Y_k \leq \frac{1}{\epsilon^2\tau_n^2}\sum_{k=1}^{\tau_n}\mathbb{E}Y_k^2$$

$$= \frac{1}{\epsilon^2\tau_n^2}\sum_{k=1}^{\tau_n}\mathbb{E}\{X_k^2\mathbf{1}_{[X_k\leq k]}\} = \frac{1}{\epsilon^2\tau_n^2}\sum_{k=1}^{\tau_n}\mathbb{E}\{X_1^2\mathbf{1}_{[X_1\leq k]}\}$$

$$\leq \frac{1}{\epsilon^2\tau_n^2}\sum_{k=1}^{\tau_n}\mathbb{E}\{X_1^2\mathbf{1}_{[X_1\leq\tau_n]}\}$$

$$= \frac{1}{\epsilon^2\tau_n^2}\tau_n\mathbb{E}\{X_1^2\mathbf{1}_{[X_1\leq\tau_n]}\}. \tag{5.4}$$

Therefore,

$$\sum_{n=1}^{\infty}P\left(\left|\frac{T_{\tau_n} - \mathbb{E}T_{\tau_n}}{\tau_n}\right| > \epsilon\right) \leq \sum_{n=1}^{\infty}\frac{1}{\epsilon^2\tau_n}\mathbb{E}\{X_1^2\mathbf{1}_{[X_1\leq\tau_n]}\} = \frac{1}{\epsilon^2}\mathbb{E}\left\{X_1^2\sum_{n=1}^{\infty}\frac{1}{\tau_n}\mathbf{1}_{[X_1\leq\tau_n]}\right\}. \tag{5.5}$$

Let $x > 0$ and let $N = \min\{n \geq 1 : \tau_n \geq x\}$. Then $\alpha^N \geq x$, and since $y \leq 2[y]$ for any $y \geq 1$,

$$\sum_{n=1}^{\infty}\frac{1}{\tau_n}\mathbf{1}_{[x\leq\tau_n]} = \sum_{\tau_n\geq x}\frac{1}{\tau_n} \leq 2\sum_{n\geq N}\alpha^{-n} = \frac{2\alpha}{\alpha - 1}\alpha^{-N} = a\alpha^{-N} \leq \frac{a}{x},$$

where $a = 2\alpha/(\alpha - 1)$. Therefore,

$$\sum_{n=1}^{\infty}\frac{1}{\tau_n}\mathbf{1}_{[X_1\leq\tau_n]} \leq \frac{a}{X_1} \qquad \text{for } X_1 > 0.$$

So

$$\sum_{n=1}^{\infty}P\left(\left|\frac{T_{\tau_n} - \mathbb{E}T_{\tau_n}}{\tau_n}\right| > \epsilon\right) \leq a\frac{\mathbb{E}[X_1]}{\epsilon^2} < \infty. \tag{5.6}$$

By the Borel–Cantelli lemma I, taking a union over positive rational values of ϵ, with probability 1, $(T_{\tau_n} - \mathbb{E}T_{\tau_n})/\tau_n \to 0$ as $n \to \infty$. Since, by (5.3),

$$\lim_{n\to\infty}\frac{1}{\tau_n}\mathbb{E}T_{\tau_n} = \mathbb{E}X_1,$$

it follows that

$$\frac{T_{\tau_n}}{\tau_n} \to \mathbb{E}X_1. \tag{5.7}$$

Furthermore, since

$$\sum_{n=1}^{\infty} P(X_n \neq Y_n) = \sum_{n=1}^{\infty} P(X_1 > n) \leq \int_0^{\infty} P(X_1 > u)\,du = \mathbb{E}X_1 < \infty,$$

we get by another application of the Borel–Cantelli lemma that, with probability 1,

$$\frac{S_n - T_n}{n} \to 0 \qquad \text{as } n \to \infty. \tag{5.8}$$

Therefore, the previous results about $\{T_n\}$ give for $\{S_n\}$ that

$$\frac{S_{\tau_n}}{\tau_n} \to \mathbb{E}X_1 \qquad \text{as } n \to \infty \tag{5.9}$$

with probability 1. If $\tau_n \leq k \leq \tau_{n+1}$, then since $X_i \geq 0$,

$$\frac{\tau_n}{\tau_{n+1}} \frac{S_{\tau_n}}{\tau_n} \leq \frac{S_k}{k} \leq \frac{\tau_{n+1}}{\tau_n} \frac{S_{\tau_{n+1}}}{\tau_{n+1}}. \tag{5.10}$$

But $\tau_{n+1}/\tau_n \to \alpha$, so that now we get with probability 1,

$$\frac{1}{\alpha}\mathbb{E}X_1 \leq \liminf_k \frac{S_k}{k} \leq \limsup_k \frac{S_k}{k} \leq \alpha\mathbb{E}X_1. \tag{5.11}$$

Letting $\alpha \downarrow 1$ via rational $\alpha > 1$, one has $\lim_{k\to\infty} S_k/k = \mathbb{E}X_1$ with probability 1. This is the strong law of large numbers (SLLN). ∎

The above proof of the SLLN is really quite remarkable, as the following observations show. First, *pairwise independence* is used only to make sure that the positive and negative parts of X_n, and their truncations Y_n remain independent and therefore *(pairwise) uncorrelated* for the calculation of the variance of T_k as the sum of the variances. Positivity can alternatively be achieved by assuming the random variables are uniformly bounded below with probability one, since one can then add a sufficiently large constant. Assuming that the random variables are square-integrable and uncorrelated it is unnecessary to truncate, so the same variance bound on the S_{τ_k} can be achieved. In fact, one may replace identical distributions with an assumption of bounded variances $Var X_j, j \geq 1$, Thus, if the random variables all are square-integrable, mean zero and a.s. uniformly bounded below, then it suffices to require that they merely be *uncorrelated* for the same proof of the a.s. convergence on the fast time scale, followed by interpolation, to go through; this is Exercise 2.

Proposition 5.5 *Let $X_1, X_2, \ldots$, be a sequence of square-integrable, mean-zero uncorrelated random variables that are a.s. uniformly bounded below, or a.s. uniformly bounded above. If, in addition, $Var X_n, n \geq 1$, is a bounded sequence, then with probability 1,*

$$\frac{X_1 + \cdots + X_n}{n} \to 0 \qquad as\ n \to \infty.$$

In particular, this holds for every bounded, mean-zero, uncorrelated sequence.

The following corollary follows directly from Theorem 5.4.

Corollary 5.6 *If $X_1, X_2, \ldots$ is an i.i.d. sequence and $\mathbb{E}X_1^+ = \infty$ and $\mathbb{E}X_1^- < \infty$, then with probability 1,*

$$\frac{X_1 + \cdots + X_n}{n} \to \infty \qquad as\ n \to \infty.$$

Similarly, if $\mathbb{E}X_1^+ < \infty$ and $\mathbb{E}X_1^- = \infty$, then the a.s. limit is $-\infty$.

Proof Assume $\mathbb{E}X_1^+ < \infty$. Then one has by Theorem 5.4 that $\frac{X_1^+ + \cdots + X_n^+}{n} \to \mathbb{E}X_1^+$ with probability one. Thus, since $X_n = X_n^+ - X_n^-, n \geq 1$, it is enough to check that $\frac{X_1^- + \cdots + X_n^-}{n} \to \infty$ a.s.. Let $Y_n^{(c)} = X_n^- \mathbf{1}_{[X_n^- \leq c]}, n \geq 1$, for a constant $c > 0$. Then, with probability one, $\frac{X_1^- + \cdots + X_n^-}{n} \geq \frac{Y_1^{(c)} + \cdots + Y_n^{(c)}}{n} \to \mathbb{E}Y_1^{(c)}$. Let $c \to \infty$ to obtain the desired result. ∎

As an obvious corollary, since a.s. convergence implies convergence in probability, one may conclude that the averages converge in probability as well. The latter statement is referred to as a **weak law of large numbers** (WLLN).

The proof of the Weierstrass approximation theorem given in Appendix B may be viewed as an application of the WLLN to a classic problem in analysis; namely, a continuous function g on $[0, 1]$ may be uniformly approximated by polynomials $h_n(x) = \sum_{k=0}^n \binom{n}{k} f(\frac{k}{n}) x^k (1 - x)^{n-k}, 0 \leq x \leq 1$, referred to as **Bernstein polynomials**.

Let us now briefly turn some attention to deviations from the law of averages. Suppose $X_1, X_2, \ldots$ is an i.i.d. sequence of random variables with mean μ. Then the WLLN implies that for any $\delta > 0$, the event that the sample average $A_n := \frac{X_1 + \cdots + X_n}{n}$ would fall outside the interval $\mu \pm \delta$, i.e., "would deviate from μ by a positive amount δ," is a rare event for large n. In fact, under suitable conditions one might expect the probability to be *exponentially small* for large n. The "large deviation theorem" below provides an important illustration of such conditions. We will need a few preliminaries to prepare for the statement and proof.

Definition 5.1 Let X be a random variable on $(\Omega, \mathcal{F}, P)$ with distribution Q. The **moment-generating function** of X (or Q) is defined by $m(h) = \mathbb{E}e^{hX} = \int_{\mathbb{R}} e^{hx} Q(dx)$. The **cumulant-generating function** is $c(h) = \ln m(h), h \in \mathbb{R}$.

Note that $m(h)$ may be infinite; see Exercise 6. The function $m(-h)$ is the **Laplace transform** of the distribution Q.

Proposition 5.7 (a) *Assume that $m(h) < \infty$ for all h in a neighborhood of $h = 0$. Then $\mathbb{E}|X|^k < \infty$ for all $k \geq 1$ and $\mathbb{E}X^k = m^{(k)}(0) \equiv \frac{d^k}{dh^k} m(0)$.* (b) *Assume that*

$m(h) < \infty$ for all h in a neighborhood of $h = r \in \mathbb{R}$. Then $\mathbb{E}|X^k e^{rX}| < \infty$ and $m^{(k)}(r) \equiv \frac{d^k}{dr^k} m(r) = \mathbb{E}X^k e^{rX}$.

Proof Since $e^{|hx|} \leq e^{hx} + e^{-hx}$, it follows from the hypothesis for (a) that $\mathbb{E}e^{hX} \leq \mathbb{E}e^{|hX|} < \infty$ for all h in a neighborhood of $h = 0$. Also, since the partial sums $\sum_{k=0}^{n} \frac{|hX|^k}{k!}$ are bounded by $e^{|hX|}$, one has by the dominated convergence theorem that $\mathbb{E}e^{hX} = \mathbb{E} \sum_{k=0}^{\infty} \frac{h^k X^k}{k!} = \sum_{k=0}^{\infty} \frac{\mathbb{E}X^k}{k!} h^k$. The assertion (a) now follows from the uniqueness of the coefficients in Taylor series expansions about the origin (Exercise 16). For part (b) consider the change of measure defined by $\tilde{Q}(dx) = \frac{e^{rx}}{m(r)} Q(dx)$. Recall that the factor $\frac{1}{m(r)}$ is the normalization of $e^{rx} Q(dx)$ to a probability. If $\tilde{X}$ is a random variable with distribution $\tilde{Q}$, then its moment-generating function is given by $\tilde{m}(h) = \frac{m(h+r)}{m(r)}$. Under hypothesis (b), $\tilde{X}$ has a moment-generating function in a neighborhood of $h = 0$, so that (a) yields $\mathbb{E}X^k e^{rX}/m(r) = m^{(k)}(r)/m(r)$. Multiplying by $m(r)$ yields the assertion (b). ∎

For the rest of this section assume that X is **non-degenerate**, i.e. Q is not a Dirac measure δ_c. Assuming that $m(h)$ is finite in a neighborhood of zero, the second derivative of $m(h)$ is obviously positive, and one may use the Cauchy–Schwarz inequality to check $c^{(2)}(h) > 0$ as well; see Exercise 7.

Corollary 5.8 *Suppose that* $m(h) = \mathbb{E}e^{hX} < \infty$ *for all* $h \in \mathbb{R}$. *Then both* $m(h)$ *and* $c(h)$ *are convex functions on* $\mathbb{R}$.

The following "change of measure"technique was used in the proof of Proposition 5.7. It provides a way in which to tilt or bias a probability distribution that is effective for analyzing deviations from the original mean. In the language of change of measure, the mean of the tilted distribution becomes the "new normal"for the sake of analysis.

Definition 5.2 Suppose that $m(b) \equiv \int e^{bx} Q(dx) < \infty$. For a fixed such parameter b, the change of measure defined by $\tilde{Q}_b(dx) = \frac{e^{bx}}{m(b)} Q(dx)$ is called an **exponential size-bias**, **Esscher** or **tilting** transformation of Q.

Let $\tilde{X}$ denote a random variable with the tilted distribution $\tilde{Q}$. The effect of the tilt on the mean is reflected in the following calculation:

$$\mathbb{E}\tilde{X} = \int_{\mathbb{R}} x \tilde{Q}_b(dx) = \frac{1}{m(b)} \int_{\mathbb{R}} x e^{bx} Q(dx) = \frac{m'(b)}{m(b)} = c'(b). \qquad (5.12)$$

The function defined by

$$c^*(x) = \sup_h (xh - c(h)), x \in \mathbb{R}, \qquad (5.13)$$

is called the **Legendre transform** of c; see Exercise 10 for more details. The maximum in (5.13) is attained at h such that $c'(h) = x$. From (5.14) it also follows that this h is also such that the size-biased distribution $\tilde{Q}_h$ has mean one. Observe that c^* is convex since for $0 \leq t \leq 1$, $x, y \in \mathbb{R}$,

$$c^*(tx + (1-t)y) = \sup_h (t(xh - tc(h)) + (1-t)(yh - (1-t)c(h))$$

$$\leq tc^*(x) + (1-t)c^*(y). \tag{5.14}$$

In the following write $A_n = \frac{X_1 + \cdots + X_n}{n}$, where $X_j (j \geq 1)$ are i.i.d. with common distribution Q, and $\tilde{A}_n = \frac{\tilde{X}_1 + \cdots + \tilde{X}_n}{n}$, where $\tilde{X}_j (j \geq 1)$ are i.i.d. with common size-biased distribution $\tilde{Q}_{h^*}$ for a suitable h^*.

Theorem 5.9 (*Cramér-Chernoff*) Suppose that $X_1, X_2, \ldots$ is an i.i.d. sequence with finite mean $\mathbb{E}X_1 = \mu$. Moreover, assume that the moment-generating function

$$m(h) := \mathbb{E}e^{hX_1}$$

is finite for all h in a neighborhood of zero. Let $c(h) := \ln m(h)$ denote the cumulant-generating function. Then for $a > \mathbb{E}X_1$ one has

$$\lim_{n \to \infty} \frac{\ln P(A_n \geq a)}{n} = I(a), \tag{5.15}$$

where $I(a) = -c^*(a)$ for

$$c^*(x) = \sup_{h \in \mathbb{R}}\{xh - c(h)\} \geq 0.$$

Proof We will first give the proof in the case that $m(h) < \infty$ for all $h \in \mathbb{R}$. The proof of the general case is outlined in Exercise 14. (and below). We may assume $P(X_1 \geq a) > 0$. For otherwise, (5.15) is trivially true, since $I(a) = -\infty$ and both sides of (5.15) are zero (Exercise 8.). To obtain the formula (5.15), first note the simple inequality

$$\mathbb{E}e^{hnA_n} \geq \mathbb{E}\{e^{hnA_n}\mathbf{1}[A_n \geq a]\} \geq e^{hn(a)}P(A_n \geq a)$$

for all $h \geq 0$. Since by independence, the moment-generating function of $nA_n \equiv X_1 + \cdots + X_n$ may be expressed as $e^{nc(h)}$, one has for any $h \geq 0$,

$$P(A_n \geq a) \leq e^{-n(ah - c(h))}.$$

Thus, one obtains an (upper) bound for the rate of decay of probability in (5.15) of the form

$$\limsup_{n \to \infty} \frac{\ln P(A_n \geq a)}{n} \leq -c^*(a).$$

It suffices to prove the reverse inequality to establish (5.15). For this it is useful to exponentially size-bias the distribution of X in such a way that the deviant event is the rule, rather than the exception. For the given deviation a, suppose that the maximum defining $c^*(a)$ is achieved at $h = h_a > 0$ (see Exercise 8), with

$$c^*(a) = ah_a - c(h_a), \qquad \frac{d}{dh}(ah - c(h))|_{h=h_a} = 0.$$

In particular, $a = \frac{d}{dh}c(h)|_{h=h_a}$. To simplify notation we will write h^* for h_a in the rest of the proof. Define a random variable $\tilde{X}$ to have the size-biased distribution given by

$$Q_{h^*}(dy) = P(\tilde{X} \in dy) = Z_\delta^{-1} e^{h^* y} P(X \in dy),$$

where $Z_\delta = m(h^*)$ normalizes $e^{h^* y} P(X \in dy)$ to a probability distribution. Now observe that

$$\mathbb{E}\tilde{X} = e^{-c(h^*)} \int_{\mathbb{R}} y e^{h^* y} P(X \in dy) = \frac{d}{dh} c(h)|_{h=h^*} = a. \tag{5.16}$$

In particular, the law of large numbers yields

$$\lim_{n \to \infty} \tilde{A}_n = \lim_{n \to \infty} \frac{\tilde{X}_1 + \cdots + \tilde{X}_n}{n} = \mathbb{E}\tilde{X} = a.$$

From here one may obtain the reverse inequality by the law of large numbers under size biasing: Namely, let $\epsilon > 0$, and consider deviations of size a (to within $\pm \epsilon$) defined by

$$D_n := \{(y_1, \ldots, y_n) \in \mathbb{R}^n : \frac{1}{n} \sum_{j=1}^{n} y_j \in (a - \epsilon, a + \epsilon)\}.$$

Note that $\exp\{-nh(a + \epsilon) + h \sum_{j=1}^{n} X_j\} \le 1$ on the event $[(X_1, \ldots, X_n) \in D_n]$. Thus one has,

$$P(A_n > a - \epsilon)$$
$$\ge P(A_n \in (a - \epsilon, a + \epsilon))$$
$$= \mathbb{E}\mathbf{1}[(X_1, \ldots, X_n) \in D_n]$$
$$\ge \mathbb{E}\mathbf{1}[(X_1, \ldots, X_n) \in D_n] \exp\{-nh^*(a + \epsilon) + h^* \sum_{j=1}^{n} X_j\}$$
$$= \exp\{-nh^*(a + \epsilon)\} Z_\delta^n \mathbb{E}\{\mathbf{1}[(X_1, \ldots, X_n) \in D_n] \prod_{j=1}^{n} Z_\delta^{-1} e^{h^* X_j}\}$$
$$= \exp\{-nh^*(a + \epsilon)\} e^{nc(h^*)} P((\tilde{X}_1, \ldots, \tilde{X}_n) \in D_n)$$
$$= \exp\{-(h^*(a + \epsilon) - c(h^*))n\} P(\tilde{A}_n \in (a - \epsilon, a + \epsilon)). \tag{5.17}$$

Now, the law of large numbers under the size-biased distribution (having mean a) makes $\tilde{A}_n \to a$ and hence $P(\tilde{A}_n \in (a - \epsilon, a + \epsilon)) \to 1$ as $n \to \infty$. In particular, it follows from (5.17) that for any $\epsilon > 0$,

$$\liminf_{n \to \infty} \frac{\ln P(A_n > a - \epsilon)}{n} \geq -c^*(a) - h^*\epsilon.$$

Since $\liminf_{n \to \infty} \frac{\ln P(A_n > a - \epsilon)}{n}$ is an increasing function of ϵ, the inequality follows. The case in which the supremum defining $c^*(a)$ is finite but not achieved is left to Exercise 16.

For the more restricted hypothesis of finiteness in a neighborhood of zero, let (a, b) be the interval on which $m(h)$ is finite, $-\infty \leq a < 0 < b < \infty$. On $(a, b)^c$, $m(h) = \infty$, and therefore, one may restrict attention to $h \in (a, b)$ in the definition of $c^*(x)$. The details are left to Exercise 14.   ∎

Remark 5.1 The inequality

$$P(A_n \geq a) \leq e^{-n \sup_h (ah - c(h))} = e^{nI(a)} \tag{5.18}$$

arrived at in the "easy-half"of the above proof is of interest in its own right and referred to as the **Chernoff inequality**. The function $I(a)$ is referred to as the **large deviation rate** and is computed here in terms of the Legendre transform of the cumulant-generating function c of the common distribution Q for the sequence of random variables.

A refinement of the Cramér–Chernoff large deviation rate that takes advantage of the central limit theorem was developed by Ranga Rao and Bahadur[2] It also yields an alternative derivation of Theorem 5.9.

Theorem 5.10 *(Bahadur and Ranga Rao)* Let $X_1, X_2, \ldots$ be an i.i.d. sequence of real-valued random variables, having finite moment-generating function $m(h) = \mathbb{E}e^{hX_1}$ for h in a neighborhood of zero. Let σ^2 denote the variance of X_1. Then for $a > \mathbb{E}X_1$,

$$p_n := P(\frac{1}{n} \sum_{j=1}^{n} X_j \geq a) = e^{nI(a)} e^{\frac{\nu_n^2}{2}} (1 - \Phi(\nu_n)] + R_n,$$

where h^* is defined as before, and

$$\frac{m'(h^*)}{m(h^*)} = a, \quad \nu_n = \sigma h^* \sqrt{n}, \quad I(a) = -ah^* + c'(h^*),$$

and $|R_n| \leq e^{nI(a)} 2c \frac{\mathbb{E}|X_1|^3}{\sigma^3 \sqrt{n}}$, for a constant $c \leq 3$.

Remark 5.2 This estimate of p_n, omitting R_n, is sharper than that given in Theorem 5.9. The error term involves the (uniform) error in normal approximation by the

---

[2]The refinement presented here is due to Bahadur, R., and Ranga Rao, R. (1960): On deviations of the sample mean, *Ann. Math. Stat.*, v31(4), 1015-1027. This work was preceded by earlier results of Blackwell, D. and Hodges, J.L. (1959): The probability in the extreme tail of a convolution, *Ann. Math. Stat.*, v30, 1113–1120.

clt. A so-called *Berry–Esseen bound* of the form $c\mathbb{E}|X_1|^3/\sigma^3\sqrt{n}$ will be proven in Chapter VI, namely Theorem 6.17, to produce Feller's constant $c = 3$. As noted there, this constant has been subsequently reduced to $c = 0.5600$ as best to date.

Proof Assume for simplicity that $m(h) < \infty$ for all $h \in \mathbb{R}$. Let $Y_j = X_j - a$, $j = 1, 2, \ldots, n$ and denote its distribution by Q_-, and its moment-generating function $\varphi(h) = \mathbb{E}e^{hY_1} = e^{-ah}m(h)$, $h \in \mathbb{R}$. Define a sequence of i.i.d. random variables $\tilde{Y}_j$, $j \geq 1$, with the size-biased distribution $\tilde{Q}_-(dy) = \varphi(\tau)^{-1}e^{\tau y}Q_-(dy)$, where h^* is chosen so that $\mathbb{E}\tilde{Y}_1 = \int y\tilde{Q}_-(dy) = 0$. Equivalently, the moment-generating function $\psi(h^*) = \mathbb{E}e^{h^*\tilde{Y}_j} = \frac{\varphi(h^*+h)}{\text{Var}(h^*)}$ has the property $\psi'(0) = 0$. Equivalently,

$$\frac{m'(h^*)}{m(h^*)} = a.$$

Also note that $\sigma^2 = \psi''(0) = \frac{\varphi''(h^*)}{\varphi(h^*)} = \frac{m'(h^*)}{m(h^*)} - a^2$. Now,

$$p_n = P\left(\sum_{j=1}^n Y_j \geq 0\right) = \int_{\{(y_1,\ldots,y_n):\sum_{j=1}^n y_j=0\}} Q_-(dy_1)\cdots Q_-(dy_n)$$

$$= \int_{\{(y_1,\ldots,y_n):\sum_{j=1}^n y_j=0\}} \varphi^n(h^*)e^{-h^*\sum_{j=1}^n y_j}\tilde{Q}_-(dy_1)\cdots\tilde{Q}_-(dy_n)$$

$$= \varphi^n(h^*)\int_{[0,\infty)} e^{-\tau y}\tilde{Q}_-^{*n}(dy) = \varphi^n(h^*)\int_{[0,\infty)} e^{-h^*\sigma\sqrt{n}x}F_n(dx),$$

where $\tilde{Q}_-^{*n}$ is the n-fold convolution of $\tilde{Q}_-$, and F_n is the distribution of $\frac{1}{\sigma\sqrt{n}}(\tilde{Y}_1 + \cdots + \tilde{Y}_n)$, for i.i.d. $\tilde{Y}_j(j \geq 1)$, having the size-biased distribution $\tilde{Q}_-$. Next apply integration by parts to obtain

$$p_n = \varphi^n(h^*)h^*\sigma\sqrt{n}\int_{[0,\infty)} e^{-h^*\sigma\sqrt{n}x}\{F_n(x) - F_n(0)\}dx.$$

By the central limit theorem, for large n and up to a (uniform) error, one can replace F_n by the standard normal distribution Φ to get

$$p_n = \varphi^n(h^*)h^*\sigma\sqrt{n}\int_{[0,\infty)} e^{-h^*\sigma\sqrt{n}x}\{\Phi(x) - \Phi(0)\}dx + R_n,$$

where, denoting the error in the central limit theorem approximation by ϵ_n, one may write $|R_n| = \varphi^n(\tau)\epsilon_n$. Using the Berry–Esseen bound noted in the remark before the proof, and using $\int_0^\infty e^{-z}dz = 1$, one has

$$|R_n| \leq e^{-nah^*}m^n(h^*)2c\frac{\mathbb{E}|X_1|^3}{\sigma^3\sqrt{n}}.$$

Integrating by parts one has, using $\Phi'(x) = \frac{1}{\sqrt{2\pi}}e^{-\frac{x^2}{2}}$ and completing the square,

$$\varphi^n(h^*)h^*\sigma\sqrt{n}\int_{[0,\infty)}e^{-h^*\sigma\sqrt{n}x}\{\Phi(x)-\Phi(0)\}dx$$

$$= \varphi^n(h^*)\int_{[0,\infty)}e^{-h^*\sigma\sqrt{n}x}\Phi'(x)$$

$$= \varphi^n(h^*)e^{\frac{1}{2}(h^*\sigma\sqrt{n})^2}\{1-\Phi(h^*\sigma\sqrt{n}\}$$

$$= e^{nI(a)}e^{\frac{v_n^2}{2}}\{1-\Phi(v_n)\}.$$

∎

We conclude this chapter with an important inequality due to Hoeffding[3] The inequality is derived from the following lemma.

Lemma 5.1 (Hoeffding's Lemma) *Let Y be a non-degenerate mean zero random variable such that $P(a \leq Y \leq b) = 1$. Then,*

$$\mathbb{E}e^{hY} \leq \exp\{\frac{h^2(b-a)^2}{8}\}, \quad \forall h \in \mathbb{R}.$$

Proof The convexity of $y \to e^{hy}$ implies that

$$e^{hY} \leq \frac{b-Y}{b-a}e^{hY} + \frac{Y-a}{b-a}e^{hb}.$$

Taking expectations one obtains

$$\mathbb{E}e^{hY} \leq \frac{b}{b-a}e^{ha} + \frac{-a}{b-a}e^{hb}$$

$$= e^{ha}(\frac{b}{b-a} + \frac{-a}{b-a}e^{h(b-a)})$$

$$= e^{ha}(1-p+pe^{h(b-a)})$$

$$= e^{-\lambda p}(1-p+pe^\lambda) = e^{g(\lambda)},$$

where $p = \frac{-a}{b-a}$, $1-p = \frac{b}{b-a}$, $\lambda = h(b-a)$, and

$$g(\lambda) = -\lambda p + \ln(1-p+pe^\lambda).$$

[3]W. Hoeffding (1963): Probability inequalities for sums of bounded random variables, *J. of the Am. Stat. Assoc.* **58**(301),1330, obtained this inequality in 1963. There are a number of related "concentration inequalities" of this type in the modern probability literature, expressing the concentration of the distribution of $\overline{X}$ near the mean.

Note that $g'(0) = 0$, $g''(p) = (1-p)pe^{\lambda}/(1-p+pe^{\lambda})^2 \leq \frac{(1-p)pe^{\lambda}}{4(1-p)pe^{\lambda}} = 1/4$. Thus, by Taylor expansion, $g(\lambda) \leq \frac{\lambda^2}{8}$. This leads to the asserted bound on $\mathbb{E}e^{hY}$. ∎

Proposition 5.11 (Hoeffding Inequality) *Suppose that X_j, $1 \leq j \leq n$, are independent non-degenerate bounded random variables, say, $P(c_j \leq X_j \leq d_j) = 1, j = 1, \ldots, n$. Let $\mu_j = \mathbb{E}X_j, \mu = \frac{1}{n}\sum_{j=1}^{n}\mu_j, \Delta^2 = \sum_{j=1}^{n}(d_j - c_j)^2$. Then for every $\delta > 0$, writing $\overline{X} = \frac{1}{n}\sum_{j=1}^{n}X_j$, one has*

$$P(\overline{X} \geq \mu + \delta) \leq \exp\{-\frac{2n^2\delta^2}{\Delta^2}\},$$

$$P(\overline{X} \leq \mu - \delta) \leq \exp\{-\frac{2n^2\delta^2}{\Delta^2}\}.$$

Proof It is clearly enough to prove the first inequality since it can then be applied to the case in which each X_j is changed to $-X_j$, $1 \leq j \leq n$. Now,

$$\begin{aligned}
P(\overline{X} \geq \mu + \delta) &= P(\overline{X} - \mu \geq \delta) \\
&\leq e^{-h\delta}\mathbb{E}e^{h(\overline{X}-\mu)} \\
&= e^{-h\delta}\prod_{j=1}^{n}\mathbb{E}e^{\frac{h}{n}(X_j-\mu_j)} \\
&\leq e^{-h\delta}\prod_{j=1}^{n}\exp\{\frac{h^2}{8n^2}(d_j - c_j)^2\} \\
&= e^{-h\delta+h^2\Delta^2/8n^2} = e^{f(h)}.
\end{aligned}$$

Using Hoeffding's lemma, the minimum of the quadratic $f(h)$ is attained at $h = 4\delta n^2/\Delta^2$, and for this value of h one has $f(h) = -2n^2\delta^2/\Delta^2$. ∎

Remark 5.3 Hoeffding's inequality may be viewed as a special case of Chernoff's inequality extended to non-identically distributed random variables, with an estimation of the Legendre transform $-I(\mu + \delta)$ provided by Hoeffding's lemma. Indeed, assuming each mgf satisfies $m_j(h) = \mathbb{E}\exp\{h(X_j - \mu)\} < \infty$ in a neighborhood $(-\epsilon, \epsilon)$ for some $\epsilon > 0$, $1 \leq j \leq n$, one has

$$P(\overline{X} - \mu > \delta) \leq e^{-\delta h}\mathbb{E}e^{h(\overline{X}-\mu)} = e^{-\delta h}\prod_{j=1}^{n}m_j(\frac{h}{n}).$$

Minimizing with respect to h (or $\frac{h}{n}$), one obtains

$$P(\overline{X} - \mu > \delta) \leq e^{nI(\delta)},$$

where

$$I(\delta) = \inf_{h \geq 0}\{-\delta\frac{h}{n} + \frac{1}{n}\sum_{j=1}^{n} c_j(\frac{h}{n})\} = \inf_{h' \geq 0}\{-\delta h' + \frac{1}{n}\sum_{j=1}^{n} c_j(h')\}.$$

Such so-called "concentration inequalities" of Chernoff and Hoeffding have many applications in areas such as risk theory and machine learning.[4]

Example 1 (*Sampling Design*) The Chernoff/Hoeffding estimate of the sample size for Bernoulli parameter estimation requires a surprisingly large value of n; namely 2, 049, Exercise 13. On the other hand the Bahadur–Ranga Rao large deviation estimate provides a reasonable reduction; Exercise 13.

Exercise Set V

1. Give a simple proof of the strong law of large numbers (SLLN) for i.i.d. random variables $Z_1, Z_2, \ldots$ having finite fourth moments. That is, for $S_n := Z_1 + \cdots + Z_n, n \geq 1, \lim_{n \to \infty} S_n/n \to \mathbb{E}Z_1$ a.s. as $n \to \infty$. [*Hint*: Use a fourth moment Chebyshev-type inequality and the Borel–Cantelli lemma I to check for each $\epsilon = 1/k, k \geq 1, P(|\frac{S_n}{n} - \mathbb{E}Z_1| > \epsilon \, i.o.) = 0$.]

2. (i) Write out the proof of Proposition 5.5 based on the remarks following the proof of Theorem 5.4. (ii) Suppose $\{X_n : n \geq 1\}$ is a sequence of mean zero uncorrelated random variables such that $\sum_{k=1}^{n} var(X_k)/n^2 \to 0$. Show that $\frac{1}{n}\sum_{k=1}^{n} X_k \to 0$ in probability.

3. (*SLLN and Transience and Recurrence of Simple Random Walk*) Let $S^x = \{S_n^x : n = 0, 1, 2, \ldots\}$, where $S_n^x = x + Z_1 + \cdots Z_n, (n \geq 1), S_0^x = x \in \mathbb{Z}$, be the simple random walk with $P(Z_1 = 1) = p, P(Z_1 = -1) = q = 1 - p$, with $0 < p < 1$. Use the SLLN to show that (i) if $p = 1/2$ then $\gamma(x) := P(S^x$ reaches every state in $\mathbb{Z}) = 1$, and (ii) if $p \neq 1/2$ then $\gamma(x) = 0$.

4. Let $X_1, X_2, \ldots$ be an i.i.d. sequence of positive random variables such that $\mathbb{E}|\ln X_1| < \infty$. (i) Calculate the a.s. limiting *geometric mean* $\lim_{n \to \infty}(X_1 \cdots X_n)^{\frac{1}{n}}$. Determine the numerical value of this limit in the case of uniformly distributed random variables on $(0, 2)$. (ii) Suppose also that $\mathbb{E}X_1 = 1$, and $P(X_1 = 1) < 1$. Show that $X_1 \cdots X_n \to 0$ a.s. as $n \to \infty$. (iii) Under the conditions of (ii), is the sequence of products $X_1, X_1 X_2, X_1 X_2 X_3, \ldots$ uniformly integrable ?

5. (*Hausdorff's Estimate*) Let $X_1, X_2, \ldots$ be an i.i.d. sequence of random variables with mean zero and moments of all orders. Let $S_n = X_1 + \cdots + X_n, n \geq 1$. Show that given any $\epsilon > 0$ the event $A := [|S_n| = O(n^{\frac{1}{2}+\epsilon})$ as $n \to \infty]$ has probability one. [*Hint*: For two sequences of numbers $\{a_n\}_n$ and $\{b_n \neq 0\}_n$ one writes $a_n = O(b_n)$ as $n \to \infty$ if and only if there is a constant $C > 0$ such that $|a_n| \leq C|b_n|$ for all n. Check that $\mathbb{E}|S_n|^{2k} \leq c_k n^k, k = 1, 2, 3, \ldots$, and use the Borel–Cantelli lemma I to prove the assertion.]

---

[4]See Mitzenmacher, M. and E. Upfal (2005) for illustrative applications in the context of machine learning.

6. Compute the moment-generating function $m(h)$ for each of the following random variables X: (i) $P(X = n) = Q(\{n\}) = \frac{c}{n^2}, n = \pm1, \pm2, \ldots$, where $c^{-1} = 2\sum_{n=1}^{\infty} \frac{1}{n^2}$. (ii) $Q(dx) = \lambda e^{-\lambda x}\mathbf{1}_{[0,\infty)}(x)dx$, where $\lambda > 0$. (iii) $Q(dx) = \frac{1}{\sqrt{2\pi}}e^{-\frac{1}{2}x^2}dx$. (iv) Show that $m(h) < \infty$ for all $h \in \mathbb{R}$ if X is a bounded random variable, i.e., $P(|X| \le B) = 1$ for some $B \ge 0$.

7. Assume $m(h) < \infty$ for all $h \in \mathbb{R}$. If X has a non-degenerate distribution, show that $c^{(2)}(h) > 0$ for all $h \in \mathbb{R}$. [*Hint*: $c^{(2)}(h) = \mathbb{E}\tilde{X}^2 - (\mathbb{E}\tilde{X})^2$, where $\tilde{X}$ has the distribution $\frac{e^{hx}Q(dx)}{m(h)}$.]

8. (i) Show that there is a maximal interval including zero on which the mgf $m(h)$ of a random variable X is finite. [*Hint*: The function $h \to m(h)$ is convex since $h \to e^{hy}$ is convex for each y.] (ii) Find examples when the maximal interval in (i) is (a) $\{0\}$, (b) $[0, \delta)$ for some $0 < \delta \le \infty$, (c) $(\delta, 0]$ for some $-\infty \le \delta < 0$, (d) $(\alpha, \beta), -\infty \le \alpha < 0 < \beta \le \infty$.

9. Suppose X is non-degenerate with a finite mgf $m(h)$ in a neighborhood of zero, and the maximal such interval is $(\alpha, \beta), -\infty \le \alpha < 0 < \beta \le \infty$. (i) Show that if $P(X > 0) > 0$, then $\lim_{h\uparrow\beta} m(h) = \infty$, and if $P(X < 0) > 0$, then $\lim_{h\downarrow\alpha} m(h) = \infty$. (ii) Prove that $m(h)$ and $c(h) = \ln m(h)$ are both strictly convex on (α, β). (iii) For every $a \in \mathbb{R}$, the mgf of $X - a$, namely, $\varphi_a(h) = e^{-ah}m(h)$, is finite on (α, β) and infinite on $(\alpha, \beta)^c$. In particular, if $a > \mu = \mathbb{E}X$, then $\lim_{h\downarrow\alpha} \varphi_a(h) = \infty$, and if $P(X > a) > 0$, then $\lim_{h\uparrow\beta} \varphi_a(h) = \infty$. (iv) Suppose $a > \mu$, and $P(X > a) > 0$. Prove that the indicated supremum $c^*(a) = \sup\{ah - c(h) : h \in \mathbb{R}\}$ is attained at some point in (α, β). [*Hint*: The function $f(h) = ah - c(h)$ satisfies $f(0) = 0$, $\lim_{h\downarrow\alpha} f(h) = -\infty = \lim_{h\uparrow\beta} f(h)$.]

10. (*Some Basic Properties of Legendre Transform*) Let u, v be twice continuously differentiable convex functions on $\mathbb{R}$ with Legendre transforms u^*, v^*, where $f^*(x) := \sup_{h\in\mathbb{R}}(xh - f(h)), x \in \mathbb{R}$.

   (i) (*Convexity*) Show that u^* is convex. [*Hint*: Directly check that $u^*(\lambda x + (1 - \lambda)y) \le \lambda u^*(x) + (1 - \lambda)u^*(y), 0 \le \lambda \le 1, x, y \in \mathbb{R}$, from the formula defining u^*. Note that u need not be convex for this property of u^*.]

   (ii) (*Involution*) Show that $u^{**} = u$. [*Hint*: Write $u^*(x) = xh(x) - u(h(x))$ and use the smoothness hypothesis on u to first obtain $\frac{d}{dx}u^*(x) = h(x)$ where $x = u'(h(x))$.]

   (iii) (*Young's Inequality*) Show that if $u = v^*$ then $xy \le u(x) + v(y)$. [*Hint*: Fix x and use the formula defining u in terms of v^* to bound $u(x)$ below by $xh - v(h)$ for all h. Replace h by y.]

   (iv) Illustrate the above relations graphically.

11. (i) Suppose that X_1 has a Gaussian distribution with mean μ and variance σ^2. Compute the large deviation rate $I(a)$. (ii) If X_1 is Poisson with mean μ, compute $I(a)$. (iii) If X_1 is Bernoulli, $P(X_1 = 0) = 1 - p, P(X_1 = 1) = p (0 < p < 1)$, compute $I(a)$. (iv) Calculate $I(a)$ for a gamma-distributed random variable with density $f(x) = \frac{\alpha^\beta}{\Gamma(\beta)}x^{\beta-1}e^{-\alpha x}\mathbf{1}_{(0,\infty)}(x)$.

12. Suppose that $X_1, X_2, \ldots$ is an i.i.d. sequence of Poisson distributed random variables with mean one, i.e., $P(X_1 = m) = \frac{1}{m!}e^{-1}, m = 0, 1, 2, \ldots$. Calculate

an exact formula for $\lim_{n\to\infty} \frac{\ln P(S_n > n(1+\epsilon))}{n}$ as a function of $\epsilon > 0$, where $S_n = X_1 + \cdots + X_n$, $n \geq 1$.

13. Determine the sample sizes for estimation of the Bernoulli parameter to within $\pm .03$ with probability at least .95 using each of the Chernoff inequality and the Bahadur–Ranga Rao large deviation estimates.

14. (i) Prove that Theorem 5.9 holds if $m(h)$ is finite in a neighborhood of zero. [*Hint*: Let (a, b) be the maximal interval on which $m(h)$ is finite, $-\infty \leq a < 0 < b < \infty$. On $(a, b)^c$, $c(h) = \infty$. Also the approach to infinity by $c(h)$ at an infinite boundary point α or β is faster than that of any linear function $h \to ah$, and therefore, one may restrict attention to $h \in (a, b)$ in the definition of $c^*(x)$.]

15. Derive the Cramér–Chernoff estimate (Theorem 5.9) from that of the Bahadur–Ranga Rao estimate (Theorem 5.10). [*Hint*: Use the Feller tail estimate Lemma 2 from Chapter IV: $1 - \Phi(x) \sim \frac{1}{x\sqrt{2\pi}} e^{-\frac{x^2}{2}}$ as $x \to \infty$.]

16. (*Interchange of the Order of Differentiation and Integration*) (i) Let $f(x, \theta)$ be a real-valued function on $S \times (c, d)$, where $(S, \mathcal{S}, \mu)$ is a measure space and (a) f is μ-integrable for all θ, (b) $\theta \mapsto f(x, \theta)$ is differentiable at $\theta = \theta_0 \in (c, d)$, and (c) $|f(x, \theta_0 + \epsilon) - f(x, \theta_0)|/|\epsilon| \leq g(x)$ for all $x \in S$ and for all ϵ such that $0 < |\epsilon| < \epsilon_0$ (for some $\epsilon_0 > 0$), and $\int_S g \, d\mu < \infty$. Then show that $\frac{d}{d\theta} \int_S f(x, \theta) \mu(dx)|_{\theta=\theta_0} = \int_S (\frac{d}{d\theta} f(x, \theta))_{\theta=\theta_0} \mu(dx)$. [*Hint*: Apply Lebesgue's dominated convergence theorem to the sequence $g_n(x) := (f(x, \theta_0 + \epsilon_n) - f(x, \theta_0))/\epsilon_n$ $(0 \neq \epsilon_n \to 0)$.] (ii) Verify the term-by-term differentiation of $m(h) = \sum_{k=0}^{\infty} \frac{h^k}{k!} \mathbb{E} X^k$ at $h = 0$ in the proof of Proposition 5.7.

17. (i) Use the definition of product probability measure via the Carathéodory construction to obtain the approximation of $A \in \mathcal{B}^{\infty}$ by finite-dimensional events $A_n = [(X_1, \ldots, X_n) \in B_n]$, $B_n \in \mathcal{B}^n$, such that $P(A \triangle A_n) \to 0$ as $n \to \infty$. [*Hint*: Given $\epsilon > 0$, obtain a cover $A \subset \cup_{m \geq 1} R_m$, with $R_m \in \sigma(X_1, \ldots, X_m)$, such that $P(\cup_{m \geq 1} R_m \backslash A) < \epsilon/2$. Use continuity of the probability from above to argue that $P(\cup_{m \geq 1} R_m \backslash \cup_{m=1}^n R_m) < \epsilon/2$ for n sufficiently large.] (ii) Show that $|P(A) - P(A_n)| \leq P(A \triangle A_n)$. (iii) Show that on a finite measure space $(S, \mathcal{S}, \mu)$, (a) $\mu(B \triangle A) = \mu(A \triangle B) \geq 0$, (b) $\mu(A \triangle A) = 0$, and (c) $\mu(A \triangle B) + \mu(B \triangle C) \geq \mu(A \triangle C)$ hold for all $A, B, C \in \mathcal{S}$. That is, $(A, B) \to \mu(A \triangle B)$ is a *pseudo-metric* on $\mathcal{S}$.

Chapter VI
Fourier Series, Fourier Transform, and Characteristic Functions

Fourier series and Fourier transform provide one of the most important tools for analysis and partial differential equations, with widespread applications to physics in particular and science in general. This is (up to a scalar multiple) a norm-preserving (i.e., isometry), linear transformation on the Hilbert space of square-integrable complex-valued functions. It turns the integral operation of convolution of functions into the elementary algebraic operation of the product of the transformed functions, and that of differentiation of a function into multiplication by its Fourier frequency.

Although beyond our scope, this powerful and elegant theory extends beyond functions on finite-dimensional Euclidean space to infinite-dimensional spaces and locally compact abelian groups.[1] From this point of view, Fourier series is the Fourier transform on the circle group.

This chapter develops the basic properties of Fourier series and the Fourier transform with applications to the central limit theorem and to transience and recurrence of random walks.

Consider a real- or complex-valued periodic function on the real line. By changing the scale if necessary, one may take the period to be 2π. Is it possible to represent f as a superposition of the periodic functions ("waves") $\cos nx$, $\sin nx$ of *frequency* n ($n = 0, 1, 2, \ldots$)? In view of **Weierstrass approximation theorem**, every continuous periodic function f of period 2π is the limit (in the sense of uniform convergence of functions) of a sequence of **trigonometric polynomials**, i.e., functions of the form

$$\sum_{n=-T}^{T} c_n e^{inx} = c_0 + \sum_{n=1}^{T} (a_n \cos nx + b_n \sin nx);$$

the Bernstein polynomials in e^{ix} illustrate one such approximation.

---

[1]Extensions of the theory can be found in the following standard references, among others: Rudin (1967), Grenander (1963), Parthasarathy (1967).

© Springer International Publishing AG 2016
R. Bhattacharya and E.C. Waymire, *A Basic Course in Probability Theory*,
Universitext, DOI 10.1007/978-3-319-47974-3_VI

As will be seen, Theorem 6.1 below gives an especially useful version of the approximation from the perspective of *Fourier series*. In a Fourier series, the coefficients of the polynomials are especially defined according to an $L^2[-\pi, \pi]$-orthogonality of the complex exponentials $e^{inx} = \cos(nx) + i \sin(nx)$ as explained below. For this special choice of coefficients the theory of Fourier series yields, among other things, that with the weaker notion of L^2-convergence the approximation holds for a wider class of functions, namely for all square-integrable functions f on $[-\pi, \pi]$; here square integrability means that f is measurable and that $\int_{-\pi}^{\pi} |f(x)|^2 \, dx < \infty$; denoted $f \in L^2[-\pi, \pi]$. It should be noted that in general, we consider integrals of complex-valued functions in this section, and the $L^p = L^p(dx)$ spaces are those of complex-valued functions (see Exercise 36 of Chapter I).

The successive coefficients c_n for this approximation are the so-called **Fourier coefficients**:

$$c_n = \frac{1}{2\pi} \int_{-\pi}^{\pi} f(x)e^{-inx} \, dx \qquad (n = 0, \pm 1, \pm 2, \ldots). \tag{6.1}$$

The main point of Theorem 6.1 in this context is to provide a tool for *uniformly approximating* continuous functions by trigonometric polynomials whose coefficients more closely approximate Fourier coefficients than alternatives such as Bernstein polynomials.

As remarked above, the functions e^{inx} $(n = 0, \pm 1, \pm 2, \ldots)$ form an **orthonormal set**:

$$\frac{1}{2\pi} \int_{-\pi}^{\pi} e^{inx} e^{-imx} \, dx = \begin{cases} 0, & \text{for } n \neq m, \\ 1 & \text{for } n = m, \end{cases} \tag{6.2}$$

so that the **Fourier series of** f is written formally, without regard to convergence for the time being, as

$$\sum_{n=-\infty}^{\infty} c_n e^{inx}. \tag{6.3}$$

As such, this is a representation of f as a superposition of orthogonal components. To make matters precise we first prove the following useful class of Fejér polynomials; see Exercise 1 for an alternative approach.

Theorem 6.1 Let f be a continuous periodic function of period 2π. Then, given $\delta > 0$, there exists a trigonometric polynomial, specifically a **Fejér average** $\sum_{n=-N}^{N} d_n e^{inx}$, where

$$d_n = (1 - \frac{|n|}{N+1}) \frac{1}{2\pi} \int_{-\pi}^{\pi} f(x)e^{-inx} dx, \, n = 0, \pm 1, \pm 2, \ldots,$$

such that

$$\sup_{x \in \mathbb{R}^1} \left| f(x) - \sum_{n=-N}^{N} d_n e^{inx} \right| < \delta.$$

Proof For each positive integer N, introduce the **Fejér kernel**

$$k_N(x) := \frac{1}{2\pi} \sum_{n=-N}^{N} \left(1 - \frac{|n|}{N+1}\right) e^{inx}. \tag{6.4}$$

This may also be expressed as

$$2\pi(N+1)k_N(x) = \sum_{0 \le j,k \le N} e^{i(j-k)x} = \left|\sum_{j=0}^{N} e^{ijx}\right|^2$$

$$= \frac{2\{1 - (\cos(N+1)x\}}{2(1 - \cos x)} = \left(\frac{\sin\{\frac{1}{2}(N+1)x\}}{\sin \frac{1}{2}x}\right)^2. \tag{6.5}$$

At $x = 2n\pi$ $(n = 0, \pm1, \pm2, \dots)$, the right side is taken to be $(N+1)^2$. The first equality in (6.5) follows from the fact that there are $N+1-|n|$ pairs (j, k) in the sum such that $j - k = n$. It follows from (6.5) that k_N is a positive continuous periodic function with period 2π. Also, k_N is a pdf on $[-\pi, \pi]$, since nonnegativity follows from (6.5) and normalization from (6.4) on integration. For every $\varepsilon > 0$ it follows from (6.5) that $k_N(x)$ goes to zero uniformly on $[-\pi, -\varepsilon] \cup [\varepsilon, \pi]$, so that

$$\int_{[-\pi,-\varepsilon]\cup[\varepsilon,\pi]} k_N(x)dx \to 0 \qquad \text{as } N \to \infty. \tag{6.6}$$

In other words, $k_N(x)dx$ converges weakly to $\delta_0(dx)$, the point mass at 0, as $N \to \infty$.
  Consider now the approximation f_N of f defined by

$$f_N(x) := \int_{-\pi}^{\pi} f(y)k_N(x-y)dy = \sum_{n=-N}^{N} \left(1 - \frac{|n|}{N+1}\right) c_n e^{inx}, \tag{6.7}$$

where c_n is the nth Fourier coefficient of f. By changing variables and using the periodicity of f and k_N, one may express f_N as

$$f_N(x) = \int_{-\pi}^{\pi} f(x-y)k_N(y)dy.$$

Therefore, writing $M = \sup\{|f(x)| : x \in \mathbb{R}^k\}$, and $\delta_\varepsilon = \sup\{|f(y) - f(y')| : |y - y'| < \varepsilon\}$, one has

$$|f(x) - f_N(x)| \le \int_{-\pi}^{\pi} |f(x-y) - f(x)|k_N(y)dy \le 2M \int_{[-\pi,-\varepsilon]\cup[\varepsilon,\pi]} k_N(y)dy + \delta_\varepsilon. \tag{6.8}$$

It now follows from (6.6) that $f - f_N$ converges to zero uniformly as $N \to \infty$. Now write $d_n = (1 - |n|/(N+1))c_n$. ∎

Remark 6.1 The representation of the approximating trigonometric polynomial for f as a convolution $f * k_N$, where k_N is a nonegative kernel such that $k_N \Rightarrow \delta_0$ is a noteworthy consequence of the proof of Theorem 6.1. The advantages of such polynomials over an approximation by Bernstein polynomials will become evident in the context of unique determination of an integrable periodic function, or even a finite measure on the circle, from its Fourier coefficients; see Proposition 6.3 and Theorem 6.4 below.

The first task is to establish the convergence of the Fourier series (6.3) to f in L^2. Here the norm $\| \cdot \|$ is $\| \cdot \|_2$ as defined by (6.10) below. If $f(x) = \sum_{n=-N}^{N} a_n e^{inx}$ is a trigonometric polynomial then the proof is immediate. The general case follows by a uniform approximation of 2π-periodic continuous function by such trigonometric polynomials, and finally the density of such continuous functions in $L^2[-\pi, \pi]$.

Theorem 6.2

a. For every f in $L^2[-\pi, \pi]$, the Fourier series of f converges to f in L^2-norm, and the identity $\| f \| = (\sum_{-\infty}^{\infty} |c_n|^2)^{1/2}$ holds for its Fourier coefficients c_n. Here $\| \cdot \|$ is defined in (6.10).
b. If (i) f is differentiable, (ii) $f(-\pi) = f(\pi)$, and (iii) f' is square-integrable, then the Fourier series of f also converges uniformly to f on $[-\pi, \pi]$.

Proof (a) Note that for every square-integrable f and all positive integers N,

$$\frac{1}{2\pi} \int_{-\pi}^{\pi} \left(f(x) - \sum_{-N}^{N} c_n e^{inx} \right) e^{-imx} dx = c_m - c_m = 0 \quad (m = 0, \pm 1, \dots, \pm N).$$
(6.9)

Therefore, if one defines the **norm** (or "length") of a function g in $L^2[-\pi, \pi]$ by

$$\| g \| = \left(\frac{1}{2\pi} \int_{-\pi}^{\pi} |g(x)|^2 dx \right)^{1/2} \equiv \| g \|_2,$$
(6.10)

then, writing $\bar{z}$ for the complex conjugate of z,

$$0 \le \| f - \sum_{-N}^{N} c_n e^{in \cdot} \|^2$$

$$= \frac{1}{2\pi} \int_{-\pi}^{\pi} \left(f(x) - \sum_{-N}^{N} c_n e^{inx} \right) \left(\bar{f}(x) - \sum_{-N}^{N} \bar{c}_n e^{-inx} \right) dx$$

$$= \frac{1}{2\pi} \int_{-\pi}^{\pi} (f(x) - \sum_{-N}^{N} c_n e^{inx}) \bar{f}(x) dx$$

$$= \|f\|^2 - \sum_{-N}^{N} c_n \bar{c}_n = \|f\|^2 - \sum_{-N}^{N} |c_n|^2. \tag{6.11}$$

This shows that $\|f - \sum_{-N}^{N} c_n e^{in\cdot}\|^2$ *decreases* as N increases and that

$$\lim_{N \to \infty} \|f - \sum_{-N}^{N} c_n e^{in\cdot}\|^2 = \|f\|^2 - \sum_{-\infty}^{\infty} |c_n|^2. \tag{6.12}$$

To prove that the right side of (6.12) vanishes, first assume that f is continuous and $f(-\pi) = f(\pi)$. Given $\varepsilon > 0$, there exists, by Theorem 6.1, a trigonometric polynomial $\sum_{-N_0}^{N_0} d_n e^{inx}$ such that

$$\max_x \left| f(x) - \sum_{-N_0}^{N_0} d_n e^{inx} \right| < \varepsilon.$$

This implies

$$\frac{1}{2\pi} \int_{-\pi}^{\pi} \left| f(x) - \sum_{-N_0}^{N_0} d_n e^{inx} \right|^2 dx < \varepsilon^2. \tag{6.13}$$

But by (6.9), $f(x) - \sum_{-N_0}^{N_0} c_n \exp\{inx\}$ is orthogonal to e^{imx} ($m = 0, \pm 1, \ldots, \pm N_0$), so that

$$\frac{1}{2\pi} \int_{-\pi}^{\pi} \left| f(x) - \sum_{-N_0}^{N_0} d_n e^{inx} \right|^2 dx$$

$$= \frac{1}{2\pi} \int_{-\pi}^{\pi} \left| f(x) - \sum_{-N_0}^{N_0} c_n e^{inx} + \sum_{-N_0}^{N_0} (c_n - d_n) e^{inx} \right|^2 dx$$

$$= \frac{1}{2\pi} \int_{-\pi}^{\pi} \left| f(x) - \sum_{-N_0}^{N_0} c_n e^{inx} \right|^2 dx$$

$$+ \frac{1}{2\pi} \int_{-\pi}^{\pi} \left| \sum_{-N_0}^{N_0} (c_n - d_n) e^{inx} \right|^2 dx. \tag{6.14}$$

Hence, by (6.13), (6.14), and (6.11),

$$\frac{1}{2\pi}\int_{-\pi}^{\pi}\left|f(x) - \sum_{-N_0}^{N_0} c_n e^{inx}\right|^2 dx < \varepsilon^2, \qquad \lim_{N\to\infty}\left\|f - \sum_{-N}^{N} c_n e^{in\cdot}\right\|^2 \leq \varepsilon^2. \quad (6.15)$$

Since $\varepsilon > 0$ is arbitrary, it follows that

$$\lim_{N\to\infty}\left\|f(x) - \sum_{-N}^{N} c_n e^{inx}\right\| = 0, \qquad (6.16)$$

and by (6.12),

$$\|f\|^2 = \sum_{-\infty}^{\infty} |c_n|^2. \qquad (6.17)$$

This completes the proof of convergence for continuous periodic f. Now it may be shown that given a square-integrable f and $\varepsilon > 0$, there exists a continuous periodic g such that $\|f - g\| < \varepsilon/2$ (Exercise 1). Also, letting $\sum a_n e^{inx}$, $\sum c_n e^{inx}$ be the Fourier series of g, f, respectively, there exists N_1 such that

$$\left\|g - \sum_{-N_1}^{N_1} a_n e^{in\cdot}\right\| < \frac{\varepsilon}{2}.$$

Hence (see (6.14))

$$\left\|f - \sum_{-N_1}^{N_1} c_n e^{in\cdot}\right\| \leq \left\|f - \sum_{-N_1}^{N_1} a_n e^{in\cdot}\right\| \leq \|f - g\| + \left\|g - \sum_{-N_1}^{N_1} a_n e^{in\cdot}\right\|$$

$$< \frac{\varepsilon}{2} + \frac{\varepsilon}{2} = \varepsilon. \qquad (6.18)$$

Since $\varepsilon > 0$ is arbitrary and $\|f(\cdot) - \sum_{-N}^{N} c_n e^{in\cdot}\|^2$ decreases to $\|f\|^2 - \sum_{-\infty}^{\infty} |c_n|^2$ as $N \uparrow \infty$ (see (6.12)), one has

$$\lim_{N\to\infty}\left\|f - \sum_{-N}^{N} c_n e^{in\cdot}\right\| = 0; \quad \|f\|^2 = \sum_{-\infty}^{\infty} |c_n|^2. \qquad (6.19)$$

To prove part (b), let f be as specified. Let $\sum c_n e^{inx}$ be the Fourier series of f, and $\sum c_n^{(1)} e^{inx}$ that of f'. Then

$$c_n^{(1)} = \frac{1}{2\pi}\int_{-\pi}^{\pi} f'(x) e^{-inx}\, dx = \frac{1}{2\pi} f(x) e^{-inx}\Big|_{-\pi}^{\pi} + \frac{in}{2\pi}\int_{-\pi}^{\pi} f(x) e^{-inx}\, dx$$

$$= 0 - inc_n = -inc_n. \qquad (6.20)$$

Since f' is square-integrable,

$$\sum_{-\infty}^{\infty} |nc_n|^2 = \sum_{-\infty}^{\infty} |c_n^{(1)}|^2 < \infty. \tag{6.21}$$

Therefore, by the Cauchy–Schwarz inequality,

$$\sum_{-\infty}^{\infty} |c_n| = |c_0| + \sum_{n \neq 0} \frac{1}{|n|} |nc_n| \leq |c_0| + \left(\sum_{n \neq 0} \frac{1}{n^2}\right)^{1/2} \left(\sum_{n \neq 0} |nc_n|^2\right)^{1/2} < \infty.$$

$$\tag{6.22}$$

But this means that $\sum c_n e^{inx}$ is uniformly absolutely convergent, since

$$\max_x \left| \sum_{|n| > N} c_n e^{inx} \right| \leq \sum_{|n| > N} |c_n| \to 0 \qquad \text{as } N \to \infty.$$

Since the continuous functions $\sum_{-N}^{N} c_n e^{inx}$ converge uniformly (as $N \to \infty$) to $\sum_{-\infty}^{\infty} c_n e^{inx}$, the latter must be a continuous function, say h. Uniform convergence to h also implies convergence in norm to h. Since $\sum_{-\infty}^{\infty} c_n e^{inx}$ also converges in norm to f, $f(x) = h(x)$ for all x. If the two continuous functions f and h are not identically equal, then

$$\int_{-\pi}^{\pi} |f(x) - h(x)|^2 dx > 0.$$

∎

Definition 6.1 For a finite measure (or a finite-signed measure) μ on the circle $[-\pi, \pi)$ (identifying $-\pi$ and π), the nth **Fourier coefficient of** μ is defined by

$$c_n = \frac{1}{2\pi} \int_{[-\pi,\pi)} e^{-inx} \mu(dx) \qquad (n = 0, \pm 1, \ldots). \tag{6.23}$$

If μ has a density f, then (6.23) is the same as the nth Fourier coefficient of f given by (6.1).

Proposition 6.3 A finite measure μ on the circle is determined by its Fourier coefficients.

Proof Approximate the measure $\mu(dx)$ by $g_N(x)\,dx$, where

$$g_N(x) := \int_{[-\pi,\pi)} k_N(x - y)\mu(dy) = \sum_{-N}^{N} \left(1 - \frac{|n|}{N+1}\right) c_n e^{inx}, \tag{6.24}$$

with c_n defined by (6.23). For every continuous periodic function h (i.e., for every continuous function on the circle),

$$\int_{[-\pi,\pi)} h(x)g_N(x)\,dx = \int_{[-\pi,\pi)} \left(\int_{[-\pi,\pi)} h(x)k_N(x-y)\,dx \right) \mu(dy). \qquad (6.25)$$

As $N \to \infty$, the probability measure $k_N(x-y)\,dx = k_N(y-x)\,dx$ on the circle converges weakly to $\delta_y(dx)$. Hence, the inner integral on the right side of (6.25) converges to $h(y)$. Since the inner integral is bounded by $\sup\{|h(y)| : y \in \mathbb{R}\}$, Lebesgue's dominated convergence theorem implies that

$$\lim_{N\to\infty} \int_{[-\pi,\pi)} h(x)g_N(x)\,dx = \int_{[-\pi,\pi)} h(y)\mu(dy). \qquad (6.26)$$

This means that μ is determined by $\{g_N : N \geq 1\}$. The latter in turn are determined by $\{c_n\}_{n\in\mathbb{Z}}$. ∎

We are now ready to answer an important question: When is a given sequence $\{c_n : n = 0, \pm1, \ldots\}$ the sequence of Fourier coefficients of a finite measure on the circle? A sequence of complex numbers $\{c_n : n = 0, \pm1, \pm2, \ldots\}$ is said to be **positive-definite** if for any finite sequence of complex numbers $\{z_j : 1 \leq j \leq N\}$, one has

$$\sum_{1\leq j,k\leq N} c_{j-k} z_j \bar{z}_k \geq 0. \qquad (6.27)$$

Theorem 6.4 *(Herglotz Theorem)* $\{c_n : n = 0, \pm1, \ldots\}$ is the sequence of Fourier coefficients of a probability measure on the circle if and only if it is positive-definite, and $c_0 = \frac{1}{2\pi}$.

Proof Necessity If μ is a probability measure on the circle, and $\{z_j : 1 \leq j \leq N\}$ a given finite sequence of complex numbers, then

$$\sum_{1\leq j,k\leq N} c_{j-k} z_j \bar{z}_k = \frac{1}{2\pi} \sum_{1\leq j,k\leq N} z_j \bar{z}_k \int_{[-\pi,\pi)} e^{i(j-k)x} \mu(dx)$$

$$= \frac{1}{2\pi} \int_{[-\pi,\pi)} \left(\sum_1^N z_j e^{ijx} \right) \left(\sum_1^N \bar{z}_k e^{-ikx} \right) \mu(dx)$$

$$= \frac{1}{2\pi} \int_{[-\pi,\pi)} \left| \sum_1^N z_j e^{ijx} \right|^2 \mu(dx) \geq 0. \qquad (6.28)$$

Also,

$$c_0 = \frac{1}{2\pi} \int_{[-\pi,\pi)} \mu(dx) = \frac{1}{2\pi}.$$

Sufficiency. Take $z_j = e^{i(j-1)x}$, $j = 1, 2, \ldots, N+1$, in (6.27) to get

$$g_N(x) := \frac{1}{N+1} \sum_{0 \le j,k \le N} c_{j-k} e^{i(j-k)x} \ge 0. \tag{6.29}$$

Again, since there are $N + 1 - |n|$ pairs (j, k) such that $j - k = n$ $(-N \le n \le N)$ it follows that (6.29) becomes

$$0 \le g_N(x) = \sum_{-N}^{N} \left(1 - \frac{|n|}{N+1}\right) e^{inx} c_n. \tag{6.30}$$

In particular, using (6.2),

$$\int_{[-\pi,\pi)} g_N(x)\,dx = 2\pi c_0 = 1. \tag{6.31}$$

Hence g_N is a pdf on $[-\pi, \pi]$. By Proposition 7.6, there exists a subsequence $\{g_{N'}\}$ such that $g_{N'}(x)\,dx$ converges weakly to a probability measure $\mu(dx)$ on $[-\pi, \pi]$ as $N' \to \infty$. Also, again using (6.2) yields

$$\int_{[-\pi,\pi)} e^{-inx} g_N(x)\,dx = 2\pi \left(1 - \frac{|n|}{N+1}\right) c_n \qquad (n = 0, \pm 1, \ldots, \pm N). \tag{6.32}$$

For each fixed n, restrict to the subsequence $N = N'$ in (6.32) and let $N' \to \infty$. Then, since for each n, $\cos(nx)$, $\sin(nx)$ are bounded continuous functions,

$$2\pi c_n = \lim_{N' \to \infty} 2\pi \left(1 - \frac{|n|}{N'+1}\right) c_n = \int_{[-\pi,\pi)} e^{-inx} \mu(dx) \qquad (n = 0, \pm 1, \ldots). \tag{6.33}$$

In other words, c_n is the nth Fourier coefficient of μ. ∎

Corollary 6.5 A sequence $\{c_n : n = 0, \pm 1, \ldots\}$ of complex numbers is the sequence of Fourier coefficients of a finite measure on the circle $[-\pi, \pi)$ if and only if $\{c_n : n = 0, \pm 1, \ldots\}$ is positive-definite.

Proof Since the measure $\mu = 0$ has Fourier coefficients $c_n = 0$ for all n, and the latter is trivially a positive-definite sequence, it is enough to prove the correspondence between nonzero positive-definite sequences and nonzero finite measures. It follows from Theorem 6.4, by normalization, that this correspondence is 1–1 between positive-definite sequences $\{c_n : n = 0, \pm 1, \ldots\}$ with $c_0 = c > 0$ and measures on the circle having total mass 2π. ∎

It is instructive to consider the **Fourier transform** $\hat{f}$ of an integrable function f on $\mathbb{R}$, defined by

$$\hat{f}(\xi) = \int_{-\infty}^{\infty} e^{i\xi y} f(y) \, dy, \qquad \xi \in \mathbb{R}. \tag{6.34}$$

as a limiting version of a Fourier series. In particular, if f is differentiable and vanishes outside a finite interval, and if f' is square-integrable, then one may use the Fourier series of f (scaled to be defined on $(-\pi, \pi]$) to obtain (see Exercise 6) the **Fourier inversion formula,**

$$f(z) = \frac{1}{2\pi} \int_{-\infty}^{\infty} \hat{f}(y) e^{-izy} \, dy. \tag{6.35}$$

Moreover, any f that vanishes outside a finite interval and is square-integrable is automatically integrable, and for such an f one has the **Plancherel identity** (see Exercise 6)

$$\|\hat{f}\|_2^2 := \int_{-\infty}^{\infty} |\hat{f}(\xi)|^2 \, d\xi = 2\pi \int_{-\infty}^{\infty} |f(y)|^2 \, dy = 2\pi \|f\|_2^2. \tag{6.36}$$

The extension of this theory relating to Fourier series and Fourier transforms in higher dimensions is straightforward along the following lines. The Fourier series of a square-integrable function f on $[-\pi, \pi) \times [-\pi, \pi) \times \cdots \times [-\pi, \pi) = [-\pi, \pi)^k$ is defined by $\sum_v c_v \exp\{iv \cdot x\}$, where the summation is over all *integral vectors* (or *multi-indices*) $v = (v^{(1)}, v^{(2)}, \ldots, v^{(k)})$, each $v^{(i)}$ being an integer. Also, $v \cdot x = \sum_{i=1}^{k} v^{(i)} x^{(i)}$ is the usual Euclidean inner product on $\mathbb{R}^k$ between two vectors $v = (v^{(1)}, \ldots, v^{(k)})$ and $x = (x^{(1)}, x^{(2)}, \ldots, x^{(k)})$. The *Fourier coefficients* are given by

$$c_v = \frac{1}{(2\pi)^k} \int_{-\pi}^{\pi} \cdots \int_{-\pi}^{\pi} f(x) e^{-iv \cdot x} \, dx. \tag{6.37}$$

The extensions of Theorems (and Proposition) 6.1–6.4 are fairly obvious. Similarly, the *Fourier transform* of an integrable function (with respect to Lebesgue measure on $\mathbb{R}^k$) f is defined by

$$\hat{f}(\xi) = \int_{-\infty}^{\infty} \cdots \int_{-\infty}^{\infty} e^{i\xi \cdot y} f(y) \, dy \qquad (\xi \in \mathbb{R}^k), \tag{6.38}$$

the Fourier inversion formula becomes

$$f(z) = \frac{1}{(2\pi)^k} \int_{-\infty}^{\infty} \cdots \int_{-\infty}^{\infty} \hat{f}(\xi) e^{-iz \cdot \xi} \, d\xi, \tag{6.39}$$

which holds when $f(x)$ and $\hat{f}(\xi)$ are integrable. The Plancherel identity (6.36) becomes

$$\int_{-\infty}^{\infty} \cdots \int_{-\infty}^{\infty} |\hat{f}(\xi)|^2 \, d\xi = (2\pi)^k \int_{-\infty}^{\infty} \cdots \int_{-\infty}^{\infty} |f(y)|^2 \, dy, \tag{6.40}$$

which holds whenever f is integrable and square-integrable, i.e., Theorem 6.7 below.

Definition 6.2 The **Fourier transform**[2] of an integrable (real- or complex-valued) function f on $\mathbb{R}^k$ is the function $\hat{f}$ on $\mathbb{R}^k$ defined by

$$\hat{f}(\xi) = \int_{\mathbb{R}^k} e^{i\xi \cdot y} f(y)\, dy, \qquad \xi \in \mathbb{R}^k. \tag{6.41}$$

As a special case, take $k = 1$, $f = \mathbf{1}_{(c,d]}$. Then,

$$\hat{f}(\xi) = \frac{e^{i\xi d} - e^{i\xi c}}{i\xi}, \tag{6.42}$$

so that $\hat{f}(\xi) \to 0$ as $|\xi| \to \infty$. Such "decay" in the Fourier transform is to be generally expected for integrable functions as follows.

Proposition 6.6 *(Riemann–Lebesgue Lemma)* The Fourier transform $\hat{f}(\xi)$ of an integrable function f on $\mathbb{R}^k$ tends to zero in the limit as $|\xi| \to \infty$.

Proof The convergence to zero as $\xi \to \pm\infty$ illustrated by (6.42) is clearly valid for arbitrary *step functions*, i.e., finite linear combinations of indicator functions of finite rectangles. Now let f be an arbitrary integrable function. Given $\varepsilon > 0$ there exists a step function f_ε such that (see Remark following Proposition 2.5)

$$\|f_\varepsilon - f\|_1 := \int_{\mathbb{R}^k} |f_\varepsilon(y) - f(y)|\, dy < \varepsilon. \tag{6.43}$$

Now it follows from (6.41) that $|\hat{f}_\varepsilon(\xi) - \hat{f}(\xi)| \le \|f_\varepsilon - f\|_1$ for all ξ. Since $\hat{f}_\varepsilon(\xi) \to 0$ as $|\xi| \to \infty$, one has $\limsup_{|\xi| \to \infty} |\hat{f}(\xi)| \le \varepsilon$. Since $\varepsilon > 0$ is arbitrary,

$$\hat{f}(\xi) \to 0 \qquad \text{as} |\xi| \to \infty.$$

∎

Let us now check that (6.35), (6.36), in fact, hold under the following more general conditions

Theorem 6.7 a. If f and $\hat{f}$ are both integrable, then the Fourier inversion formula (6.35) holds.

  b. If f is integrable as well as square-integrable, then the Plancherel identity (6.36) holds.

---

[2]There are several different ways in which Fourier transforms can be parameterized and/or normalized by extra constant factors and/or a different sign in the exponent. The definition given here follows the standard conventions of probability theory.

Proof (a) Let f, $\hat{f}$ be integrable. Assume for simplicity that f is continuous. Note that this assumption is innocuous since the inversion formula yields a continuous (version of) f (see Exercise 7(i) for the steps of the proof without this a priori continuity assumption for f). Let φ_{ε^2} denote the pdf of the Gaussian distribution with mean zero and variance $\varepsilon^2 > 0$. Then writing Z to denote a standard normal random variable,

$$f * \varphi_{\varepsilon^2}(x) = \int_{\mathbb{R}} f(x - y)\varphi_{\varepsilon^2}(y)dy = \mathbb{E}f(x - \varepsilon Z) \to f(x), \qquad (6.44)$$

as $\varepsilon \to 0$. On the other hand (see Exercise 3),

$$\begin{aligned}
f * \varphi_{\varepsilon^2}(x) &= \int_{\mathbb{R}} f(x - y)\varphi_{\varepsilon^2}(y)dy = \int_{\mathbb{R}} f(x - y)\left\{\frac{1}{2\pi}\int_{\mathbb{R}} e^{-i\xi y}e^{-\varepsilon^2\xi^2/2}d\xi\right\}dy \\
&= \frac{1}{2\pi}\int_{\mathbb{R}} e^{-\varepsilon^2\xi^2/2}\left\{\int_{\mathbb{R}} e^{i\xi(x-y)}f(x - y)dy\right\}e^{-i\xi x}d\xi \\
&= \frac{1}{2\pi}\int_{\mathbb{R}} e^{-i\xi x}e^{-\varepsilon^2\xi^2/2}\hat{f}(\xi)d\xi \to \frac{1}{2\pi}\int_{\mathbb{R}} e^{-i\xi x}\hat{f}(\xi)d\xi \qquad (6.45)
\end{aligned}$$

as $\varepsilon \to 0$. The inversion formula (6.35) follows from (6.44), (6.45). For part (b) see Exercise 7(ii). ∎

Remark 6.2 Since $L^1(\mathbb{R}, dx) \cap L^2(\mathbb{R}, dx)$ is dense in $L^2(\mathbb{R}, dx)$ in the L^2-metric, the Plancheral identity (6.36) may be extended to all of $L^2(\mathbb{R}, dx)$, extending in this process the definition of the Fourier transform $\hat{f}$ of $f \in L^2(\mathbb{R}, dx)$. However, we do not make use of this extension in this text.

Suppose $k = 1$ to start. If f is continuously differentiable and f, f' are both integrable, then integration by parts yields (Exercise 2(b))

$$\widehat{f'}(\xi) = -i\xi\hat{f}(\xi). \qquad (6.46)$$

The boundary terms in deriving (6.46) vanish, if f' is integrable (as well as f) then $f(x) \to 0$ as $x \to \pm\infty$. More generally, if f is r-times continuously differentiable and $f^{(j)}, 0 \le j \le r$, are all integrable, then one may repeat the relation (6.46) to get by induction (Exercise 2(b))

$$\widehat{f^{(r)}}(\xi) = (-i\xi)^r \hat{f}(\xi). \qquad (6.47)$$

In particular, (6.47) implies that if f, f', f'' are integrable then $\hat{f}$ is integrable. Similar formulae are readily obtained for dimensions $k > 1$ using integration by parts. From this and the Riemann–Lebesgue lemma one may therefore observe a clear sense in which the *smoothness* of the function f is related to the *rate of decay* of the Fourier transform at ∞. The statements of smoothness in higher dimensions use the multi-index notation for derivatives: For a k-tuple of positive integers $\alpha = (\alpha_1, \ldots, \alpha_k)$ $|\alpha| = \sum_{j=1}^{k} \alpha_j$, $\partial^\alpha = \frac{\partial^{\alpha_1}}{\partial x_1^{\alpha_1}} \cdots \frac{\partial^{\alpha_k}}{\partial x_k^{\alpha_k}}$.

Theorem 6.8

a. Suppose f in $L^1(\mathbb{R}^k)$. For $|\alpha| \leq m$, $\hat{f} \in C^m$, and $\partial^\alpha \hat{f} = (ix)^\alpha \hat{f})$.
b. If (i) $x^\alpha f \in C^m$, (ii) $\partial^\alpha f \in L^1$ for $\alpha \leq m$, and (iii) $\partial^\alpha f \in C_0$ for $|\alpha| \leq m - 1$, then $\widehat{\partial^\alpha f}(\xi) = (i\xi)^\alpha \hat{f}(\xi)$.

Proof To establish part (i) requires differentiation under the integral and induction on $|\alpha|$. The differentiation is justified by the dominated convergence theorem. Integration by parts yields part (ii) in the case $|\alpha| = 1$, as indicated above. The result then follows by induction on $|\alpha|$.   ∎

Definition 6.3 The **Fourier transform** $\hat{\mu}$ of a finite measure μ on $\mathbb{R}^k$, with Borel σ-field $\mathcal{B}^k$, is defined by

$$\hat{\mu}(\xi) = \int_{\mathbb{R}^k} e^{i\xi \cdot x} \, d\mu(x). \tag{6.48}$$

If μ is a *finite-signed measure*, i.e., $\mu = \mu_1 - \mu_2$ where μ_1, μ_2 are finite measures, then also one defines $\hat{\mu}$ by (6.48) directly, or by setting $\hat{\mu} = \hat{\mu}_1 - \hat{\mu}_2$. In particular, if $\mu(dx) = f(x) \, dx$, where f is real-valued and integrable, then $\hat{\mu} = \hat{f}$. If μ is a probability measure, then $\hat{\mu}$ is also called the **characteristic function** of μ, or of any random vector $X = (X_1, \ldots, X_k)$ on $(\Omega, \mathcal{F}, P)$ whose distribution is $\mu = P \circ X^{-1}$. In this case, by the change of variable formula, one has the equivalent definition

$$\hat{\mu}(\xi) = \mathbb{E}e^{i\xi \cdot X}, \, \xi \in \mathbb{R}^k. \tag{6.49}$$

In the case that $\hat{Q} \in L^1(\mathbb{R}^k)$ the Fourier inversion formula yields a density function for $Q(dx)$, i.e., integrability of $\hat{Q}$ implies absolute continuity of Q with respect to Lebesgue measure.

We next consider the **convolution** of two integrable functions f, g:

$$f * g(x) = \int_{\mathbb{R}^k} f(x - y)g(y) \, dy \qquad (x \in \mathbb{R}^k). \tag{6.50}$$

Since by the Tonelli part of the Fubini–Tonelli theorem,

$$\int_{\mathbb{R}^k} |f * g(x)| \, dx = \int_{\mathbb{R}^k} \int_{\mathbb{R}^k} |f(x - y)||g(y)| \, dy \, dx$$
$$= \int_{\mathbb{R}^k} |f(x)| \, dx \int_{-\infty}^{\infty} |g(y)| \, dy, \tag{6.51}$$

$f * g$ is integrable. Its Fourier transform is

$$(f * g)\hat{\ }(\xi) = \int_{\mathbb{R}^k} e^{i\xi \cdot x} \left(\int_{\mathbb{R}^k} f(x - y)g(y) \, dy \right) dx$$

$$= \int_{\mathbb{R}^k} \int_{\mathbb{R}^k} e^{i\xi \cdot (x-y)} e^{i\xi \cdot y} f(x - y)g(y) \, dy \, dx$$

$$= \int_{\mathbb{R}^k} \int_{\mathbb{R}^k} e^{i\xi \cdot z} e^{i\xi \cdot y} f(z)g(y) \, dy \, dz = \hat{f}(\xi)\hat{g}(\xi), \qquad (6.52)$$

a result of importance in both probability and analysis. By iteration, one defines the *n-fold convolution* $f_1 * \cdots * f_n$ of n integrable functions $f_1, \ldots, f_n$ and it follows from (6.52) that $(f_1 * \cdots * f_n)\hat{\ } = \hat{f}_1 \hat{f}_2 \cdots \hat{f}_n$. Note also that if f, g are real-valued integrable functions and one defines the measures μ, ν by $\mu(dx) = f(x) \, dx$, $\nu(dx) = g(x) \, dx$, and $\mu * \nu$ by $(f * g)(x) \, dx$, then

$$(\mu * \nu)(B) = \int_B (f * g)(x) \, dx = \int_{\mathbb{R}^k} \left(\int_B f(x - y) \, dx \right) g(y) \, dy$$

$$= \int_{\mathbb{R}^k} \mu(B - y)g(y) \, dy \int_{\mathbb{R}^k} \mu(B - y) d\nu(y), \qquad (6.53)$$

for every interval (or, more generally, for every Borel set) B. Here $B - y$ is the *translate* of B by $-y$, obtained by subtracting from each point in B the number y. Also $(\mu * \nu)\hat{\ } = (f * g)\hat{\ } = \hat{f}\hat{g} = \hat{\mu}\hat{\nu}$. In general (i.e., whether or not finite-signed measures μ and/or ν have densities), the last expression in (6.53) defines the *convolution* $\mu * \nu$ of finite-signed measures μ and ν. The Fourier transform of this finite-signed measure is still given by $(\mu * \nu)\hat{\ } = \hat{\mu}\hat{\nu}$. Recall that if X_1, X_2 are independent k-dimensional random vectors on some probability space $(\Omega, \mathcal{A}, P)$ and have distributions Q_1, Q_2, respectively, then the distribution of $X_1 + X_2$ is $Q_1 * Q_2$. The characteristic function (i.e., Fourier transform) may also be computed from

$$(Q_1 * Q_2)\hat{\ }(\xi) = \mathbb{E}e^{i\xi \cdot (X_1 + X_2)} = \mathbb{E}e^{i\xi \cdot X_1} \mathbb{E}e^{i\xi \cdot X_2} = \hat{Q}_1(\xi)\hat{Q}_2(\xi). \qquad (6.54)$$

This argument extends to finite-signed measures, and is an alternative way of thinking about (or deriving) the result $(\mu * \nu)\hat{\ } = \hat{\mu}\hat{\nu}$.

Theorem 6.9 *(Uniqueness)* Let Q_1, Q_2 be probabilities on the Borel σ-field of $\mathbb{R}^k$. Then $\hat{Q}_1(\xi) = \hat{Q}_2(\xi)$ for all $\xi \in \mathbb{R}^k$ if and only if $Q_1 = Q_2$.

Proof For each $\xi \in \mathbb{R}^k$, one has by definition of the characteristic function that $e^{-i\xi \cdot x}\hat{Q}_1(\xi) = \int_{\mathbb{R}^k} e^{i\xi(y-x)} Q_1(dy)$. Thus, integrating with respect to Q_2, one obtains the duality relation

$$\int_{\mathbb{R}^k} e^{-i\xi \cdot x} \hat{Q}_1(\xi) Q_2(d\xi) = \int_{\mathbb{R}^k} \hat{Q}_2(y - x) Q_1(dy). \qquad (6.55)$$

Let $\varphi_{1/\sigma^2}(x) = \frac{\sigma}{\sqrt{2\pi}} e^{-\frac{\sigma^2 x^2}{2}}$, $x \in \mathbb{R}$, denote the Gaussian pdf with variance $1/\sigma^2$ centered at 0, and take $Q_2(dx) \equiv \Phi_{1/\sigma^2}(dx) := \prod_{j=1}^{k} \varphi_{1/\sigma^2}(x_j) dx_1 \cdots dx_k$ in (6.55). Then $\hat{Q}_2(\xi) = \hat{\Phi}_{1/\sigma^2}(\xi) = e^{-\sum_{j=1}^{k} \frac{\xi_j^2}{2\sigma^2}} = (\sqrt{2\pi\sigma^2})^k \prod_{j=1}^{k} \varphi_{\sigma^2}(\xi_j)$ so that the right-hand side may be expressed as $(\sqrt{2\pi\sigma^2})^k$ times the pdf of $\Phi_{\sigma^2} * Q_1$. In particular, one has

$$\frac{1}{2\pi} \int_{\mathbb{R}^j} e^{-i\xi \cdot x} \hat{Q}_1(\xi) e^{-\sum_{j=1}^{k} \frac{\sigma^2 \xi_j^2}{2}} d\xi_j = \int_{\mathbb{R}^k} \prod_{j=1}^{k} \varphi_{\sigma^2}(y_j - x_j) Q_1(dy).$$

The right-hand side may be viewed as the pdf of the distribution of the sum of independent random vectors $X_{\sigma^2} + Y$ with respective distributions Φ_{σ^2} and Q_1. Also, by the Chebyshev inequality, $X_{\sigma^2} \to 0$ in probability as $\sigma^2 \to 0$. Thus the distribution of $X_\sigma^2 + Y$ converges weakly to Q_1. Equivalently, the pdf of $X_{\sigma^2} + Y$ is given by the expression on the left side, involving Q_1 only through $\hat{Q}_1$. In this way $\hat{Q}_1$ uniquely determines Q_1.   ∎

Remark 6.3 Equation (6.55) may be viewed as a form of **Parseval's relation**.

The following version of **Parseval relation** is easily established by an application of the Fubini–Tonelli theorem and definition of characteristic function.

Proposition 6.10 *(Parseval Relation)* Let Q_1 and Q_2 be probabilities on $\mathbb{R}^k$ with characteristic functions $\hat{Q}_1$ and $\hat{Q}_2$, respectively. Then

$$\int_{\mathbb{R}^k} \hat{Q}_1(\xi) Q_2(d\xi) = \int_{\mathbb{R}^k} \hat{Q}_2(\xi) Q_1(d\xi).$$

At this point we have established that the map $Q \in \mathcal{P}(\mathbb{R}^k) \to \hat{Q} \in \widehat{\mathcal{P}}(\mathbb{R}^k)$ is one to one, and transforms convolution as pointwise multiplication. Some additional basic properties of this map are presented in the exercises. We next consider important special cases of an *inversion formula* for absolutely continuous finite (signed) measures $\mu(dx) = f(x)dx$ on $\mathbb{R}^k$. This is followed by a result on the *continuity* of the map $Q \to \hat{Q}$ for respectively the weak topology on $\mathcal{P}(\mathbb{R}^k)$ and the topology of pointwise convergence on $\widehat{\mathcal{P}}(\mathbb{R}^k)$. Finally the identification of the *range* of the Fourier transform of finite positive measures is provided. Such results are of notable theoretical and practical value.

Next we will see that the correspondence $Q \mapsto \hat{Q}$, on the set of probability measures with the weak topology onto the set of characteristic functions with the topology of pointwise convergence is *continuous*, thus providing a basic tool for obtaining weak convergence of probabilities on the finite-dimensional space $\mathbb{R}^k$.

Theorem 6.11 *(Cramér–Lévy Continuity Theorem)* Let $P_n (n \geq 1)$ be probability measures on $(\mathbb{R}^k, \mathcal{B}^k)$.

a. If P_n converges weakly to P, then $\hat{P}_n(\xi)$ converges to $\hat{P}(\xi)$ for every $\xi \in \mathbb{R}^k$.

b. If for some continuous function φ one has $\hat{P}_n(\xi) \to \varphi(\xi)$ for every ξ, then φ is the characteristic function of a probability P, and P_n converges weakly to P.

Proof (a) Since $\hat{P}_n(\xi)$, $\hat{P}(\xi)$ are the integrals of the bounded continuous function $\exp\{i\xi \cdot x\}$ with respect to P_n and P, it follows from the definition of weak convergence that $\hat{P}_n(\xi) \to \hat{P}(\xi)$. (b) We will show that $\{P_n : n \geq 1\}$ is tight. First, let $k = 1$. For $\delta > 0$ one has

$$
\frac{1}{2\delta} \int_{-\delta}^{\delta} (1 - \hat{P}_n(\xi)) d\xi = \frac{1}{2\delta} \int_{\mathbb{R}} \left\{ \int_{-\delta}^{\delta} (1 - e^{i\xi x}) d\xi \right\} P_n(dx)
$$
$$
= \frac{1}{2\delta} \int_{\mathbb{R}} (2\delta - \xi[\frac{\sin(\xi x)}{\xi x} \mid_{-\delta}^{\delta}) P_n(dx)
$$
$$
= \frac{1}{2\delta} \int_{\mathbb{R}} \left(2\delta - 2\delta \frac{\sin(\delta x)}{\delta x} \right) P_n(dx)
$$
$$
= \int_{\mathbb{R}} \left(1 - \frac{\sin(\delta x)}{\delta x} \right) P_n(dx)
$$
$$
\geq \frac{1}{2} P_n(\{x : |\delta x| \geq 2\}) = \frac{1}{2} P_n \left(\left\{ x : |x| \geq \frac{2}{\delta} \right\} \right).
$$

Hence, by assumption,

$$
P_n \left(\left\{ x : |x| \geq \frac{2}{\delta} \right\} \right) \leq \frac{2}{2\delta} \int_{-\delta}^{\delta} (1 - \hat{P}_n(\xi)) d\xi \to \frac{2}{2\delta} \int_{-\delta}^{\delta} (1 - \varphi(\xi)) d\xi,
$$

as $n \to \infty$. Since φ is continuous and $\varphi(0) = 1$, given any $\varepsilon > 0$ one may choose $\delta > 0$ such that $(1 - \varphi(\xi)) \leq \varepsilon/4$ for $|\xi| \leq \delta$. Then the limit in (6.56) is no more than $\varepsilon/2$, proving tightness. For $k > 1$, consider the distribution $P_{j,n}$ under P_n of the one-dimensional projections $x = (x_1, \ldots, x_k) \mapsto x_j$ for each $j = 1, \ldots, k$. Then $\hat{P}_{j,n}(\xi_j) = \hat{P}_n(0, \ldots, 0, \xi_j, 0, \ldots, 0) \to \varphi_j(\xi_j) := \varphi(0, \ldots, 0, \xi_j, 0, \ldots, 0)$ for all $\xi_j \in \mathbb{R}^1$. The previous argument shows that $\{P_{j,n} : n \geq 1\}$ is a tight family for each $j = 1, \ldots, k$. Hence there is a $\delta > 0$ such that $P_n(\{x \in \mathbb{R}^k : |x_j| \leq 2/\delta, j = 1, \ldots, k\}) \geq 1 - \sum_{j=1}^{k} P_{j,n}(\{x_j : |x_j| \geq 2/\delta\}) \geq 1 - k\varepsilon/2$ for all sufficiently large n, establishing the desired tightness. By Prohorov's Theorem (Theorem 7.11), there exists a subsequence of $\{P_n\}_{n=1}^{\infty}$, say $\{P_{n_m}\}_{m=1}^{\infty}$, that converges weakly to some probability P. By part (a), $\hat{P}_{n_m}(\xi) \to \hat{P}(\xi)$, so that $\hat{P}(\xi) = \varphi(\xi)$ for all $\xi \in \mathbb{R}^k$. Since the limit characteristic function $\varphi(\xi)$ is the same regardless of the subsequence $\{P_{n_m}\}_{m=1}^{\infty}$, it follows that P_n converges weakly to P as $n \to \infty$. ∎

The **law of rare events**, or **Poisson approximation to the binomial distribution**, provides a simple illustration of the Cramér–Lévy continuity Theorem 6.11.

Proposition 6.12 *(Law of Rare Events)* For each $n \geq 1$, suppose that $X_{n,1}, \ldots, X_{n,n}$ is a sequence of n i.i.d. 0 or 1-valued random variables with $p_n = P(X_{n,k} = 1)$, $q_n = P(X_{n,k} = 0)$, where $\lim_{n \to \infty} np_n = \lambda > 0, q_n = 1 - p_n$. Then $Y_n = \sum_{k=1}^{n} X_{n,k}$ converges in distribution to Y, where Y is distributed by the Poisson law

$$P(Y = m) = \frac{\lambda^m}{m!} e^{-\lambda},$$

$m = 0, 1, 2, \ldots.$

Proof Using the basic fact that $\lim_{n \to \infty}(1 + \frac{a_n}{n})^n = e^{\lim_n a_n}$ whenever $\{a_n\}_{n=1}^{\infty}$ is a sequence of complex numbers such that $\lim_n a_n$ exists, one has by independence, and in the limit as $n \to \infty$,

$$\mathbb{E}e^{i\xi Y_n} = (q_n + p_n e^{i\xi})^n = \left(1 + \frac{np_n(e^{i\xi} - 1)}{n}\right)^n \to \exp(\lambda(e^{i\xi} - 1)), \quad \xi \in \mathbb{R}.$$

One may simply check that this is the characteristic function of the asserted limiting Poisson distribution.   ∎

The development of tools for Fourier analysis of probabilities is concluded with an application of the Herglotz theorem (Theorem 6.4) to identify the **range** of the Fourier transform of finite positive measures.

Definition 6.4 A complex-valued function φ on $\mathbb{R}^k$ is said to be **positive-definite** if for every positive integer n and finite sequences $\{\xi_1, \xi_2, \ldots, \xi_n\} \subset \mathbb{R}^k$ and $\{z_1, z_2, \ldots, z_n\} \subset \mathbb{C}$ (the set of complex numbers), one has

$$\sum_{1 \le j, k \le n} z_j \bar{z}_k \varphi(\xi_j - \xi_k) \ge 0. \tag{6.56}$$

Theorem 6.13 *(Bochner's theorem)* A function φ on $\mathbb{R}^k$ is the Fourier transform of a finite measure on $\mathbb{R}^k$ if and only if it is positive-definite and continuous.

Proof We give the proof in the case $k = 1$ and leave $k > 1$ to the reader. The proof of necessity is entirely analogous to (6.28). It is sufficient to consider the case $\varphi(0) = 1$. For each positive integer N, $c_{j,N} := \varphi(-j2^{-N}))$, $j = 0, \pm 1, \pm 2, \ldots$, is positive-definite in the sense of (6.27). Hence, by the Herglotz theorem, there exists a probability γ_N on $[-\pi, \pi)$ such that $c_{j,N} = (2\pi)^{-1} \int_{[-\pi, \pi)} e^{-ijx} \gamma_N(dx)$ for each j. By the change of variable $x \to 2^N x$, one has $\varphi(j2^{-N}) = (2\pi)^{-1} \int_{[-2^N \pi, 2^N \pi)} e^{ij2^{-N}x} \mu_N(dx)$ for some probability $\mu_N(dx)$ on $[-2^N \pi, 2^N \pi)$. The characteristic function $\hat{\mu}_N(\xi) := \int_{\mathbb{R}^1} e^{i\xi x} \mu_N(dx)$ agrees with φ at all dyadic rational points $j2^{-N}$, $j \in \mathbb{Z}$, dense in $\mathbb{R}$. To conclude the proof we note that one may use the continuity of $\varphi(\xi)$ to see that the family of functions $\hat{\mu}_N(\xi)$ is equicontinuous by the lemma below. With this it will follow by the Arzelà–Ascoli theorem (Appendix B) that there is a subsequence that converges pointwise to a continuous function g on $\mathbb{R}$. Since g and φ agree on a dense subset of $\mathbb{R}$, it follows that $g = \varphi$.   ∎

Lemma 1 *(An Equicontinuity Lemma)*

a. Let φ_N, $N \ge 1$, be a sequence of characteristic functions of probabilities μ_N. If the sequence is equicontinuous at $\xi = 0$ then it is equicontinuous at all $\xi \in \mathbb{R}$.

b. In the notation of the above proof of Bochner's theorem, let μ_N be the probability on $[-2^N\pi, 2^N\pi]$ with characteristic function $\varphi_N = \hat{\mu}_N$, where $\varphi_N(\xi) = \varphi(\xi)$ for $\xi = j2^{-N}, j \in \mathbb{Z}$. Then, (i) for $h \in [-1, 1]$, $0 \le 1 - \operatorname{Re}\varphi_N(h2^{-N}) \le 1 - \operatorname{Re}\varphi(2^{-N})$. (ii) φ_N is equicontinuous at 0, and hence at all points of $\mathbb{R}$ (by (i)).

Proof For the first assertion (a) simply use the Cauchy–Schwarz inequality to check that $|\varphi_N(\xi) - \varphi_N(\xi+\eta)|^2 \le 2|\varphi_N(0) - \operatorname{Re}\varphi_N(\eta)|$.

For (i) of the second assertion (b), write the formula and note that $1 - \cos(hx) \le 1 - \cos(x)$ for $-\pi \le x \le \pi$, $0 \le h \le 1$. For (ii), given $\varepsilon > 0$ find $\delta > 0$, ($0 < \delta < 1$) such that $|1 - \varphi(\theta)| < \varepsilon$ for all $|\theta| < \delta$. Now express each such θ as $\theta = (h_N + k_N)2^{-N}$, where $k_N = [2^N\theta]$ is the integer part of $2^N\theta$, and $h_N = 2^N\theta - [2^N\theta] \in [-1, 1]$. Using the inequality $|a+b|^2 \le 2|a|^2 + 2|b|^2$ together with the inequality in the proof of (a), one has that $|1 - \varphi_N(\theta)|^2 = |1 - \varphi_N((h_N+k_N)2^{-N})|^2 \le 2|1 - \varphi(k_N2^{-N})|^2 + 4|1 - \operatorname{Re}\varphi(2^{-N})| \le 2\varepsilon^2 + 4\varepsilon$. ∎

We will illustrate the use of characteristic functions in two probability applications. For the first, let us recall the general random walk on $\mathbb{R}^k$ from Chapter II. A basic consideration in the probabilistic analysis of the long-run behavior of a stochastic evolution involves frequencies of visits to specific states.

Let us consider the random walk $S_n := Z_1 + \cdots + Z_n$, $n \ge 1$, starting at $S_0 = 0$. The state 0 is said to be **neighborhood recurrent** if for every $\varepsilon > 0$, $P(S_n \in B_\varepsilon \text{ i.o.}) = 1$, where $B_\varepsilon = \{x \in \mathbb{R}^k : |x| < \varepsilon\}$. It will be convenient for the calculations to use the **rectangular norm** $|x| := \max\{|x_j| : j = 1, \ldots, k\}$, for $x = (x_1, \ldots, x_k)$. All finite-dimensional norms being equivalent, there is no loss of generality in this choice.

Observe that if 0 is not neighborhood recurrent, then for some $\varepsilon > 0$, $P(S_n \in B_\varepsilon \text{ i.o.}) < 1$, and therefore by the Hewitt–Savage 0-1 law, $P(S_n \in B_\varepsilon \text{ i.o.}) = 0$. Much more may be obtained with regard to recurrence dichotomies, expected return times, nonrecurrence, etc., which is postponed to a fuller treatment of stochastic processes. However, the following lemma is required for the result given here. As a warm-up, note that by the Borel–Cantelli lemma I, if $\sum_{n=1}^\infty P(S_n \in B_\varepsilon) < \infty$ for some $\varepsilon > 0$ then 0 cannot be neighborhood recurrent. In fact one has the following basic result.

Lemma 2 (*Chung–Fuchs*) 0 is neighborhood recurrent if and only if for all $\varepsilon > 0$, $\sum_{n=1}^\infty P(S_n \in B_\varepsilon) = \infty$.

Proof As noted above, if for some $\varepsilon > 0$, $\sum_{n=1}^\infty P(S_n \in B_\varepsilon) < \infty$, then with probability one, S_n will visit B_ε at most finitely often by the Borel–Cantelli lemma I. So it suffices to show that if $\sum_{n=1}^\infty P(S_n \in B_\varepsilon) = \infty$ for every $\varepsilon > 0$ then S_n will visit any given neighborhood of zero infinitely often with probability one. The proof is based on establishing the following two calculations:

$$\text{(A)} \quad \sum_{n=1}^{\infty} P(S_n \in B_\varepsilon) = \infty \Rightarrow P(S_n \in B_{2\varepsilon} \ i.o.) = 1,$$

$$\text{(B)} \quad \sum_{n=1}^{\infty} P(S_n \in B_\varepsilon) \geq \frac{1}{(2m)^k} \sum_{n=1}^{\infty} P(S_n \in B_{m\varepsilon}), \quad m \geq 2.$$

In particular, $\sum_{n=0}^{\infty} P(S_n \in B_\varepsilon) = \infty$ for some $\varepsilon > 0$, then from (B), $\sum_{n=0}^{\infty} P(S_n \in B_{\varepsilon'}) = \infty$ for all $\varepsilon' < \varepsilon$. In view of (A) this would make 0 neighborhood recurrent. To prove (A), let $N_\varepsilon := card\{n \geq 0 : S_n \in B_\varepsilon\}$ count the number of visits to B_ε. Also let $T_\varepsilon := \sup\{n : S_n \in B_\varepsilon\}$ denote the (possibly infinite) time of the last visit to B_ε. To prove (A) we will show that if $\sum_{m=0}^{\infty} P(S_m \in B_\varepsilon) = \infty$, then $P(T_{2\varepsilon} = \infty) = 1$. Let r be an arbitrary positive integer. One has

$$P(|S_m| < \varepsilon, |S_n| \geq \varepsilon, \forall n \geq m+r)$$
$$= P(m \leq T_\varepsilon < m+r)$$
$$= P(T_\varepsilon = m) + P(T_\varepsilon = m+1) + \cdots + P(T_\varepsilon = m+r-1).$$

Hence,

$$\sum_{m=1}^{\infty} P(|S_m| < \varepsilon, |S_n| \geq \varepsilon, \forall n \geq m+r) = \sum_{m=1}^{\infty} P(T_\varepsilon = m) + \cdots + \sum_{m=1}^{\infty} P(T_\varepsilon = m+r-1) \leq r.$$

Thus,

$$r \geq \sum_{m=0}^{\infty} P(S_m \in B_\varepsilon, |S_n| \geq \varepsilon \, \forall \, n \geq m+r)$$

$$\geq \sum_{m=0}^{\infty} P(S_m \in B_\varepsilon, |S_n - S_m| \geq 2\varepsilon \, \forall \, n \geq m+r)$$

$$= \sum_{m=0}^{\infty} P(S_m \in B_\varepsilon) P(|S_n| \geq 2\varepsilon \, \forall \, n \geq r). \tag{6.57}$$

Assuming $\sum_{m=0}^{\infty} P(S_m \in B_\varepsilon) = \infty$, one must therefore have $P(T_{2\varepsilon} \leq r) \leq P(|S_n| \geq 2\varepsilon \, \forall \, n \geq r) = 0$. Thus $P(T_{2\varepsilon} < \infty) = 0$. For the proof of (B), let $m \geq 2$ and for $x = (x_1, \ldots, x_k) \in \mathbb{R}^k$, define $\tau_x = \inf\{n \geq 0 : S_n \in R_\varepsilon(x)\}$, where $R_\varepsilon(x) := [0, \varepsilon)^k + x := \{y \in \mathbb{R}^k : 0 \leq y_i - x_i < \varepsilon, i = 1, \ldots, k\}$ is the translate of $[0, \varepsilon)^k$ by x, i.e., "square with lower left corner at x of side lengths ε." For arbitrary fixed $x \in \{-m\varepsilon, -(m-1)\varepsilon, \ldots, (m-1)\varepsilon\}^k$,

$$\sum_{n=0}^{\infty} P(S_n \in R_\varepsilon(x)) = \sum_{m=0}^{\infty} \sum_{n=m}^{\infty} P(S_n \in R_\varepsilon(x), \tau_x = m)$$

$$\leq \sum_{m=0}^{\infty} \sum_{n=m}^{\infty} P(|S_n - S_m| < \varepsilon, \tau_x = m)$$

$$= \sum_{m=0}^{\infty} P(\tau_x = m) \sum_{j=0}^{\infty} P(S_j \in B_\varepsilon)$$

$$\leq \sum_{j=0}^{\infty} P(S_j \in B_\varepsilon).$$

Thus, it now follow that

$$\sum_{n=0}^{\infty} P(S_n \in B_{m\varepsilon}) \leq \sum_{n=0}^{\infty} \sum_{x \in \{-m\varepsilon, -(m-1)\varepsilon, \ldots, (m-1)\varepsilon\}^k} P(S_n \in R_\varepsilon(x))$$

$$= \sum_{x \in \{-m\varepsilon, -(m-1)\varepsilon, \ldots, (m-1)\varepsilon\}^k} \sum_{n=0}^{\infty} P(S_n \in R_\varepsilon(x))$$

$$\leq (2m)^k \sum_{n=0}^{\infty} P(S_n \in B_\varepsilon).$$

∎

Remark 6.4 On a countable state space such as $\mathbb{Z}^d$, the topology is discrete and $\{j\}$ is an open neighborhood of j for every state j. Hence neighborhood recurrence is equivalent to *point recurrence*. Using the so-called *strong Markov property* discussed in Chapter XI, one may show that if a state i of a Markov chain on a countable state space is point recurrent, then the probability of reaching a state j, starting from i, is *one*, provided that the n-step transition probability from i to j, $p_{ij}^{(n)}$, is nonzero for some n; see Example 1 below, and Exercise 5 of Chapter XI.

Example 1 *(Polya's Theorem)* The simple symmetric random walk $\{S_n : n = 0, 1, 2, \ldots\}$ on $\mathbb{Z}^k$ starting at $\mathbf{S}_0 = \mathbf{0}$ is defined by the random walk with the discrete displacement distribution $Q(\{\mathbf{e}_j\}) = Q(\{-\mathbf{e}_j\}) = \frac{1}{2k}, j = 1, 2, \ldots, k$, where $\mathbf{e}_j$ is the jth standard basis vector, i.e., jth column of the $k \times k$ identity matrix. For $k = 1$ the recurrence follows easily from Lemma 2 by the combinatorial identity $P(S_{2n} = 0) = \binom{2n}{n} 2^{-2n}$ and Stirling's formula. For $k = 2$, one may rotate the coordinate axis by $\pi/4$ to map the simple symmetric two-dimensional random walk onto a random walk on the rotated lattice having *independent* one-dimensional simple symmetric random walk coordinates. It then follows for the two-dimensional walk that $P(S_{2n} = 0) = (\binom{2n}{n} 2^{-2n})^2$, from which the point recurrence also follows in two dimensions. Combinatorial arguments for the transience in three or more dimensions are also possible, but quite a bit more involved. An alternative approach by Fourier analysis is given below.

We turn now to conditions on the distribution of the displacements for neighborhood recurrence in terms of Fourier transforms. If, for example, $\mathbb{E}Z_1$ exists and is nonzero, then it follows from the strong law of large numbers that a.s. $|S_n| \to \infty$. The following is a complete characterization of neighborhood recurrence in terms of the distribution of the displacements. A simpler warm-up version for random walks on the integer lattice is given in Exercise 25. In the following theorem $Re(z)$ refers to the *real part* of a complex number z.

Theorem 6.14 (*Chung–Fuchs Recurrence Criterion*) Let $Z_1, Z_2, \ldots$ be an i.i.d. sequence of random vectors in $\mathbb{R}^k$ with common distribution Q. Let $\{S_n = Z_1 + \cdots + Z_n : n \geq 1\}$, $S_0 = 0$, be a random walk on $\mathbb{R}^k$ starting at 0. Then 0 is a neighborhood-recurrent state if and only if for every $\varepsilon > 0$,

$$\sup_{0<r<1} \int_{B_\varepsilon} \mathrm{Re}\left(\frac{1}{1 - r\hat{Q}(\xi)}\right) d\xi = \infty.$$

Proof First observe that the "triangular probability density function" $\hat{f}(\xi) = (1 - |\xi|)^+$, $\xi \in \mathbb{R}$, has the characteristic function $f(x) = 2\frac{1-\cos(x)}{x^2}$, $x \in \mathbb{R}$, and therefore, $\frac{1}{2\pi} f(x)$ has characteristic function $\hat{f}(\xi)$ (Exercise 23). One may also check that $f(x) \geq 1/2$ for $|x| \leq 1$ (Exercise 23). Also $\mathbf{f}(\mathbf{x}) := \prod_{j=1}^k f(x_j)$, $\mathbf{x} = (x_1, \ldots, x_k)$, has characteristic function $\hat{\mathbf{f}}(\xi) = \prod_{j=1}^k \hat{f}(\xi_j)$, and $\hat{\mathbf{f}}$ has characteristic function $(2\pi)^k \mathbf{f}$. In view of Parseval's relation (Proposition 6.10), one may write

$$\int_{\mathbb{R}^k} \mathbf{f}\left(\frac{\mathbf{x}}{\lambda}\right) Q^{*n}(d\mathbf{x}) = \lambda^k \int_{\mathbb{R}^k} \hat{\mathbf{f}}(\lambda\xi)\hat{Q}^n(\xi)d\xi,$$

for any $\lambda > 0$, $n \geq 1$. Using the Fubini–Tonelli theorem one therefore has for $0 < r < 1$ that

$$\int_{\mathbb{R}^k} \mathbf{f}(\frac{\mathbf{x}}{\lambda}) \sum_{n=0}^{\infty} r^n Q^{*n}(d\mathbf{x}) = \lambda^k \int_{\mathbb{R}^k} \frac{\hat{\mathbf{f}}(\lambda\xi)}{1 - r\hat{Q}(\xi)}d\xi.$$

Also, since the integral on the left is real, the right side must also be a real integral. For what follows note that when an indicated integral is real, one may replace the integrand by its respective real part. Suppose that for some $\varepsilon > 0$,

$$\sup_{0<r<1} \int_{B_{\frac{1}{\varepsilon}}} \mathrm{Re}\left(\frac{1}{1 - r\hat{Q}(\xi)}\right) d\xi < \infty.$$

Then, it follows that

$$\sum_{n=1}^{\infty} P(S_n \in B_\varepsilon) = \sum_{n=1}^{\infty} Q^{*n}(B_\varepsilon) \le 2^k \int_{\mathbb{R}^k} \mathbf{f}(\frac{\mathbf{x}}{\varepsilon}) \sum_{n=0}^{\infty} Q^{*n}(d\mathbf{x})$$

$$\le 2^k \varepsilon^k \sup_{0<r<1} \int_{\mathbb{R}^k} \frac{\hat{\mathbf{f}}(\varepsilon\xi)}{1 - r\hat{Q}(\xi)} d\xi$$

$$\le 2^k \varepsilon^k \sup_{0<r<1} \int_{B_{\frac{1}{\varepsilon}}} \mathrm{Re}\left(\frac{1}{1 - r\hat{Q}(\xi)}\right) d\xi < \infty.$$

Thus, in view of of Borel–Cantelli I, 0 cannot be neighborhood recurrent.

For the converse, suppose that 0 is not neighborhood recurrent. Then, by Lemma 2, one must have for any $\varepsilon > 0$ that $\sum_{n=1}^{\infty} Q^{*n}(B_\varepsilon) < \infty$.

Let $\varepsilon > 0$. Then, again using the Parseval relation with $(2\pi)^k \hat{\mathbf{f}}$ as the Fourier transform of $\mathbf{f}$,

$$\sup_{0<r<1} \int_{B_\varepsilon} \mathrm{Re}\left(\frac{1}{1-r\hat{Q}(\xi)}\right) d\xi \le 2^k \sup_{0<r<1} \int_{B_\varepsilon} \mathrm{Re}\left(\frac{\mathbf{f}(\frac{\mathbf{x}}{\varepsilon})}{1-r\hat{Q}(\mathbf{x})}\right) d\mathbf{x}$$

$$\le 2^k (2\pi)^k \varepsilon^k \sup_{0<r<1} \int_{\mathbb{R}^k} \hat{\mathbf{f}}(\varepsilon\mathbf{x}) \sum_{n=0}^{\infty} r^n Q^{*n}(d\mathbf{x})$$

$$\le 2^k (2\pi)^k \varepsilon^k \int_{B_{\varepsilon^{-1}}} \hat{\mathbf{f}}(\varepsilon\mathbf{x}) \sum_{n=0}^{\infty} Q^{*n}(d\mathbf{x})$$

$$\le (4\varepsilon\pi)^k \sum_{n=1}^{\infty} Q^{*n}(B_{\varepsilon^{-1}}) < \infty.$$

∎

Corollary 6.15 If $\int_{B_\varepsilon} \mathrm{Re}\left(\frac{1}{1-\hat{Q}(\xi)}\right) d\xi = \infty$ for $\varepsilon > 0$, then the random walk with displacement distribution Q is neighborhood recurrent.[3] [*Hint*: Pass to the limit as $r \to 1$ in $0 \le \mathrm{Re}\left(\frac{1}{1-r\hat{Q}(\xi)}\right)$, using the Chung–Fuchs criterion]

Example 2 *(Gaussian Random Walk)* Suppose that Q is the k-dimensional standard normal distribution. Then $\hat{Q}(\xi) = e^{-\frac{|\xi|^2}{2}}, \xi \in \mathbb{R}^k$.

We now turn to a hallmark application of Theorem 6.11 in probability to prove the celebrated Theorem 6.16 below. First, we need an estimate on the error in the Taylor polynomial approximation to the exponential function. The following lemma exploits the special structure of the exponential to obtain two bounds: a "good small x bound" and a "good large x bound", each of which is valid for all x.

---

[3] That the converse is also true was independently established in Stone, C. J. (1969): On the potential operator for one-dimensional recurrent random walks, *Trans. AMS*, **136** 427–445, and Ornstein, D. (1969): Random walks, *Trans. AMS*, **138**, 1–60.

Lemma 3 *(Taylor Expansion of Characteristic Functions)* Suppose that X is a random variable defined on a probability space $(\Omega, \mathcal{F}, P)$ such that $\mathbb{E}|X|^m < \infty$. Then

$$\left| \mathbb{E}e^{i\xi X} - \sum_{k=0}^{m} \frac{(i\xi)^k}{k!} \mathbb{E}X^k \right| \le \mathbb{E} \min \left\{ \frac{|\xi|^{m+1}|X|^{m+1}}{(m+1)!}, 2\frac{|\xi|^m|X|^m}{m!} \right\}, \qquad \xi \in \mathbb{R}.$$

Proof Let $f_m(x) = e^{ix} - \sum_{j=0}^{m} \frac{(ix)^j}{j!}$. Note that $f_m(x) = i \int_0^x f_{m-1}(y)dy$. Iteration yields a succession of $m-1$ iterated integrals with integrand of modulus $|f_0(y_{m-1})| = |e^{iy_{m-1}} - 1| \le 2$. The iteration of the integrals is therefore at most $2\frac{|x|^m}{m!}$. To obtain the other bound note the following integration by parts identity:

$$\int_0^x (x-y)^m e^{iy}dy = \frac{x^{m+1}}{m+1} + \frac{i}{m+1} \int_0^x (x-y)^{m+1} e^{iy}dy.$$

This defines a recursive formula that by induction leads to the expansion

$$e^{ix} = \sum_{j=0}^{m} \frac{(ix)^j}{j!} + \frac{i^{m+1}}{m!} \int_0^x (x-y)^m e^{iy}dy. \tag{6.58}$$

For $x \ge 0$, bound the modulus of the integrand by $|x-y|^m \le y^m$ to get the bound on the modulus of the integral term by $\frac{|x|^{m+1}}{(m+1)!}$. Similarly for $x < 0$. Since both bounds hold for all x, the smaller of the two also holds for all x. Now replace x by $|\xi X|$ and take expected values to complete the proof. ∎

Theorem 6.16 *(The Classical Central Limit Theorem)* Let $\mathbf{X}_n, n \ge 1$, be i.i.d. k-dimensional random vectors with (common) mean μ and a finite covariance matrix D. Then the distribution of $(\mathbf{X}_1 + \cdots + \mathbf{X}_n - n\mu)/\sqrt{n}$ converges weakly to Φ_D, the normal distribution on $\mathbb{R}^k$ with mean zero and covariance matrix D.

Proof It is enough to prove the result for $\mu = \mathbf{0}$ and $D = I$, the $k \times k$ identity matrix I, since the general result then follows by an affine linear (and hence continuous) transformation. First, consider the case $k = 1, \{X_n : n \ge 1\}$ i.i.d. $\mathbb{E}X_n = 0, \mathbb{E}X_n^2 = 1$. Let φ denote the (common) characteristic function of X_n. Then the characteristic function, say φ_n, of $(X_1 + \cdots + X_n)/\sqrt{n}$ is given at a fixed ξ by

$$\varphi_n(\xi) = \varphi^n(\xi/\sqrt{n}) = \left(1 - \frac{\xi^2}{2n} + o\left(\frac{1}{n}\right) \right)^n, \tag{6.59}$$

where $no(\frac{1}{n}) = o(1) \to 0$ as $n \to \infty$. The limit of (6.59) is $e^{-\frac{\xi^2}{2}}$, the characteristic function of the standard normal distribution, which proves the theorem for the case $k = 1$, using Theorem 6.11(b).

For $k > 1$, let $\mathbf{X}_n, n \ge 1$, be i.i.d. with mean zero and covariance matrix I. Then for each fixed $\xi \in \mathbb{R}^k, \xi \ne \mathbf{0}, Y_n = \xi \cdot \mathbf{X}_n, n \ge 1$, defines an i.i.d. sequence of real-valued random variables with mean zero and variance $\sigma_\xi^2 = \xi \cdot \xi$. Hence by the preceding,

$Z_n := (Y_1 + \cdots + Y_n)/\sqrt{n}$ converges in distribution to the one-dimensional normal distribution with mean zero and variance $\xi \cdot \xi$, so that the characteristic function of Z_n converges to the function $\eta \mapsto \exp\{-(\xi \cdot \xi)\eta^2/2\}, \eta \in \mathbb{R}$. In particular, at $\eta = 1$, the characteristic function of Z_n is

$$\mathbb{E}e^{i\xi \cdot (\mathbf{X}_1 + \cdots + \mathbf{X}_n)/\sqrt{n}} \to e^{-\xi \cdot \xi/2}. \tag{6.60}$$

Since (6.60) holds for every $\xi \in \mathbb{R}^k$, the proof is complete by the Cramér–Lévy continuity theorem.   ∎

Let us now establish the Berry–Esseen bound on the rate of convergence first noted in Chapter IV.[4]

Theorem 6.17 *(Berry–Esseen Convergence Rate)* Let $X_1, X_2, \ldots$ be an i.i.d. sequence of random variables having finite third moments $\rho = \mathbb{E}|X_1|^3 < \infty$, with mean μ and variance σ^2. Then, for $S_n = X_1 + \cdots + X_n, n \geq 1$, one has

$$\sup_{x \in \mathbb{R}} |P(\frac{S_n - n\mu}{\sigma\sqrt{n}} \leq x) - \Phi(x)| \leq \frac{3\mathbb{E}|X_1|^3}{\sigma^3\sqrt{n}}.$$

The proof rests on the following lemma[5] exploiting the fact that for any $T > 0$, the clearly integrable function $\omega_T(\xi) := 1 - \frac{|\xi|}{T}, |\xi| \leq T$, and zero on $|\xi| \geq T$, is by Bochner's theorem the characteristic function of a probability distribution. In fact, one can exhibit this distribution as $v_T(x) := \frac{1}{\pi}\frac{1-\cos(Tx)}{Tx^2}, x \in \mathbb{R}$.

Lemma 4 Let F be a distribution function on $\mathbb{R}$, and G any function on $\mathbb{R}$ such that $\lim_{x \to -\infty} G(x) = 0$, $\lim_{x \to \infty} G(x) = 1$, and having bounded derivative $|G'(x)| \leq m < \infty$. Then, for $T > 0$,

$$\sup_{y \in \mathbb{R}} |\int_{\mathbb{R}} (F(y-x) - G(y-x))\frac{1}{\pi}\frac{1 - \cos(Tx)}{Tx^2} dx| \geq \frac{1}{2}\sup_{x \in \mathbb{R}} |F(x) - G(x)| - \frac{12m}{\pi T}.$$

Proof Let $\Delta(x) = F(x) - G(x), x \in \mathbb{R}$. Since G is continuous and F has left and right limits at any point $x \in \mathbb{R}$, so does $\Delta(x)$. Also $\Delta(x) \to 0$ as $x \to \pm\infty$. So there is an x_0 such that either $|\Delta(x_0^+)|$ or $|\Delta(x_0^-)|$ takes the maximum value $\eta = \sup_{x \in \mathbb{R}} |\Delta(x)|$. Say $|\Delta(x_0)| = \eta$. We take $\Delta(x_0) = \eta$, by changing $F - G$ to $G - F$ in the desired inequality, if necessary. Since F is nondecreasing $|G'(x)|$ is bounded by m, $\Delta(x_0 + s) \geq \eta - ms, s > 0$. Taking $h = \eta/2m$, $y = x_0 + h$, $x = h - s$, for $|x| \leq h$ one has

$$\Delta(y - x) \geq \frac{\eta}{2} + mx.$$

---

[4] A comprehensive account of errors of normal approximation for the clt in general multidimensions may be found in Bhattacharya, R. and R. Ranga Rao (2010).

[5] The proof given here follows that given in Feller, W. (1971), vol 2. Feller refers to this particular estimate, attributed to A.C. Berry, as the *smoothing inequality*.

For $|x| > h$, $\Delta(y - x) \geq -\eta$. This, and the properties that v_T is symmetric about $x = 0$, and $\int_{|x|>h} v_T(x)dx \leq \frac{4}{\pi Th}$, provides the asserted bounds as follows:

$$
\sup_{y\in\mathbb{R}} \left| \int_{\mathbb{R}} (F(y-x) - G(y-x)) \frac{1}{\pi} \frac{1 - \cos(Tx)}{Tx^2} dx \right|
$$

$$
\geq \int_{\mathbb{R}} \Delta(y-x) v_T(x) dx
$$

$$
\geq \frac{\eta}{2}(1 - \frac{4}{\pi Th}) - \eta \frac{4}{\pi Th}. \tag{6.61}
$$

This is the asserted lower bound. ∎

Proof of Berry–Esseen theorem Let $Q(dx)$ denote the distribution of X_1. Apply Lemma 4 to $F(x) = F_n(x) = P(\frac{S_n - n\mu}{\sigma\sqrt{n}} \leq x)$, $x \in \mathbb{R}$, and $G(x) = \Phi(x)$, $x \in \mathbb{R}$, with, using Liapounov inequality,

$$
T = \frac{4}{3}\frac{\sigma^3}{\rho}\sqrt{n} \leq \frac{4}{3}\sqrt{n}.
$$

The integral on the left side of Lemma 4 is the distribution function of the signed measure $(F_n - \Phi) * v_T$ whose density is given by Fourier inversion as

$$
\frac{1}{2\pi}\int_{\mathbb{R}} e^{-i\xi x}(\varphi^n(\frac{\xi}{\sigma\sqrt{n}}) - e^{-\frac{\xi^2}{2}})\hat{v}_T(\xi)d\xi = \frac{d}{dx}\int_{\mathbb{R}} \frac{e^{-i\xi x}}{-i\xi}(\varphi^n(\frac{\xi}{\sigma\sqrt{n}}) - e^{-\frac{\xi^2}{2}})\hat{v}_T(\xi)d\xi.
$$

Thus the integral on the right equals the integral on the left in the lemma. Since $|\Phi'(x)| = m < 2/5$, the smoothing lemma now yields

$$
\pi|F_n(x) - \Phi(x)| \leq \int_{-T}^{T} |\varphi^n(\frac{\xi}{\sigma n}) - e^{-\frac{\xi^2}{2}}| \frac{d\xi}{|\xi|} + \frac{9.6}{T}. \tag{6.62}
$$

Recall (6.58) from which it follows that $|e^{ix} - \sum_{j=0}^{n-1}\frac{(ix)^j}{j!}| \leq \frac{x^n}{n!}$, $x > 0$, $n = 1, 2, \ldots$. Thus,

$$
|\varphi(x) - 1 + \frac{1}{2}\sigma^2 x^2| = |\int_{\mathbb{R}}(e^{ixy} - 1 - ixy + \frac{1}{2}y^2 x^2)Q(dy)| \leq \frac{1}{6}\rho|x|^3. \tag{6.63}
$$

Since $e^{-x} - 1 + x \leq \frac{x^2}{2}$, $x > 0$, one has

$$\left|\varphi\left(\frac{\xi}{\sigma\sqrt{n}}\right) - e^{-\frac{\xi^2}{2n}}\right| \le \left|\varphi\left(\frac{\xi}{\sigma\sqrt{n}}\right) - 1 + \frac{\xi^2}{2n}\right|$$

$$+ \left|1 - \frac{\xi^2}{2n} - e^{-\frac{\xi^2}{2n}}\right|$$

$$\le \frac{1}{6\sigma^3 n^{\frac{3}{2}}}|\xi|^3 + \frac{|\xi|^4}{8n^2}. \tag{6.64}$$

Also from (6.63), $|\varphi(x)| \le 1 - \frac{1}{2}\sigma^2 x^2 + \frac{\rho}{6}|x|^3$, for $\frac{1}{2}\sigma^2 x^2 \le 1$. So for $|\xi| \le T$ one has

$$\left|\varphi\left(\frac{\xi}{\sigma\sqrt{n}}\right)\right| \le 1 - \frac{1}{2n}\xi^2 + \frac{\rho}{6\sigma^3 n^{\frac{3}{2}}}|\xi|^3 \le 1 - \frac{5}{18n}\xi^2 \le e^{-\frac{5}{18n}\xi^2}.$$

Since $\sigma^3 < \rho$, assume $n \ge 10$; otherwise the theorem is clearly true for $\sqrt{n} \le 3$. In this case, $|\varphi(\frac{\xi}{\sigma\sqrt{n}})|^{n-1} \le e^{-\frac{1}{4}\xi^2}$. These estimates can be used to bound the integrand on the right side of (6.62) based on the simple inequality $|a^n - b^n| \le n|a - b|c^{n-1}$, for $|a| \le c, |b| \le c$, with $a = \varphi(\frac{\xi}{\sigma\sqrt{n}}), b = e^{-\frac{\xi^2}{2n}}, c = e^{-\frac{\xi^2}{4}}$. In particular, for $\sqrt{n} > 3$, one obtains using this inequality that

$$\left|\varphi^n\left(\frac{\xi}{\sigma n}\right) - e^{-\frac{\xi^2}{2}}\right| \le \frac{1}{T}\left(\frac{2}{9}\xi^2 + \frac{1}{18}|\xi|^3\right)e^{-\frac{1}{4}\xi^2}.$$

Inserting this (integrable) bound on the integrand in (6.62) and integrating by parts, yields

$$\pi|F_n(x) - \Phi(x)| \le \frac{8}{9}\sqrt{\pi} + \frac{98}{99}.$$

The assertion follows since $\sqrt{\pi} < \frac{9}{5}$ making the right side smaller than 4π. ∎

Remark 6.5 After a rather long succession of careful estimates, the constant $c = 3$ in Feller's bound $c\rho/\sigma^3\sqrt{n}$ has been reduced[6] to $c = 0.5600$ as best to date.

Definition 6.5 A nondegenerate distribution Q on $\mathbb{R}$, i.e., $Q \ne \delta_{\{c\}}$, is said to be *stable* if for every integer n there is a centering constant c_n and a scaling *index* $\alpha > 0$ such that $n^{-\frac{1}{\alpha}}(X_1 + \cdots + X_n - c_n)$ has distribution Q whenever $X_j, j \ge 1$, are i.i.d. with distribution Q.

It is straightforward to check that the normal distribution and Cauchy distribution are both stable with respective indices $\alpha = 2$ and $\alpha = 1$. Notice also that it follows directly from the definition that every stable distribution Q is **infinitely divisible** in the sense that for any integer $n \ge 1$, there is a probability distribution Q_n such that Q may be exposed as an n-fold convolution $Q = Q_n^{*n}$.

---

[6]See Shevtsova, I. G. (2010): An Improvement of Convergence Rate Estimates in the Lyapunov Theorem, Doklady Math. **82**(3), 862–864.

The following example[7] illustrates a general framework in which symmetric stable laws arise naturally.

Example 3 *(One-dimensional Holtzmark problem)* Consider $2n$ points (eg., masses or charges) $X_1, \ldots, X_{2n}$ independently and uniformly distributed within an interval $[-n, n]$ so that the density of points is one. Suppose that there is a fixed point (mass, charge) at the origin that exerts an *inverse rth power force* on the randomly distributed points, where $r > 1/2$. That is, the force exerted by the point at the origin on a mass at location x is $-sgn(x)|x|^{-r}$. Let $F_n = -\sum_{j=1}^{2n} \frac{sgn(X_j)}{|X_j|^r}$ denote the total force exerted by the origin on the $2n$ points. The characteristic function of the limit distribution Q_r of F_n as $n \to \infty$ may be calculated as follows: For $\xi > 0$, using an indicated change of variable,

$$
\mathbb{E}e^{i\xi F_n} = \left(\mathbb{E} \cos(\frac{\xi \, sgn(X_1)}{|X_1|^r}) \right)^{2n}
$$

$$
= \left(1 - \frac{\xi^{\frac{1}{r}}}{nr} \int_{\xi(\frac{1}{n})^r}^{\infty} (1 - \cos(y)) y^{-\frac{r+1}{r}} \, dy \right)^{2n}
$$

$$
\to e^{-a\xi^\alpha},
$$

where $\alpha = 1/r$. This calculation uses the fact that $|1 - \cos(y)| \le 2$ to obtain integrability on $[1, \infty)$. Also $\frac{1-\cos(y)}{y^2} \to 1$ as $y \downarrow 0$ on $(0, 1)$. So one has $0 < a < \infty$ for $0 < \frac{1}{r} < 2$. Similar calculation holds for $\xi < 0$ to obtain $e^{-a|\xi|^{\frac{1}{r}}}$. In particular Q_r is a so-called **symmetric stable distribution** with index $\alpha = \frac{1}{r} \in (0, 2)$ in the following sense: If $F_1^{(\infty)}, F_2^{(\infty)}, \ldots$ are i.i.d. with distribution Q_r, then $m^{-r}(F_1^{(\infty)} + \cdots + F_m^{(\infty)})$ has distribution Q_r. This example includes all such one-dimensional symmetric stable distributions with the notable exception of $\alpha = 2$, corresponding to the normal distribution. The case $\alpha = 2$ represents a different phenomena covered by the central limit theorem in Chapter IV and to be expanded upon in the next chapter.

Exercise Set VI

1. Prove that given $f \in L^2[-\pi, \pi]$ and $\varepsilon > 0$, there exists a continuous function g on $[-\pi, \pi]$ such that $g(-\pi) = g(\pi)$ and $\|f - g\| < \varepsilon$, where $\|\|$ is the L^2-norm defined by (6.10). [*Hint:* By Proposition 2.6 in Appendix A, there exists a continuous function h on $[-\pi, \pi]$ such that $\|f - h\| < \frac{\varepsilon}{2}$. If $h(-\pi) \ne h(\pi)$, modify it on $[\pi - \delta, \pi]$ by a linear interpolation with a value $h(\pi - \delta)$ at $\pi - \delta$ and a value $h(-\pi)$ at π, where $\delta > 0$ is suitably small.]
2. (a) Prove that if $\mathbb{E}|X|^r < \infty$ for some positive integer r, then the characteristic function $\varphi(\xi)$ of X has a continuous rth order derivative $\varphi^{(r)}(\xi) =$

---

[7]For a more elaborate treatment of the physics of the Holtsmark distribution in higher dimensions see S. Chandreskhar (1943): Stochastic problems in physics and astronomy, *Reviews of Modern Physics*, 15(3), reprinted in Wax (1954). The treatment provided here was inspired by Lamperti (1996).

$i^r \int_{\mathbb{R}} x^r e^{i\xi x} P_X(dx)$, where P_X is the distribution of X. In particular, $\varphi^{(r)}(0) = i^r \mathbb{E}X^r$. (b) Prove (6.47) assuming that f and $f^{(j)}$, $1 \leq j \leq r$, are integrable. [*Hint*: Prove (6.46) and use induction.] (c) If $r \geq 2$ in (b), prove that $\hat{f}$ is integrable.

3. This exercise concerns the *normal* (or *Gaussian*) distribution.

   (i) Prove that for every $\sigma \neq 0$, $\varphi_{\sigma^2,\mu}(x) = (2\pi\sigma^2)^{-\frac{1}{2}} e^{-\frac{(x-\mu)^2}{2\sigma^2}}$, $-\infty < x < \infty$, is a probability density function (pdf). The probability on $(\mathbb{R}, \mathcal{B}(\mathbb{R}))$ with this pdf is called the *normal* (or *Gaussian*) *distribution with mean* μ *variance* σ^2, denoted by $\Phi_{\sigma^2,\mu}$. [*Hint*: Let $c = \int_{-\infty}^{\infty} e^{-x^2/2} dx$. Then $c^2 = \int_{\mathbb{R}^2} e^{-(x^2+y^2)/2} dx dy = \int_0^{\infty} \int_0^{2\pi} r e^{-r^2/2} d\theta dr = 2\pi$.]

   (ii) Show that $\int_{-\infty}^{\infty} x \varphi_{\sigma^2,\mu}(x) dx = \mu$, $\int_{-\infty}^{\infty} (x-\mu)^2 \varphi_{\sigma^2,\mu}(x) dx = \sigma^2$. [*Hint*: $\int_{-\infty}^{\infty} (x-\mu)\varphi_{\sigma^2,\mu}(x) dx = 0$, $\int_{-\infty}^{\infty} x^2 e^{-x^2/2} dx = 2 \int_0^{\infty} x(-de^{-x^2/2}) = 2 \int_0^{\infty} e^{-x^2/2} dx = \sqrt{2\pi}$.]

   (iii) Write $\varphi = \varphi_{1,0}$, the *standard normal density*. Show that its odd-order moments vanish and the even-order moments are given by $\mu_{2n} = \int_{-\infty}^{\infty} x^{2n} \varphi(x) dx = (2n-1)\cdot(2n-3)\cdots 3 \cdot 1$ for $n = 1, 2, \ldots$. [*Hint*: Use integration by parts to prove the recursive relation $\mu_{2n} = (2n-1)\mu_{2n-2}$, $n = 1, 2 \ldots$, with $\mu_0 = 1$.]

   (iv) Show $\hat{\Phi}_{\sigma^2,\mu}(\xi) = e^{i\xi\mu - \sigma^2\xi^2/2}$, $\hat{\varphi}(\xi) = e^{-\xi^2/2}$. [*Hint*: $\hat{\varphi}(\xi) = \int_{-\infty}^{\infty} (\cos(\xi x)) \varphi(x) dx$. Expand $\cos(\xi x)$ in a power series and integrate term by term using (iii).]

   (v) (*Fourier Inversion for* $\varphi_{\sigma^2} \equiv \varphi_{\sigma^2,0}$). Show $\varphi_{\sigma^2}(x) = (2\pi)^{-1} \int_{-\infty}^{\infty} e^{-i\xi x} \hat{\varphi}_{\sigma^2}(\xi) d\xi$. [*Hint*: $\hat{\varphi}_{\sigma^2}(\xi) = \sqrt{\frac{2\pi}{\sigma^2}} \varphi_{\frac{1}{\sigma^2}}(\xi)$. Now use (iv).]

   (vi) Let $\mathbf{Z} = (Z_1, \ldots, Z_k)$ be a random vector where $Z_1, Z_2, \ldots, Z_k$ are i.i.d. random variables with standard normal density φ. Then $\mathbf{Z}$ is said to have the *k-dimensional standard normal distribution*. Its pdf (with respect to Lebesgue measure on $\mathbb{R}^k$) is $\varphi_I(\mathbf{x}) = \varphi(x_1) \cdots \varphi(x_k) = (2\pi)^{-\frac{k}{2}} e^{-\frac{|x|^2}{2}}$, for $\mathbf{x} = (x_1, \ldots, x_k)$. If Σ is a $k \times k$ positive-definite symmetric matrix and $\mu \in \mathbb{R}^k$, then the *normal* (or *Gaussian*) *distribution* $\Phi_{\Sigma,\mu}$ *with mean* μ *and dispersion* (or *covariance*) *matrix* Σ has pdf $\varphi_{\Sigma,\mu}(x) = (2\pi)^{-\frac{k}{2}} (\det \Sigma)^{-\frac{1}{2}} \exp\{-\frac{1}{2}(x-\mu) \cdot \Sigma^{-1}(x-\mu)\}$, where $\cdot$ denotes the inner (dot) product on $\mathbb{R}^k$. (a) Show that $\hat{\varphi}_{\Sigma,\mu}(\xi) = \exp\{i\xi \cdot \mu - \frac{1}{2}\xi \cdot \Sigma\xi\}$, $\xi \in \mathbb{R}^k$. (Note that the characteristic function of any absolutely continuous distribution is the Fourier transform of its pdf). (b) If A is a $k \times k$ matrix such that $AA' = \Sigma$, show that for standard normal $\mathbf{Z}$, $A\mathbf{Z} + \mu$ has the distribution $\Phi_{\Sigma,\mu}$. (c) Prove the *inversion formula* $\varphi_{\Sigma,\mu}(x) = (2\pi)^{-k} \int_{\mathbb{R}^k} \hat{\varphi}_{\Sigma,\mu}(\xi) e^{-i\xi \cdot x} d\xi$, $x \in \mathbb{R}^k$.

   (vii) If $(X_1, \ldots, X_k)$ has a k-dimensional Gaussian distribution, show $\{X_1, \ldots, X_k\}$ is a collection of independent random variables if and only if they are uncorrelated.

4. Suppose that $\{P_n\}_{n=1}^{\infty}$ is a sequence of Gaussian probability distributions on $(\mathbb{R}^k, \mathcal{B}^k)$ with respective mean vectors $m^{(n)} = (m_1^{(n)}, \ldots, m_k^{(n)})$ and variance–

covariance matrices $\Gamma^{(n)} = ((\gamma_{i,j}^{(n)}))_{1\leq i,j\leq k}$. (i) Show that if $m^{(n)} \to m$ and $\Gamma^{(n)} \to \Gamma$ (componentwise) as $n \to \infty$, then $P_n \Rightarrow P$, where P is Gaussian with mean vector m and variance–covariance matrix Γ. [*Hint*: Apply the continuity theorem for characteristic functions. Note that in the case of nonsingular Γ one may apply Scheffé's theorem, or apply Fatou's lemma to $P_n(G)$, G open.] (ii) Show that if $P_n \Rightarrow P$, then P must be Gaussian. [*Hint*: Consider the case $k = 1$, $m_n = 0$, $\sigma_n^2 = \int_{\mathbb{R}} x^2 P_n(dx)$. Use the continuity theorem and observe that if σ_n^2 ($n \geq 1$) is unbounded, then $\hat{P}_n(\xi) \equiv e^{-\frac{\sigma_n^2}{2}\xi^2}$ does not converge to a continuous limit at $\xi = 0$.]

5. (*Change of Location/Scale/Orientation*) Let $\mathbf{X}$ be a k-dimensional random vector and compute the characteristic function of $\mathbf{Y} = A\mathbf{X} + \mathbf{b}$, where A is a $k \times k$ matrix and $\mathbf{b} \in \mathbb{R}^k$.

6. (*Fourier Transform, Fourier Series, Inversion, and Plancherel*) Suppose f is differentiable and vanishes outside a finite interval, and f' is square-integrable. Derive the inversion formula (6.35) by justifying the following steps. Define $g_N(x) := f(Nx)$, vanishing outside $(-\pi, \pi)$. Let $\sum c_{n,N}e^{inx}$, $\sum c_{n,N}^{(1)}e^{inx}$ be the Fourier series of g_N and its derivative g_n', respectively.

(i) Show that $c_{n,N} = \frac{1}{2N\pi}\hat{f}\left(-\frac{n}{N}\right)$.

(ii) Show that $\sum_{n=-\infty}^{\infty}|c_{n,N}| \leq \frac{1}{2\pi}\left|\int_{-\infty}^{\infty}g_N(x)\,dx\right| + A\left(\frac{1}{2\pi}\int_{-\pi}^{\pi}|g_N'(x)|^2\,dx\right)^{1/2}$ $< \infty$, where $A = (2\sum_{n=1}^{\infty}n^{-2})^{1/2}$. [*Hint*: Split off $|c_{0,N}|$ and apply Cauchy–Schwarz inequality to $\sum_{n\neq 0}\frac{1}{|n|}(|nc_{n,N}|)$. Also note that $|c_{n,N}^{(1)}|^2 = |nc_{n,N}|^2$.]

(iii) Show that for all sufficiently large N, the following convergence is uniform: $f(z) = g_N\left(\frac{z}{N}\right) = \sum_{n=-\infty}^{\infty}c_{n,N}e^{inz/N} = \sum_{n=-\infty}^{\infty}\frac{1}{2N\pi}\hat{f}\left(-\frac{n}{N}\right)e^{inz/N}$.

(iv) Show that (6.35) follows by letting $N \to \infty$ in the previous step if $\hat{f} \in L^1(\mathbb{R},dx)$.

(v) Show that for any f that vanishes outside a finite interval and is square-integrable, hence integrable, one has, for all sufficiently large N, $\frac{1}{N}\sum_{n=-\infty}^{\infty}\left|\hat{f}\left(\frac{n}{N}\right)\right|^2 = 2\pi\int_{-\infty}^{\infty}|f(y)|^2\,dy$. [*Hint*: Check that $\frac{1}{2\pi}\int_{-\pi}^{\pi}|g_N(x)|^2\,dx = \frac{1}{2N\pi}\int_{-\infty}^{\infty}|f(y)|^2\,dy$, and $\frac{1}{2\pi}\int_{-\pi}^{\pi}|g_N(x)|^2\,dx = \sum_{n=-\infty}^{\infty}|c_{n,N}|^2 = \frac{1}{4N^2\pi^2}\sum_{n=-\infty}^{\infty}\left|\hat{f}\left(\frac{n}{N}\right)\right|^2$.] Show that the Plancherel identity (6.36) follows in the limit as $N \to \infty$.

7. (*General Inversion Formula and Plancherel Identity*)

(i) Prove (6.35) assuming only that f, $\hat{f}$ are integrable. [*Hint: Step 1.* Continuous functions with compact support are dense in $L^1 \equiv L^1(\mathbb{R},dx)$. *Step 2.* Show that *translation* $y \to g(\cdot + y)(\equiv g(x+y), x \in \mathbb{R})$, is continuous on $\mathbb{R}$ into L^1, for any $g \in L^1$. For this, given $\delta > 0$, find continuous h with compact support such that $\|g - h\|_1 < \delta/3$. Then find $\varepsilon > 0$ such that $\|h(\cdot+y)-h(\cdot+y')\|_1 < \delta/3$ if $|y-y'| < \varepsilon$. Then use $\|g(\cdot+y)-g(\cdot+y')\|_1 \leq \|g(\cdot+y)-h(\cdot+y)\|_1+\|h(\cdot+y)-h(\cdot+y')\|_1+\|h(\cdot+y')-g(\cdot+y')\|_1 < \delta$, noting that the *Lebesgue integral (measure) is translation invariant. Step 3.*

Use Step 2 to prove that $\mathbb{E}f(x + \varepsilon Z) \to f(x)$ in L^1 as $\varepsilon \to 0$, where Z is standard normal. *Step 4.* Use (6.45), which does not require f to be continuous, and Step 3, to show that the limit in (6.45) is equal a.e. to f.]

(ii) *(Plancherel Identity)*. Let $f \in L^1 \cap L^2$. Prove (6.36). [*Hint:* Let $\tilde{f}(x) := \overline{f(-x)}$, $g = f * \tilde{f}$. Then $g \in L^1$, $|g(x)| \le \|f\|_2^2$, $g(0) = \|f\|_2^2$. Also $g(x) = \langle f_x, f \rangle$, where $f_x(y) = f(x + y)$. Since $x \to f_x$ is continuous on $\mathbb{R}$ into L^2 (using arguments similar to those in Step 2 of part (i) above), and $\langle , \rangle$ is continuous on $L^2 \times L^2$ into $\mathbb{R}$, g is continuous on $\mathbb{R}$. Apply the inversion formula (in part (i)) to get $\|f\|_2^2 = g(0) = \frac{1}{2\pi}\int \hat{g}(\xi)d\xi \equiv \frac{1}{2\pi}\int |\hat{f}(\xi)|^2 d\xi$.]

8. *(Smoothing Property of Convolution)* (a) Suppose μ, ν are probabilities on $\mathbb{R}^k$, with ν absolutely continuous with pdf f; $\nu(dx) = f(x)dx$. Show that $\mu * \nu$ is absolutely continuous and calculate its pdf. (b) If $f, g \in L^1(\mathbb{R}^k, dx)$ and if g is bounded and continuous, show that $f * g$ is continuous. (c) If $f, g \in L^1(\mathbb{R}^k, dx)$, and if g and its first r derivatives $g^{(j)}$, $j = 1, \ldots, r$ are bounded and continuous, show that $f * g$ is r times continuously differentiable. [*Hint:* Use induction.]

9. Suppose $f, \hat{f}$ are integrable on $(\mathbb{R}, dx)$. Show $\hat{\hat{f}}(x) = 2\pi f(-x)$.

10. Let $Q(dx) = \frac{1}{2}\mathbf{1}_{[-1,1]}(x)dx$ be the uniform distribution on $[-1, 1]$.

   (i) Find the characteristic functions of Q and $Q^{*2} \equiv Q * Q$.
   (ii) Show that the probability with pdf $c \sin^2 x/x^2$, for appropriate normalizing constant c, has a characteristic function with compact support and compute this characteristic function. [*Hint:* Use Fourier inversion for $f = \hat{Q}^2$.]

11. Derive the multidimensional extension of the Fourier inversion formula.

12. Show that if Q is a stable distribution symmetric about 0 with exponent α, then $c_n = 0$ and $0 < \alpha \le 2$. [*Hint:* $\hat{Q}(\xi)$ must be real by symmetry, and positivity follows from the case $n = 2$ in the definition.]

13. Show that

   (i) The Cauchy distribution with pdf $(\pi(1 + x^2))^{-1}$, $x \in \mathbb{R}$, has characteristic function $e^{-|\xi|}$.
   (ii) The characteristic function of the double-sided exponential distribution $\frac{1}{2}e^{-|x|}dx$ is $(1 + \xi^2)^{-1}$. [*Hint:* Use integration by parts twice to show $\int_{-\infty}^{\infty} e^{i\xi x}(\frac{1}{2}e^{-|x|})dx \equiv \int_0^{\infty} e^{-x}\cos(\xi x)dx = (1 + \xi^2)^{-1}$.]

14. (i) Give an example of a pair of *dependent* random variables X, Y such that the distribution of their sum is the convolution of their distributions. [*Hint:* Consider the Cauchy distribution with $X = Y$.] (ii) Give an example of a non-Gaussian bivariate distribution such that the marginals are Gaussian. [*Hint:* Extend the proof of Theorem 6.7.]

15. Show that if φ is the characeristic function of a probability then φ must be uniformly continuous on $\mathbb{R}$.

16. *(Symmetric Distributions)* (i) Show that the characteristic function of $\mathbf{X}$ is real-valued if and only if $\mathbf{X}$ and $-\mathbf{X}$ have the same distribution. (ii) A *symmetrization* of (the distribution of) a random variable $\mathbf{X}$ may be defined by (the distribution of) $\mathbf{X} - \mathbf{X}'$, where $\mathbf{X}'$ is an independent copy of $\mathbf{X}$, i.e., independent of $\mathbf{X}$ and

having the same distribution as **X**. Express symmetrization of a random variable in terms of its characteristic function.

17. (*Multidimensional Gaussian characterization*) Suppose that $\mathbf{X} = (X_1, \ldots, X_k)$ is a k-dimensional random vector having a positive pdf $f(x_1, \ldots, x_k)$ on $\mathbb{R}^k (k \geq 2)$. Assume that (a) f is differentiable, (b) $X_1, \ldots, X_k$ are independent, and (c) have an *isotropic density*, i.e., $f(x_1, \ldots, x_k)$ is a function of $\|x\|^2 = x_1^2 + \cdots + x_k^2, (x_1, \ldots, x_k) \in \mathbb{R}^k$. Show that $X_1, \ldots, X_k$ are i.i.d. normal with mean zero and common variance. [*Hint*: Let f_j denote the marginal pdf of X_j and argue that $\frac{f_j'}{2x_j f_j}$ must be a constant.]

18. (i) Show that the functions $\{e_\xi : \xi \in \mathbb{R}^k\}$ defined by $e_\xi(x) := \exp(i\boldsymbol{\xi} \cdot \mathbf{x}), \mathbf{x} \in \mathbb{R}^k$ constitute a measure-determining class for probabilities on $(\mathbb{R}^k, \mathcal{B}^k)$.[*Hint*: Given two probabilities P, Q for which the integrals of the indicated functions agree, construct a sequence by $P_n = P \; \forall \, n = 1, 2, \ldots$ whose characteristic functions will obviously converge to that of Q.]
    (ii) Show that the closed half-spaces of $\mathbb{R}^k$ defined by $F_a := \{x \in \mathbb{R}^k : x_j \leq a_j, 1 \leq j \leq k\}, a = (a_1, \ldots, a_k)$ constitute a measure-determining collection of Borel subsets of $\mathbb{R}^k$. [*Hint*: Use a trick similar to that above.]

19. Compute the distribution with characteristic function $\varphi(\xi) = \cos^2(\xi), \xi \in \mathbb{R}^1$.

20. (*Fourier Inversion for Lattice Random Variables*)
    (i) Let $p_j, j \in \mathbb{Z}$, be a probability mass function (pmf) of a probability distribution Q on the integer lattice $\mathbb{Z}$. Show that the Fourier transform $\hat{Q}$ is periodic with period 2π, and derive the inversion formula $p_j = (2\pi)^{-1} \int_{(-\pi, \pi]} e^{-ij\xi} \hat{Q}(\xi) d\xi$.
    (ii) Let Q be a *lattice distribution of span $h > 0$*, i.e., for some $a_0, Q(\{a_0 + jh : j = 0, \pm 1, \pm 2, \ldots\}) = 1$. Show that $\hat{Q}$ is periodic with period $2\pi/h$ and write down an inversion formula. (iii) Extend (i), (ii) to the multidimensional lattice distributions with $\mathbb{Z}^k$ in place of $\mathbb{Z}$.

21. (*Parseval's Relation*) Let $f, g, \in L^2([-\pi, \pi))$, with Fourier coefficients $\{c_n\}$, $\{d_n\}$, respectively. Prove that $\sum_n c_n \overline{d_n} = \frac{1}{2\pi} \int_{(-\pi, \pi]} f(x) \overline{g}(x) dx \equiv \langle f, g \rangle$.
    (ii) Let $f, g \in L^2(\mathbb{R}^k, dx)$ with Fourier transforms $\hat{f}, \hat{g}$. Prove that $\langle \hat{f}, \hat{g} \rangle = 2\pi \langle f, g \rangle$. [*Hint*: Use (a) the Plancherel identity and (b) the *polar identity* $4\langle f, g \rangle = \|f + g\|^2 - \|f - g\|^2$.]

22. (i) Let φ be continuous and positive-definite on $\mathbb{R}$ in the sense of Bochner, and $\varphi(0) = 1$. Show that the sequence $\{c_j \equiv \varphi(j) : j \in \mathbb{Z}\}$ is positive-definite in the sense of Herglotz (6.27). (ii) Show that there exist distinct probability measures on $\mathbb{R}$ whose characteristic functions agree at all integer points.

23. Show that the "triangular function" $\hat{f}(\xi) = (1 - |\xi|)^+$ is the characteristic function of $f(x) = 2\frac{1 - \cos(x)}{x^2}, x \in \mathbb{R}$. [*Hint*: Consider the characteristic function of the convolution of two uniform distributions on $[-1/2, 1/2]$ and Fourier inversion.] Also show that $1 - \cos(x) \geq x^2/4$ for $|x| \leq \pi/3$. [*Hint*: Use $\cos(y) \geq 1/2$ and $\sin(y) \geq y$ for $0 < y < \pi/3$ in the formula $1 - \cos(x) = \int_0^x \sin(y) dy$.]

24. (*Chung–Fuchs*) For the one-dimensional random walk show that if $\frac{S_n}{n} \to 0$ in probability as $n \to \infty$, i.e., WLLN holds, then 0 is neighborhood recurrent. [*Hint:* Using the lemma for the proof of Chung–Fuchs, for any positive integer m and $\delta, \varepsilon > 0$, $\sum_{n=0}^{\infty} P(S_n \in B_\varepsilon) \geq \frac{1}{2m} \sum_{n=0}^{\infty} P(S_n \in B_{m\varepsilon}) \geq \frac{1}{2m} \sum_{n=0}^{m\delta^{-1}} P(S_n \in B_{\delta\varepsilon})$, using monotonicity of $r \to P(S_n \in B_r)$. Let $m \to \infty$ to obtain for the indicated Cesàro average, using $\lim_{n\to\infty} P(S_n \in B_{\delta\varepsilon}) = 1$ from the WLLN hypothesis, that $\sum_{n=0}^{\infty} P(S_n \in B_\varepsilon) \geq \frac{1}{2\delta}$. Let $\delta \to 0$ and apply the Lemma 2.]

25. This exercise provides a version of the Chung–Fuchs Fourier analysis criteria for the case of random walks on the integer lattice. Show that

   (i) $P(S_n = 0) = \frac{1}{(2\pi)^k} \int_{[-\pi,\pi)^k} \varphi(\xi)d\xi$, where $\varphi(\xi) = \mathbb{E}e^{i\xi \cdot X_1}$. [*Hint:* Use Fourier inversion formula.]

   (ii) $\sum_{n=0}^{\infty} r^n P(S_n = 0) = \frac{1}{(2\pi)^k} \int_{[-\pi,\pi)^k} Re(\frac{1}{1-r\varphi(\xi)})d\xi$.

   (iii) The lattice random walk $\{S_n : n = 0, 1, 2, \dots\}$ is recurrent if and only if $\lim_{r\uparrow 1} \int_{[-\pi,\pi)^k} Re(\frac{1}{1-r\varphi(\xi)})d\xi = \infty$. [*Hint:* Justify passage to the limit $r \uparrow 1$ and use Borel–Cantelli lemma.]

26. (i) Use the Chung–Fuchs criteria, in particular Corollary 6.15 and its converse, to determine whether the random walk with symmetric Cauchy displacement distribution is recurrent or transient. (ii) Extend this to symmetric stable displacement distributions with exponent $0 < \alpha \leq 2$.[8]

27. Show that 0 is neighborhood recurrent for the random walk if and only if $\sum_{n=0}^{\infty} P(S_n \in B_1) = \infty$.

28. Prove that the set of trigonometric polynomials is dense in $L^2([-\pi, \pi), \mu)$, where μ is a finite measure on $[-\pi, \pi)$.

29. Establish the formula $\int_{\mathbb{R}} g(x)\mu * \nu(dx) = \int_{\mathbb{R}} \int_{\mathbb{R}} g(x + y)\mu(dx)\nu(dy)$ for any bounded measurable function g.

---

[8] Surprisingly, recurrence and heavy tails may coexist, see Shepp, L (1964): Recurrent random walks with arbitrarily large steps, *Bull. Amer. Math. Soc.*, v. 70, 540–542; Grey, D.R. (1989): Persistent random walks may have arbitrarily large tails, *Adv. Appld. Probab.* **21**, 229–230.

Chapter VII
Weak Convergence of Probability Measures on Metric Spaces

Let (S, ρ) be a metric space and let $\mathcal{P}(S)$ be the set of all probability measures on $(S, \mathcal{B}(S))$. In this chapter we consider a general formulation of convergence in $\mathcal{P}(S)$, referred to as **weak convergence** or **convergence in distribution**.

For motivation, suppose that $Y_1, Y_2, \ldots$ is an i.i.d. sequence of square-integrable random variables with values in $S = \mathbb{R}$, defined on a probability space $(\Omega, \mathcal{F}, P)$. Assume that the Y_j are centered and scaled to have mean zero and variance one. A well-known *central limit theorem*, already proven in Chapter IV, asserts that

$$\lim_{n \to \infty} P(\sqrt{n}\frac{S_n}{n} \leq y) = \int_{-\infty}^{y} \frac{1}{\sqrt{2\pi}} e^{-\frac{x^2}{2}} dx, \quad \forall y \in \mathbb{R}, \tag{7.1}$$

where $S_n = Y_1 + \cdots + Y_n, n \geq 1$. Letting $Q_n, n \geq 1$, denote the distribution of $X_n = \sqrt{n}\frac{S_n}{n} \equiv \frac{S_n}{\sqrt{n}}$, and $Q(dx) = \frac{1}{\sqrt{2\pi}} e^{-\frac{x^2}{2}} dx$, the convergence from Q_n to Q may be expressed as

$$Q_n(A) \to Q(A), \quad A = (-\infty, y], y \in \mathbb{R}.$$

While this formulation is somewhat too special for a general formulation of limits of probability measures on a metric space S, we will also see that it has an equivalent formulation of the form

$$\lim_{n \to \infty} \int_{\mathbb{R}} f(x) Q_n(dx) = \int_{\mathbb{R}} f(x) Q(dx), \tag{7.2}$$

for all bounded continuous functions f on $S = \mathbb{R}$. Equivalently,

$$\lim_{n \to \infty} \mathbb{E}f(X_n) = \mathbb{E}f(Z), \tag{7.3}$$

where Z is the standard normal random variable with distribution Q.

© Springer International Publishing AG 2016
R. Bhattacharya and E.C. Waymire, *A Basic Course in Probability Theory*,
Universitext, DOI 10.1007/978-3-319-47974-3_VII

This formulation of convergence in distribution is sufficiently general to accommodate the convergence of any sequence of probability measures $Q_n, n \geq 1$, to a probability measure Q on $(S, \mathcal{B}(S))$ as in Definition 7.1 below.

Another example is obtained by consideration of the distribution of polygonal paths $X_n = \{X_n(t) : 0 \leq t \leq 1\} \in S = C[0, 1]$ of a random walk defined by linear interpolation between the values $(\frac{k}{n}, \frac{S_n}{\sqrt{n}}), k = 0, 1, \ldots, n$, given by,

$$
X_n(t) = \begin{cases} \frac{S_k}{\sqrt{n}} & \text{if } t = \frac{k}{n}, \quad k = 0, 1, 2, \ldots n \\ \sqrt{n} Y_{k+1}(t - \frac{k}{n}) + \frac{S_k}{\sqrt{n}} & \text{if } \frac{k}{n} < t < \frac{k+1}{n}, \ k = 0, 1, \ldots n - 1. \end{cases} \tag{7.4}
$$

Then, viewing X_n as a random path with values in the metric space $S = C[0, 1]$ for the uniform metric, the convergence of X_n in distribution to a continuous parameter stochastic process $Z = \{Z(t) : 0 \leq t \leq 1\} \in C[0, 1]$, known as the standard Brownian motion, enjoys the same formulation as above, with the metric space $S = \mathbb{R}$ replaced by $S = C[0, 1]$.

For yet another motivation, recall the total variation metric (distance) for $\mathcal{P}(S)$ that emerged in the context of Scheffe's theorem. It will be seen that the convergence in total variation metric is stronger than convergence in distribution.

To fix ideas one may regard a sequence of probabilities $Q_n \in \mathcal{P}(S), n \geq 1$, as respective distributions of random maps $X_n, n \geq 1$, defined on some probability space and taking values in the metric space S.

A topology may be defined on $\mathcal{P}(S)$ by the following neighborhood system: For $Q_0 \in \mathcal{P}(S), \delta > 0$, and $f_i (1 \leq i \leq m)$ real-valued bounded continuous functions on S, define an open neighborhood of $Q_0 \in \mathcal{P}(S)$ as

$$
N(Q_0 : f_1, f_2, \ldots, f_m; \delta) := \{Q \in \mathcal{P}(S) : \left| \int_S f_i \, dQ - \int_S f_i \, dQ_0 \right| < \delta \ \forall \ i \leq m\}. \tag{7.5}
$$

Here all $\delta > 0$, $m \geq 1$, and $f_i \in C_b(S)$ (the *set of all real-valued bounded continuous functions on S*), $1 \leq i \leq m$, are allowed. An **open set** of $\mathcal{P}(S)$ is defined to be a set U such that every Q_0 in U has an open neighborhood of the form (7.5) contained in U. Since the neighborhoods (7.5) are taken to be open, the topology is the collection of all unions of such sets. The topology (i.e., the collection of open sets) so defined is called the **weak topology**[1] of $\mathcal{P}(S)$.

Definition 7.1 A sequence of probabilities $\{Q_n : n \geq 1\}$ is said to **converge weakly** to a probability Q if $\int_S f \, dQ_n \rightarrow \int_S f \, dQ \ \forall \ f \in C_b(S)$. Denote this convergence by $Q_n \Rightarrow Q$.

---

[1]Billingsley (1968) provides a detailed exposition and comprehensive account of the weak convergence theory. From the point of view of functional analysis, weak convergence is actually convergence in the weak* topology. However the abuse of terminology has become the convention in this context.

Recall that the collection $C_b(S)$ is a measure-determining class of functions. Thus the limit Q of $\{Q_n\}_{n=1}^{\infty}$ is uniquely defined by weak convergence (also see Remark 7.2 below).

Note that if Q_n, Q are viewed as distributions of random maps X_n, X, respectively, defined on some probability space, then the definition of weak convergence, equivalently **convergence in distribution**, takes the form

$$\lim_n \mathbb{E} f(X_n) = \mathbb{E} f(X) \qquad \forall f \in C_b. \tag{7.6}$$

There are a number of equivalent formulations of weak convergence that are useful in various contexts. We will need the following topological notions. Recall that a point belongs to the **closure** of a set A if it belongs to A or if every neighborhood of the point intersects both A and A^c. On the other hand, a point belongs to the **interior** of A if there is an open set contained in A that includes the point. Denoting the closure of a set A by A^- and the interior by A°, one defines the **boundary** by $\partial A = A^- \backslash A^\circ$. A set A in $\mathcal{B}$ whose boundary ∂A satisfies $P(\partial A) = 0$ is called a P**-continuity set**. Since ∂A is closed, it clearly belongs to the σ-field $\mathcal{S} = \mathcal{B}(S)$.

Theorem 7.1 *(Alexandrov Theorem)* Let $Q_n, n \geq 1$, Q be probability measures on $(S, \mathcal{B}(S))$. The following are equivalent:

  (i) $Q_n \Rightarrow Q$.
 (ii) $\lim_n \int_S f \, dQ_n = \int_S f \, dQ$ for all bounded, uniformly continuous real f.
(iii) $\lim \sup_n Q_n(F) \leq Q(F)$ for all closed F.
 (iv) $\lim \inf_n Q_n(G) \geq Q(G)$ for all open G.
  (v) $\lim_n Q_n(A) = Q(A)$ for all Q-continuity sets A.

Proof The plan is to first prove (i) implies (ii) implies (iii) implies (i), and hence that (i), (ii), and (iii) are equivalent. We then directly prove that (iii) and (iv) are equivalent and that (iii) and (v) are equivalent.

(i) implies (ii): This follows directly from the definition.

(ii) implies (iii): Let F be a closed set and $\delta > 0$. For a sufficiently small but fixed value of ε, $G_\varepsilon = \{x : \rho(x, F) < \varepsilon\}$ satisfies $Q(G_\varepsilon) < Q(F) + \delta$, by continuity of the probability measure Q from above, since the sets G_ε decrease to $F = \cap_{\varepsilon \downarrow 0} G_\varepsilon$. Adopt the construction from the proof of Proposition 1.6 that $C_b(S)$ is measure-determining to produce a uniformly continuous function h on S such that $h(x) = 1$ on F, $h(x) = 0$ on the complement G_ε^c of G_ε, and $0 \leq h(x) \leq 1$ for all x. In view of (ii) one has $\lim_n \int_S h \, dQ_n = \int_S h \, dQ$. In addition,

$$Q_n(F) = \int_F h \, dQ_n \leq \int_S h \, dQ_n$$

and

$$\int_S h \, dQ = \int_{G_\varepsilon} h \, dQ \leq Q(G_\varepsilon) < Q(F) + \delta.$$

Thus

$$\limsup_n Q_n(F) \leq \lim_n \int_S h \, dQ_n = \int_S h \, dQ < Q(F) + \delta.$$

Since δ was arbitrary this proves (iii).

(iii) implies (i): Let $f \in C_b(S)$. It suffices to prove

$$\limsup_n \int_S f \, dQ_n \leq \int_S f \, dQ. \tag{7.7}$$

For then one also gets $\liminf_n \int_S f \, dQ_n \geq \int_S f \, dQ$, and hence (i), by replacing f by $-f$. But in fact, for (7.7) it suffices to consider $f \in C_b(S)$ such that $0 < f(x) < 1, x \in S$, since the more general $f \in C_b(S)$ can be reduced to this by translating and rescaling f. Fix an integer k and let F_i be the closed set $F_i = [f \geq i/k] \equiv \{x \in S : f(x) \geq i/k\}$, $i = 0, 1, \dots, k$. Then taking advantage of $0 < f < 1$, one has

$$\sum_{i=1}^k \frac{i-1}{k} Q\left([\frac{i-1}{k} \leq f < \frac{i}{k}]\right) \leq \int_S f \, dQ \leq \sum_{i=1}^k \frac{i}{k} Q\left([\frac{i-1}{k} \leq f < \frac{i}{k}]\right).$$

Noting that $F_0 = S$, $F_k = \emptyset$, the sum on the right telescopes as

$$\sum_{i=1}^k \frac{i}{k}[Q(F_{i-1}) - Q(F_i)] = \frac{1}{k} + \frac{1}{k}\sum_{i=1}^k Q(F_i),$$

while the sum on the left is smaller than this by $1/k$. Hence

$$\frac{1}{k}\sum_{i=1}^k Q(F_i) \leq \int_S f \, dQ < \frac{1}{k} + \frac{1}{k}\sum_{i=1}^k Q(F_i). \tag{7.8}$$

In view of (iii), $\limsup_n Q_n(F_i) \leq Q(F_i)$ for each i. So, using the upper bound in (7.8) with Q_n in place of Q and the lower bound with Q, it follows that

$$\limsup_n \int_S f \, dQ_n \leq \frac{1}{k} + \frac{1}{k}\sum_{i=1}^k Q(F_i) \leq \frac{1}{k} + \int_S f \, dQ.$$

Now let $k \to \infty$ to obtain the asserted inequality (7.7) to complete the proof of (i) from (iii).

(iii) iff (iv): This is simply due to the fact that open and closed sets are complementary.

(iii) implies (v): Let A be a Q-continuity set. Since (iii) implies (iv) one has

$$Q(A^-) \geq \limsup_n Q_n(A^-) \geq \limsup_n Q_n(A)$$

$$\geq \liminf_n Q_n(A) \geq \liminf_n Q_n(A^\circ) \geq Q(A^\circ). \tag{7.9}$$

Since $Q(\partial A) = 0$, $Q(A^-) = Q(A^\circ)$, so that the inequalities squeeze down to $Q(A)$ and $\lim_n Q_n(A) = Q(A)$ follows.

(v) implies (iii): Let F be a closed set. The idea is to observe that F may be expressed as the limit of a decreasing sequence of Q-continuity sets as follows. Since $\partial\{x : \rho(x, F) \leq \delta\} \subset \{x : \rho(x, F) = \delta\}$, these boundaries are disjoint for distinct δ, (Exercise 1). Thus at most countably many of them can have positive Q-measure (Exercise 1), all other, therefore, being Q-continuity sets. In particular, there is a sequence of positive numbers $\delta_k \downarrow 0$ such that the sets $F_k = \{x : \rho(x, F) \leq \delta_k\}$ are Q-continuity sets. From (v) one has $\limsup_n Q_n(F) \leq \lim_n Q_n(F_k) = Q(F_k)$ for each k. Since F is closed one also has $F_k \downarrow F$, so that (iii) follows from continuity of the probability Q from above. This completes the proof of the theorem. ∎

Definition 7.2 Let S be a metric space. A family $\mathcal{F} \subset \mathcal{B}(S)$ is said to be **convergence-determining** if $P_n(B) \to P(B)$ for all P-continuity sets $B \in \mathcal{F}$ implies $P_n \Rightarrow P$.

Proposition 7.2 Let S be a separable metric space. Suppose $\mathcal{F} \subset \mathcal{B}(S)$ has the properties: (i) it is closed under finite intersections, (ii) every open set is a finite or countable union of P-continuous sets in $\mathcal{F}$. Then $\mathcal{F}$ is convergence-determining.

Proof Let $P_n(F) \to P(F)$ for all P-continuity sets $F \in \mathcal{F}$. Note that if A, B are P-continuity sets in $\mathcal{F}$, then so is $A \cap B$, since $\partial(A \cap B) \subset \partial(A) \cup \partial(B)$. Let G be open and a finite union, $G = \cup_{1 \leq i \leq k} G_i$ with each G_i P-continuous. Then, by the inclusion–exclusion formula, $P_n(\cup_{1 \leq i \leq k} G_i) = \sum_{1 \leq i \leq k} P_n(G_i) - \sum_{1 \leq i_1 < i_2 \leq k} P_n(G_{i_1} \cap G_{i_2}) \pm \cdots + (-1)^{k+1} P_n(G_1 \cap G_2 \cap \cdots \cap G_k) \to \sum_{1 \leq i \leq k} P(G_i) - \sum_{1 \leq i_1 < i_2 \leq k} P(G_{i_1} \cap G_{i_2}) \pm \cdots + (-1)^{k+1} P(G_1 \cap G_2 \cap \cdots \cap G_k) = P(G)$. If G is open and a countable union of P-continuous sets $G_i \in \mathcal{F}$, $G = \cap_{i \geq 1} G_i$, then given any $\varepsilon > 0$ there exists k such that $P(\cup_{1 \leq i \leq k} G_i) \geq P(G) - \varepsilon$, and, therefore, $\liminf P_n(G) \geq \liminf P_n(\cup_{1 \leq i \leq k} G_i) = P(\cup_{1 \leq i \leq k} G_i) \geq P(G) - \varepsilon$. This being true for every $\varepsilon > 0$, one has $\liminf P_n(G) \geq P(G)$ for all open G. By Alexandrovs Theorem, $P_n \Rightarrow P$. ∎

As an example, it follows that in $\mathbb{R}^\infty$ the class $\mathcal{F}$ of finite-dimensional Borel sets is convergence-determining (see Chapter I, Exercises 19, 20), i.e., finite-dimensional distributions are convergence-determining.

Weak convergence is often referred to as an **integral limit theorem**, referring to convergence of the distribution functions. In contrast, a **local limit theorem** can be a precursor to weak convergence as follows. Consider a sequence of discrete random variables X_n ($n \geq 1$). In fact one may permit possibly defective random variables, i.e., allow $P(X_n \in \mathbb{R}) \leq 1$. Let the set of values of X_n be contained in a discrete set $L_n = \{x_i^{(n)} : i \in \mathcal{I}_n\}$, where $\mathcal{I}_n$ is a countable index set. Write $p_i^{(n)} = P(X_n = x_i^{(n)})$. Assume that there exist nonoverlapping intervals $A_i^{(n)}$ of lengths $|A_i^{(n)}| > 0$, $i \in \mathcal{I}_n$, which partition an interval $J \subset \mathbb{R}$ such that (i) $x_i^{(n)} \in A_i^{(n)}$, (ii) $\delta_n := \sup\{|A_i^{(n)}| : i \in \mathcal{I}_n\} \to 0$ as $n \to \infty$, (iii) for every $x \in J$ outside a set of Lebesgue measure zero, and with the index $i = i(n, x)$ such that $x \in A_i^{(n)}$, one has

$$p_{i(n,x)}^{(n)} / |A_{i(n,x)}^{(n)}| \longrightarrow f(x) \qquad \text{as } n \to \infty, \tag{7.10}$$

and (iv) $1 \geq \alpha_n := \sum_{i \in \mathcal{I}_n} p_i^{(n)} \to \alpha := \int_J f(x)dx > 0$.

Proposition 7.3 (*Local Limit Theorem*) Under the assumptions (i)–(iv) above, $\sum_{\{i \in \mathcal{I}_n : x_i^{(n)} \leq x\}} p_i^{(n)} \to \int_{J \cap (-\infty, x]} f(y)dy$ for every $x \in J$. In particular if X_n are proper random variables, i.e., $\sum_{i \in \mathcal{I}_n} p_i^{(n)} = 1$, then X_n converges in distribution to the law with density f.

Proof First assume $\alpha_n = 1 = \alpha$ for all n. On J define the density function $f_n(x) = p_i^{(n)}/|A_i^{(n)}|$ if $x \in A_i^{(n)}$ ($x \in J$). By assumption (iv) and Scheffé's theorem, $\int_J |f_n(y) - f(y)|dy \to 0$. On the other hand, since $p_i^{(n)} = \int_{A_i^{(n)}} f_n(y)dy$ for all $i \in \mathcal{I}_n$,

$$\sum_{\{i \in \mathcal{I}_n : x_i^{(n)} \leq x\}} p_i^{(n)} = \int_{J \cap (-\infty, x]} f_n(y)dy.$$

For the general case of $\alpha_n \to \alpha > 0$, rescale using $p_i^{(n)}/\alpha_n$ and $f(x)/\alpha$ to get the displayed equality above. Then compare the sum on the left with that divided by α. ∎

Remark 7.1 Note also that this extends to higher dimensions $\mathbb{R}^k$, where $A_i^{(n)}$ is a rectangle of positive k-dimensional volume $|A_i^{(n)}|$.

Example 1 For a very simple illustrative example, suppose that for each fixed $n = 1, 2, \ldots, P(T_n = i) = (1 - \theta_n)^{i-1}\theta_n, i = 1, 2, \ldots$ is a *geometrically distributed* random variable with parameter $\theta_n \in (0, 1)$. Let $X_n = \frac{T_n}{n}$. Then $L_n = \{\frac{1}{n}, \frac{2}{n}, \ldots, \}$. Define $A_i(n) = (\frac{2i-1}{2n}, \frac{2i+1}{2n}], i = 0, \ldots,$ and $\delta_n = \frac{1}{n}$. Also $p_i^{(n)} = (1 - \theta_n)^{ni-1}\theta_n, i = \frac{1}{n}, \frac{2}{n}, \ldots,$. Assume that $n\theta_n \to \theta > 0$ as $n \to \infty$. Then, for any $x \in J = [0, \infty)$ and sequence $i(n, x)$ such that $\frac{2i(n,x)-1}{2n} \leq x < \frac{2i(n,x)+1}{2n}$, one has

$$np_{i(n,x)}^{(n)} = (n\theta_n)(1 - \frac{n\theta_n}{n})^{ni(n,x)}(1 - \theta_n)^{-1} \to \theta e^{-\theta x}$$

as $n \to \infty$, from which convergence in distribution of X_n to the *exponential distribution* with parameter $\theta > 0$ follows.

The following result provides a useful tool for tracking weak convergence in a variety of settings. Note that in the case that h is continuous it follows immediately from the definition of weak convergence since compositions of bounded continuous functions with h are bounded and continuous.

Theorem 7.4 (*Mann–Wald*) Let S_1, S_2 be a pair of metric spaces and $h : S_1 \to S_2$ a Borel-measurable map. Suppose that $\{Q_n\}_{n=1}^{\infty}, Q$ are probabilities on the Borel σ-field of S_1 such that $Q_n \Rightarrow Q$. If h is Q-a.s. continuous, then $Q_n \circ h^{-1} \Rightarrow Q \circ h^{-1}$.

Proof Let F be a closed subset of S_2. Then, letting $F_h = h^{-1}(F)$, it follows from Alexandrov conditions that $\limsup_n Q_n(F_h) \leq \limsup_n Q_n(F_h^-) \leq Q(F_h^-)$. But $Q(F_h^-) = Q(F_h)$ since $F_h^- \subset D_h \cup F_h$, where D_h denotes the set of discontinuities of h (Exercise 1) and, by hypothesis, $Q(D_h) = 0$. ∎

Remark 7.2 It follows from Theorem 7.1, in particular, that if (S, ρ) is a metric space, $Q_1, Q_2 \in \mathcal{P}(S)$, then $\int_S f \, dQ_1 = \int_S f \, dQ_2 \forall f \in C_b(S)$ implies $Q_1 = Q_2$. Note that by a simple rescaling this makes $\{f \in C_b(S) : \|f\|_\infty \leq 1\}$ measure-determining as well. The same is true for the set $UC_b(S)$ of all bounded uniformly continuous real-valued functions on S in place of $C_b(S)$.

Using a technique from the proof of Theorem 7.1 one may also obtain the following equivalent specification of the weak topology.

Proposition 7.5 The weak topology is defined by the system of open neighborhoods of the form (7.5) with $f_1, f_2, \ldots, f_m$ bounded and uniformly continuous.

Proof Fix $Q_0 \in \mathcal{P}(S)$, $f \in C_b(S)$, $\varepsilon > 0$. We need to show that the set $\{Q \in \mathcal{P}(S) : |\int_S f \, dQ - \int_S f \, dQ_0| < \varepsilon\}$ contains a set of the form (7.5), but with f_i's that are uniformly continuous and bounded. Without essential loss of generality, assume $0 < f < 1$. As in the proof of Theorem 7.1, (iii) implies (i); see the relations (7.8), choose and fix a large integer k such that $1/k < \varepsilon/4$ and consider the F_i in that proof. Next, as in the proof of (ii) implies (iii) of Theorem 7.1, there exist uniformly continuous functions g_i, $0 \leq g_i \leq 1$, such that $g_i = 1$ on F_i and $|\int_S g_i \, dQ_0 - Q_0(F_i)| < \varepsilon/4$, $1 \leq i \leq k$. Then on the set $\{Q : |\int_S g_i \, dQ - \int_S g_i \, dQ_0| < \varepsilon/4, 1 \leq i \leq k\}$, one has (see (7.5))

$$\int_S f \, dQ \leq \frac{1}{k} \sum_{i=1}^k Q(F_i) + \frac{1}{k} \leq \frac{1}{k} \sum_{i=1}^k \int_S g_i \, dQ + \frac{1}{k}$$

$$< \frac{1}{k} \sum_{i=1}^k \int_S g_i \, dQ_0 + \frac{\varepsilon}{4} + \frac{1}{k} \leq \frac{1}{k} \sum_{i=1}^k Q_0(F_i) + \frac{2\varepsilon}{4} + \frac{1}{k}$$

$$< \int_S f \, dQ_0 + \frac{2\varepsilon}{4} + \frac{1}{k} \leq \int_S f \, dQ_0 + \varepsilon. \qquad (7.11)$$

Similarly, replacing f by $1 - f$ in the above argument, one may find uniformly continuous h_i, $0 \leq h_i \leq 1$, such that on the set $\{Q : |\int_S h_i \, dQ - \int_S h_i \, dQ_0| < \varepsilon/4, 1 \leq i \leq k\}$, one has $\int_S (1 - f) dQ < \int_S (1 - f) dQ_0 + \varepsilon$. Therefore

$$\{Q \in \mathcal{P}(S) : |\int_S f \, dQ - \int_S f \, dQ_0| < \varepsilon\}$$

$$\supset \left\{Q : \left|\int_S g_i \, dQ - \int_S g_i \, dQ_0\right| < \varepsilon/4, \left|\int_S h_i \, dQ - \int_S h_i \, dQ_0\right| < \varepsilon/4, 1 \leq i \leq k\right\}.$$

By taking intersections over m such sets, it follows that a neighborhood $N(Q_0)$, say, of Q_0 of the form (7.5) (with $f_i \in C_b(S)$, $1 \leq i \leq m$) contains a neighborhood of Q_0 defined with respect to bounded uniformly continuous functions. In particular, $N(Q_0)$ is an open set defined by the latter neighborhood system. Since the latter neighborhood system is a subset of the system (7.5), the proof is complete. ∎

Two points of focus for the remainder of this section are metrizability and (relative) compactness in the weak topology. Compactness in a metric space may be equivalently viewed as the existence of a limit point for any sequence in the space.

In the case that (S, ρ) is a compact metric space, $C(S) \equiv C_b(S)$ is a *complete separable metric space* under the **"sup" norm** $\|f\| := \max\{|f(x)| : x \in S\}$, i.e., under the metric $d(f, g) := \|f - g\| \equiv \max\{|f(x) - g(x)| : x \in S\}$ (see Appendix B). In this case the weak topology is metrizable, i.e., $\mathcal{P}(S)$ is a metric space with the metric

$$d_W(Q_1, Q_2) := \sum_{n=1}^{\infty} 2^{-n} \left| \int_S f_n \, dQ_1 - \int_S f_n \, dQ_2 \right|, \qquad (7.12)$$

where $\{f_n : n \geq 1\}$ is a dense subset of $\{f \in C(S) : \|f\| \leq 1\}$. Using Cantor's diagonal procedure and the Riesz representation theorem (for bounded linear functionals on $C(S)$ in Appendix A), one may check that every sequence $\{Q_n : n \geq 1\}$ has a convergent subsequence. In other words, one has the following result (Exercise 4).

Proposition 7.6 If (S, ρ) is compact, then $\mathcal{P}(S)$ is a compact metric space under the weak topology, with a metric given by (7.12).

A slightly weaker form of convergence is sometimes useful to consider within the general theme of this section, for example in analyzing the nature of certain failures of weak convergence (see Exercise 3). A function $f \in C_b(S)$ is said to **vanish at infinity** if for each $\varepsilon > 0$ there is a compact subset K_ε such that $|f(x)| < \varepsilon$ for all $x \in K_\varepsilon^c$. Let $C_b^0(S)$ denote the collection of all such functions on S.

Definition 7.3 A sequence of probability measures $\{Q_n\}_{n=1}^{\infty}$ on $(S, \mathcal{B}(S))$ is said to **converge vaguely** to a finite measure $Q^{(0)}$, not necessarily a probability, if $\lim_n \int_S f \, dQ_n = \int_S f \, dQ^{(0)}$ for all $f \in C_b^0(S)$.

Corollary 7.7 *(Helly Selection Principle)* Every sequence of probabilities $\mu_n, n \geq 1$, on $(\mathbb{R}, \mathcal{B})$ has a vaguely convergent subsequence.

Proof Let $\varphi : \mathbb{R} \to (-1, 1)$ be given by $\varphi(x) = \frac{2}{\pi} \tan^{-1}(x), x \in \mathbb{R}$, and define a probability ν_n supported on $(-1, 1)$ by $\nu_n(A) = \mu_n(\{x : \varphi(x) \in A\})$ for Borel subsets A of $(-1, 1)$. One may regard ν_n as a probability on the compact interval $[-1, 1]$ (supported on the open interval). Thus, by Proposition 7.6, there is a probability ν on $[-1, 1]$ and a subsequence $\{\nu_{n_m} : m \geq 1\}$ such that $\nu_{n_m} \Rightarrow \nu$ as $m \to \infty$. Define $\tilde{\nu}(A) = \nu(A)$ for Borel subsets A of $(-1, 1)$. Then $\tilde{\nu}$ is a measure on $(-1, 1)$ with $\tilde{\nu}(-1, 1) \leq 1$. Let $\mu(B) = \tilde{\nu}(\{y \in (-1, 1) : \varphi^{-1}(y) \in B\})$ for Borel subsets B of $\mathbb{R}$. Since for $f \in C_b^0(\mathbb{R})$, the map $g := f \circ \varphi^{-1}$ is in $C_b([-1, 1])$, where $g(1) = g(-1) = f(\varphi^{-1}(\pm 1)) := 0$, one has, using the change of variable formula,

$$\int_{\mathbb{R}} f(x)\mu(dx) = \int_{(-1,1)} f(\varphi^{-1}(y))\mu \circ \varphi^{-1}(dy) = \int_{[-1,1]} g(y)\tilde{\nu}(dy)$$

$$= \lim_{m \to \infty} \int_{[-1,1]} g(y)\nu_{n_m}(dy)$$

$$= \lim_{m \to \infty} \int_{\mathbb{R}} f(x)\mu_{n_m}(dx),$$

where the change of variable formula is again used to write the last equality.   ∎

The following theorem is typically proven using Fourier transforms, as will be presented here.[2]

Proposition 7.8 *(Cramér–Wold Device)* A sequence of k-dimensional random vectors $\mathbf{X}_n (n \geq 1)$ converges in distribution to (the distribution of a random vector) $\mathbf{X}$ if and only if all linear functions $\mathbf{c} \cdot \mathbf{X}_n \equiv \sum_{j=1}^{k} c_j X_n^{(j)}$ converge in distribution to $\mathbf{c} \cdot \mathbf{X}$ for all $\mathbf{c} = (c_1, \ldots, c_k) \in \mathbb{R}^k$.

Proof Certainly if one has convergence in distribution of the sequence $\mathbf{X}_n (n \geq 1)$ to $\mathbf{X}$, then the linear functions of $\mathbf{X}_n$ must converge in distribution since for any choice of $\mathbf{c}$, the function $\mathbf{x} \to \mathbf{c} \cdot \mathbf{x}$ is continuous. For if g is a continuous, bounded function then so is the composition $g(\mathbf{c} \cdot \mathbf{x})$, and $\mathbb{E}g(c \cdot \mathbf{X}_n) \to \mathbb{E}g(c \cdot \mathbf{X})$. For the converse suppose that $\mathbf{c} \cdot \mathbf{X}_n \equiv \sum_{j=1}^{k} c_j X_n^{(j)}$ converges in distribution to $\mathbf{c} \cdot \mathbf{X}$ for all $\mathbf{c} = (c_1, \ldots, c_k) \in \mathbb{R}^k$. Then one has for any $\theta \in \mathbb{R}^k$, $\mathbb{E}e^{i\theta \cdot \mathbf{X}_n} \to \mathbb{E}e^{i\theta \cdot \mathbf{X}}$. Now use the Cramér–Lévy continuity theorem (Theorem 6.11).   ∎

As a simple application one may obtain the classical one-dimensional central limit theorem for sums of i.i.d. random variables having finite second moments to $\mathbb{R}^k$. Namely,

Theorem 7.9 *(Multivariate Classical CLT)* Let $\{\mathbf{X}_n : n = 1, 2, \ldots\}$ be a sequence of i.i.d. random vectors with values in $\mathbb{R}^k$. Let $\mathbb{E}\mathbf{X}_1 = \mu \in \mathbb{R}^k$ (defined componentwise) and assume that the dispersion matrix (i.e., variance–covariance matrix) $\mathbf{D}$ of $\mathbf{X}_1$ is finite. Then as $n \to \infty$, $n^{-1/2}(\mathbf{X}_1 + \cdots + \mathbf{X}_n - n\mu)$ converges in distribution to the Gaussian probability measure with mean zero and dispersion matrix $\mathbf{D}$.

Proof For each $\xi \in \mathbb{R}^k \backslash \{0\}$ apply Corollary 4.3, to the sequence $\xi \cdot \mathbf{X}_n, n \geq 1$. Then use the Cramér–Wold device.   ∎

For our next result we need the following lemma. Let $H = [0, 1]^{\mathbb{N}}$ be the space of all sequences in $[0, 1]$ with the product topology, referred to as the **Hilbert cube**

Lemma 1 *(Hilbert Cube Embedding)* Let (S, ρ) be a separable metric space. There exists a map h on S into the Hilbert cube $H \equiv [0, 1]^{\mathbb{N}}$ with the product topology, such that h is a homeomorphism of S onto $h(S)$, in the relative topology of $h(S)$.

---

[2]A non-Fourier analytic proof was found by Guenther Walther (1997): On a conjecture concerning a theorem of Cramér and Wold, *J. Multivariate Anal.*, **63**, 313–319, resolving some serious doubts about whether it would be possible.

Proof Without loss of generality, assume $\rho(x, y) \leq 1 \forall x, y \in S$. Let $\{z_k : k = 1, 2, \dots\}$ be a dense subset of S. Define the map

$$h(x) = (\rho(x, z_1), \ \rho(x, z_2), \dots, \rho(x, z_k), \dots) \quad (x \in S). \qquad (7.13)$$

If $x_n \to x$ in S, then $\rho(x_n, z_k) \to \rho(x, z_k) \ \forall k$, so that $h(x_n) \to h(x)$ in the (metrizable) product topology (of pointwise convergence) on $h(S)$. Also, h is one-to-one. For if $x \neq y$, one may find z_k such that $\rho(x, z_k) < \frac{1}{3}\rho(x, y)$, and hence $\rho(y, z_k) \geq \rho(y, x) - \rho(z_k, x) > \frac{2}{3}\rho(x, y)$, so that $\rho(x, z_k) \neq \rho(y, z_k)$. Finally, let $\tilde{a}_n \equiv (a_{n1}, a_{n2}, \dots) \to \tilde{a} = (a_1, a_2, \dots)$ in $h(S)$, and let $x_n = h^{-1}(\tilde{a}_n)$, $x = h^{-1}(\tilde{a})$. One then has $(\rho(x_n, z_1), \ \rho(x_n, z_2), \dots) \to (\rho(x, z_1), \ \rho(x, z_2), \dots)$. Hence $\rho(x_n, z_k) \to \rho(x, z_k) \ \forall k$, implying $x_n \to x$, since $\{z_k : k \geq 1\}$ is dense in S. ∎

Theorem 7.10 Let (S, ρ) be a separable metric space. Then $\mathcal{P}(S)$ is a separable metric (i.e., metrizable) space under the weak topology.

Proof By Lemma 1, S may be replaced by its homeomorphic image $S_h \equiv h(S)$ in $[0, 1]^{\mathbb{N}}$ which is compact under the product topology by Tychonov's theorem (Appendix B), and is metrizable with the metric

$$d(\tilde{a}, \tilde{b}) := \sum_{n=1}^{\infty} 2^{-n} |a_n - b_n| (\tilde{a} = (a_1, a_2, \dots), \ \tilde{b} = (b_1, b_2, \dots)).$$

We shall consider uniform continuity of functions on S_h with respect to this metric d. Every uniformly continuous (bounded) f on S_h has a unique extension $\bar{f}$ to $\bar{S}_h$ ($\equiv$ closure of S_h in $[0, 1]^{\mathbb{N}}$) : $\bar{f}(\tilde{a}) := \lim_{k \to \infty} f(\tilde{a}^k)$, where $\tilde{a}^k \in S_h$, $\tilde{a}^k \to \tilde{a}$. Conversely, the restriction of every $g \in C(\bar{S}_h)$ is a uniformly continuous bounded function on S_h. In other words, the space $UC_b(S_h)$ of all bounded uniformly continuous functions on S_h may be identified with $C(\bar{S}_h)$ as sets and as metric spaces under the supremum distance d_∞ between functions. Since $\bar{S}_h$ is compact, $C_b(\bar{S}_h) \equiv C(\bar{S}_h)$ is a separable metric space under the supremum distance d_∞, and therefore, so is $UC_b(S_h)$. Letting $\{f_n : n = 1, 2, \dots\}$ be a dense subset of $UC_b(S_h)$, one now defines a metric d_W on $\mathcal{P}(S_h)$ as in (7.12). This proves metrizability of $\mathcal{P}(S_h)$.

To prove separability of $\mathcal{P}(S_h)$, for each $k = 1, 2, \dots$, let $D_k := \{x_{ki} : i = 1, 2, \dots, n_k\}$ be a finite $(1/k)$-net of S_h (i.e., every point of S_h is within a distance $1/k$ from some point in this net). This is possible since $\bar{S}_h$ is a compact metric space. Let $D = \{x_{ki} : i = 1, \dots, n_k, k \geq 1\} = \cup_{k=1}^{\infty} D_k$. Consider the set $\mathcal{E}$ of all probabilities with finite support contained in D and having rational mass at each point of support. Then $\mathcal{E}$ is countable and is dense in $\mathcal{P}(S_h)$. To prove this last assertion, fix $Q_0 \in \mathcal{P}(S_h)$. Consider the partition generated by the set of open balls $\{x \in S_h : d(x, x_{ki}) < \frac{1}{k}\}$, $1 \leq i \leq n_k$. Let Q_k be the probability measure defined by letting the mass of Q_0 on each nonempty set of the partition be assigned to a singleton $\{x_{ki}\}$ in D_k that is at a distance of at most $1/k$ from the set. Now construct $\tilde{Q}_k \in \mathcal{E}$, where $\tilde{Q}_k$ has the same support as Q_k but the point masses of $\tilde{Q}_k$ are rational and are such that the sum of the absolute differences between these masses of Q_k

and the corresponding ones of $\tilde{Q}_k$ is less than $1/k$. Then it is simple to check that $d_W(Q_0, \tilde{Q}_k) \to 0$ as $k \to \infty$, that is, $\int_{S_h} g \, d\tilde{Q}_k \to \int_{S_h} g \, dQ_0$ for every uniformly continuous and bounded g on S_h.   ∎

The next result is of considerable importance in probability. To state it we need a notion called "tightness."

Definition 7.4 A subset Λ of $\mathcal{P}(S)$ is said to be **tight** if for every $\varepsilon > 0$, there exists a compact subset K_ε of S such that

$$Q(K_\varepsilon) \geq 1 - \varepsilon \quad \forall Q \in \Lambda. \tag{7.14}$$

Theorem 7.11 *(Prohorov's Theorem)* (a) Let (S, ρ) be a separable metric space. If $\Lambda \subset \mathcal{P}(S)$ is tight then its weak closure $\bar{\Lambda}$ is compact (metric) in the weak topology. (b) If (S, ρ) is Polish, then the converse is true: For a set Λ to be conditionally compact (i.e., $\bar{\Lambda}$ compact) in the weak topology, it is necessary that Λ be tight.

Proof We begin with a proof of part (a). Suppose $\Lambda \subset \mathcal{P}(S)$ is tight. Let $\tilde{S} = \cup_{j=1}^{\infty} K_{1/j}$, where $K_{1/j}$ is a compact set determined from (7.14) with $\varepsilon = 1/j$. Then $P(\tilde{S}) = 1 \, \forall \, Q \in \Lambda$. Accordingly, since $\tilde{S}$ is σ-compact, so is it's image $\tilde{S}_h = \cup_{j=1}^{\infty} h(K_{1/j})$ under the map h (appearing in the proofs of Lemma 1 and Theorem 7.10). That is, since the image of a compact set under a continuous map is compact, we see that $\tilde{S}_h$ is also expressible as a countable union of compact sets, i.e., σ-compact.

In particular, $\tilde{S}_h$ is a Borel subset of $[0, 1]^{\mathbb{N}}$ and therefore of $\bar{\tilde{S}}_h$. Let Λ_h be the image of Λ in $\tilde{S}_h$ under h, i.e., $\Lambda_h = \{P \circ h^{-1} : P \in \Lambda\} \subset \mathcal{P}(\tilde{S}_h)$. In view of the homeomorphism $h : \tilde{S} \to \tilde{S}_h$, it is enough to prove that Λ_h is conditionally compact as a subset of $\mathcal{P}(\tilde{S}_h)$.

Since $\tilde{S}_h$ is a Borel subset of $\bar{\tilde{S}}_h$, one may take $\mathcal{P}(\tilde{S}_h)$ as a subset of $\mathcal{P}(\bar{\tilde{S}}_h)$, extending Q in $\mathcal{P}(\tilde{S}_h)$ by setting $Q(\bar{\tilde{S}}_h \backslash \tilde{S}_h) = 0$. Thus $\Lambda_h \subset \mathcal{P}(\tilde{S}_h) \subset \mathcal{P}(\bar{\tilde{S}}_h)$. By Proposition 7.6, $\mathcal{P}(\bar{\tilde{S}}_h)$ is compact metric (in the weak topology). Hence every sequence $\{Q_n : n = 1, 2, \ldots\}$ in Λ_h has a subsequence $\{Q_{n_k} : k = 1, 2, \ldots\}$ converging weakly to some $Q \in \mathcal{P}(\bar{\tilde{S}}_h)$. We need to show that $Q \in \mathcal{P}(\tilde{S}_h)$, that is, $Q(\tilde{S}_h) = 1$. By Theorem 7.1, $Q(h(K_{1/j})) \geq \limsup_{k\to\infty} Q_{n_k}(h(K_{1/j})) \geq 1 - 1/j$. (By hypothesis, $Q(h(K_{1/j})) \geq 1 - 1/j \, \forall \, Q \in \Lambda_h$). Letting $j \to \infty$, one gets $Q(\tilde{S}_h) = 1$. Finally, note that if Λ is conditionally compact when considered as a subset of $\mathcal{P}(\tilde{S})$, it is also conditionally compact when regarded as a subset of $\mathcal{P}(S)$ (Exercise 8).

For part (b) suppose that (S, ρ) is separable and complete and let Λ be relatively compact in the weak topology. We will first show that given any nondecreasing sequence $G_n, n \geq 1$, of open subsets of S such that $\cup_n G_n = S$ and given any $\varepsilon > 0$, there is an $n = n(\varepsilon)$ such that $Q(G_{n(\varepsilon)}) \geq 1 - \varepsilon$ for all $Q \in \Lambda$. For suppose this is not true. Then there are an $\varepsilon > 0$ and $Q_1, Q_2, \ldots$ in Λ such that $Q_n(G_n) < 1 - \varepsilon$ for all $n \geq 1$. But by the assumed compactness, there is a subsequence $Q_{n(k)}$ that converges weakly to some probability $Q \in \mathcal{P}(S)$. By Alexandrov's theorem this implies, noting $G_n \subset G_{n(k)}$ for $n \leq n(k)$, that $Q(G_n) \leq \liminf_{k\to\infty} Q_{n(k)}(G_n) \leq$

$\liminf_{k\to\infty} Q_{n(k)}(G_{n(k)}) \leq 1 - \varepsilon$, for $n \geq 1$. This leads to the contradiction $1 = Q(S) = \lim_{n\to\infty} Q(G_n) \leq 1 - \varepsilon$. Now to prove that Λ is tight, fix $\varepsilon > 0$. By separability of S for each $k \geq 1$ there is a sequence of open balls $B_{n,k}, n \geq 1$, having radii smaller than $1/k$ and such that $\cup_{n\geq 1} B_{n,k} = S$. Let $G_{n,k} := \cup_{m=1}^{n} B_{m,k}$. Using the first part of this proof of (b), it follows that for each k there is an $n = n(k)$ such that $Q(G_{n(k),k}) \geq 1 - 2^{-k}\varepsilon$ for all $Q \in \Lambda$. Define $G := \cap_{k=1}^{\infty} G_{n(k),k}$. Then its closure $\overline{G}$ is totally bounded, since for each k there is a finite cover of $\overline{G}$ by $n(k)$ closed balls $\overline{B}_{n,k}$ of diameter smaller than $1/k$. Thus completeness of S implies that $\overline{G}$ is compact (see Appendix B, Lemma 4). But $Q(\overline{G}) \geq Q(G) \geq 1 - \sum_{k=1}^{\infty} 2^{-k}\varepsilon = 1 - \varepsilon$ for all $Q \in \Lambda$. ∎

We state the following proposition without proof[3].

Proposition 7.12 Let (S, ρ) be a Polish space. Then $\mathcal{P}(S)$ is Polish for the weak topology.

Corollary 7.13 Let (S, ρ) be a Polish space. Then any finite collection Λ of probabilities on $(S, \mathcal{B}(S))$ is tight.

Remark 7.3 The compactness asserted in part (a) of Theorem 7.11 remains valid without the requirement of separability for the metric space (S, ρ). To see this, simply note that the set $\tilde{S} = \cup_{j=1}^{\infty} K_{1/j}$ is σ-compact metric whether S is separable or not. However, in this case $\mathcal{P}(S)$ may not be metric under the weak topology. Nonetheless, the relative weak topology on Λ (and $\bar{\Lambda}$) is metrizable.

In applications one might have $\Lambda = \{Q_n\}_{n=1}^{\infty}$, where $Q_n = P \circ X_n^{-1}$ is the distribution of a random map X_n. If X_n is real valued, for example, then one might try to check tightness by a Chebyshev-type inequality, see, for example, Exercise 5.

The following definition and proposition provide a frequently used metrization in weak convergence theory.

Definition 7.5 The **Prohorov metric** d_π on $\mathcal{P}(S)$ is defined by

$$d_\pi(Q_1, Q_2) := \inf\{\varepsilon > 0 : Q_1(A) \leq Q_2(A^\varepsilon) + \varepsilon, Q_2(A) \leq Q_1(A^\varepsilon) + \varepsilon, \forall A \in \mathcal{B}(S)\}.$$

Remark 7.4 Essentially using the symmetry that $A \subset S\backslash B^\varepsilon$ if and only if $B \subset S\backslash A^\varepsilon$, one may check that if $Q_1(A) \leq Q_2(A^\varepsilon) + \varepsilon$ for all $A \in \mathcal{B}(S)$ then $d_\pi(Q_1, Q_2) \leq \varepsilon$. That is it suffices to check that one of the inequalities holds for all $A \in \mathcal{B}(S)$ to get the other. For if the first inequality holds for all A, taking $B = S\backslash A^\varepsilon$, one has $Q_1(A^\varepsilon) = 1 - Q_1(B) \geq 1 - Q_2(B^\varepsilon) - \varepsilon = Q_2(S\backslash B^\varepsilon) - \varepsilon \geq Q_2(A) - \varepsilon$.

Proposition 7.14 Let (S, ρ) be a separable metric space. Then d_π metrizes the weak topology on $\mathcal{P}(S)$ in the sense that:

---

[3]See Parthasarathy, K.R. (1967), Theorem 6.5, pp. 46–47, or Bhattacharya and Majumdar (2007), Theorem C11.6, p. 237.

(i) d_π defines a metric on $\mathcal{P}(S)$.

(ii) If $d_\pi(Q_n, Q) \to 0$ as $n \to \infty$ then $Q_n \Rightarrow Q$.

(iii) If $Q_n \Rightarrow Q$, then $d_\pi(Q_n, Q) \to 0$ as $n \to \infty$.

Proof Suppose that $d_\pi(Q_1, Q_2) = 0$. Then from the definition of d_π one arrives for all closed sets F, letting $\varepsilon \downarrow 0$ with $A = F$ in the definition, at $Q_1(F) \leq Q_2(F)$ and $Q_2(F) \leq Q_1(F)$. Symmetry and nonnegativity are obvious. For the triangle inequality let $d_\pi(Q_i, Q_{i+1}) = \varepsilon_i, i = 1, 2$. Then $Q_1(A) \leq Q_2(A^{\varepsilon_1'}) + \varepsilon_1' \leq Q_3((A)^{\varepsilon_1'})^{\varepsilon_2'}) + \varepsilon_1' + \varepsilon_2'$, for all $\varepsilon_i' > \varepsilon_i, i = 1, 2$. Thus $d_\pi(Q_1, Q_3) \leq \varepsilon_1' + \varepsilon_2'$ since $(A^{\varepsilon_1'})^{\varepsilon_2'} \subset A^{\varepsilon_1 + \varepsilon_2}$. Since this is true for all $\varepsilon_i' > \varepsilon_i, i = 1, 2$, the desired triangle inequality follows. Next suppose that $d_\pi(Q_n, Q) \to 0$ as $n \to \infty$. Let $\varepsilon_n \to 0$ be such that $d_\pi(Q_n, Q) < \varepsilon_n$. Then, by definition, $Q_n(F) \leq Q(F^{\varepsilon_n}) + \varepsilon_n$ for all closed F. Thus $\limsup_n Q_n(F) \leq Q(F)$ for all closed F, and weak convergence follows from Alexandrov's conditions. For the converse, fix an $\varepsilon > 0$. In view of the remark following the definition of d_π it suffices to show that for all n sufficiently large, say $n \geq n_0$, one has for any Borel set A that $Q(A) \leq Q_n(A^\varepsilon) + \varepsilon$. By separability, S is the union of countably many open balls $B_i, i \geq 1$, of diameter smaller than ε. Choose N such that $Q(S\setminus \cup_{m=1}^N B_m) \leq Q(\cup_{m \geq N+1} B_m) < \varepsilon$. Now by Alexandrov's conditions, $Q_n \Rightarrow Q$ implies that for any of the finitely many open sets of the form $G := B_{i_1} \cup \cdots \cup B_{i_m}, 1 \leq i_1 < \cdots < i_m \leq N$, there is an n_0 such that $Q_n(G) > Q(G) - \varepsilon$ for all $n \geq n_0$. For $A \in \mathcal{B}(S)$ let $\hat{A} = \cup_{i=1}^N \{B_i : B_i \cap A \neq \emptyset\}$. Then consider the open set $\hat{A}^\varepsilon := \{x \in S : \rho(x, \hat{A}) < \varepsilon\}$. In particular, there exists $n > n_0$ that $Q(A) \leq Q(\hat{A}) + Q(\cup_{i>N} B_i) \leq Q(\hat{A}) + \varepsilon < Q_n(\hat{A}) + 2\varepsilon \leq Q_n(\hat{A}^\varepsilon) + 2\varepsilon \leq Q_n(A^{2\varepsilon}) + 2\varepsilon$, since $\hat{A} \subset A^\varepsilon$, so that $\hat{A}^\varepsilon \subset A^{2\varepsilon}$. Thus $d_\pi(Q_n, Q) \leq 2\varepsilon$ for all $n \geq n_0$. ∎

Example 2 *(Functional Central Limit Theorem (FCLT, Invariance Principle))* Suppose $\{Z_m : m = 1, 2, \ldots\}$ is an i.i.d. sequence with $\mathbb{E}Z_1 = 0$ and variance $\mathbb{E}Z_1^2 = \sigma^2 > 0$ defined on a probability space $(\Omega, \mathcal{F}, P)$. To simplify calculations for this example we will assume $\mathbb{E}Z_1^4 < \infty$. This result will be proved by another method in the full generality of finite second moments in Chapter XI.

For a positive integer n, define a polygonal path process $B^{(n)}$ with values in the metric space $S = C[0, \infty)$, with the topology of uniform convergence on compacts, by

$$B^{(n)}(t) = \frac{S_{[nt]}}{\sqrt{n}} = \frac{S_m}{\sqrt{n}} = \frac{Z_1 + \cdots + Z_m}{\sqrt{n}}, \quad t = \frac{m}{n}, m = 1, 2, \ldots, \quad (7.15)$$

with linear interpolation between $(mn, \frac{S_m}{\sqrt{n}})$. Note that by Corollary 4.7 from the classical central limit theory, one has $(B^{(n)}(1), \ldots, B^{(n)}(kn)) \Rightarrow (B(1), \ldots, B(k))$ as $n \to \infty$, where $(B(1), \cdots, B(k))$ is a k-dimensional Gaussian random vector with mean zero and variance–covariance $\mathbb{E}B(i)B(j) = \sigma^2 i \wedge j, 1 \leq i, j \leq k$. From this point onwards take $\sigma^2 = 1$ for without loss of generality.

The functional central limit theorem provides the existence of a limit distribution W on $C[0, \infty)$, referred to as **Wiener measure**. Equivalently, this is the existence of a stochastic process B with values in $C[0, \infty)$ having distribution W, referred to as **standard Brownian motion**, such that $B^{(n)} \Rightarrow B$ as $n \to \infty$. In particular it will follow that in general the limit process depends on the distribution of the displacements Z only through the variance $\sigma^2 > 0$, and is given by the rescaling σB. This invariance of the distribution is the reason for the alternative reference to *invariance principle*. The following definition provides a complete description of the stochastic process obtained in the limit when $\sigma^2 = 1$.

Definition 7.6 The one-dimensional standard Brownian motion starting at 0 defined on a probability space $(\Omega, \mathcal{F}, P)$ is the stochastic process $\mathbf{B} := \{\mathbf{B}_t : t \geq 0\}$ having a.s. continuous paths, $B_0 = 0$, with the *independent increments property*: For any $0 = t_0 < t_1 < t_2 < \cdots < t_k$, $B_{t_{i+1}} - B_{t_i}, i = 0, 1, \ldots k - 1$, are independent. The increments have a Gaussian distribution with mean zero and variances $t_{i+1} - t_i$, respectively. The probability measure $W = P \circ \mathbf{B}^{-1}$ on the Borel σ-field of $C[0, \infty)$ defined by the distribution of standard Brownian motion starting at 0 is called the *Wiener measure*. If B is standard Brownian motion then $X_t = \sigma B_t + \mu t, t \geq 0$, $\sigma > 0, \mu \in \mathbb{R}$, is referred to as Brownian motion with *drift coefficient* μ, and *diffusion coefficient* σ^2. [4]

Note that the property of independent Gaussian increments of standard Brownian motion may be equivalently formulated as the property that for any $0 = t_0 < t_1 < t_2 < \cdots < t_k$, $(B_{t_1}, \ldots, B_{t_k})$ is Gaussian random vector with mean zero and variance–covariance matrix $((\mathbb{E}B_{t_i} B_{t_j}))_{0 \leq i, j \leq k} = ((t_i \wedge t_j))_{0 \leq i, j \leq k}$.

Theorem 7.15 (*The Functional Central Limit Theorem (Invariance Principle)*) Suppose $\{Z_m : m = 1, 2, \ldots\}$ is an i.i.d. sequence with $\mathbb{E}Z_m = 0$ and variance $\sigma^2 > 0$. Then as $n \to \infty$ the stochastic processes $\{\tilde{X}_t^{(n)} : t \geq 0\}$ converge in distribution to a Brownian motion starting at the origin with zero drift and diffusion coefficient σ^2.

For the proof we will require a few lemmas of general utility.

First to check tightness in the context of probabilities on $C[0, 1]$ we appeal to the Arzela–Ascoli theorem from Appendix B for a description of the (relatively) compact subsets of $C[0, 1]$. Accordingly, a subset A of functions in $C[0, 1]$ has compact closure if and only if

i $\sup\limits_{\omega \in A} |\omega_0| < \infty$,

ii $\lim\limits_{\delta \to 0} \sup\limits_{\omega \in A} \nu_\omega(\delta) = 0$, where $\nu_\omega(\delta)$ is the *oscillation at scale of resolution* δ in $\omega \in C[0, 1]$ defined by $\nu_\omega(\delta) = \sup_{|s-t|<\delta} |\omega_s - \omega_t|$.

The condition (ii) refers to the *equicontinuity* of the functions in A in the sense that given any $\varepsilon > 0$ there is a common $\delta > 0$ such that for all functions $\omega \in A$ we have $|\omega_t - \omega_s| < \varepsilon$ if $|t - s| < \delta$. Conditions (i) and (ii) together imply that A is *uniformly bounded* in the sense that there is a number B for which

---

[4]The notations $X_t, B_t, X(t), B(t)$ are all common and used freely in this text.

$$\|\omega\| := \sup_{0 \le t \le 1} |\omega(t)| \le B \qquad \text{for all } \omega \in A.$$

This is because for N sufficiently large we have $\sup_{\omega \in A} \nu_\omega(1/N) < 1$ and, therefore, for each $0 \le t \le 1$

$$|\omega_t| \le |\omega_0| + \sum_{i=1}^{N} |\omega_{it/N} - \omega_{(i-1)t/N}| \le \sup_{\omega \in A} |\omega_0| + N \sup_{\omega \in A} \nu_\omega(1/N) = B.$$

Combining this with the Prohorov theorem gives the following criterion for tightness of probability measures $\{Q_n\}_{n \ge 1}$ on $S = C[0, 1]$.

Lemma 2 Let $\{Q_n : n \ge 1\}$ be a sequence of probability measures on $C[0, 1]$. Then $\{Q_n : n \ge 1\}$ is tight if and only if the following two conditions hold.

i For each $\eta > 0$ there is a number B such that

$$Q_n(\{\omega \in C[0, 1] : |\omega_0| > B\}) \mathbb{E} \eta, \qquad n = 1, 2, \dots.$$

ii For each $\varepsilon > 0$, $\eta > 0$, there is a $0 < \delta < 1$ such that

$$Q_n(\{\omega \in C[0, 1] : \nu_\omega(\delta) \mathbb{E} \varepsilon\}) \mathbb{E} \eta, \qquad n \ge 1.$$

Proof If $\{Q_n : n \ge 1\}$ is tight, then given $\eta > 0$ there is a compact K such that $Q_n(K) > 1 - \eta$ for all n. By the Arzela–Ascoli theorem, if $B > \sup_{\omega \in K} |\omega_0|$ then

$$Q_n(\{\omega \in C[0, 1] : |\omega_0| \ge B\}) \le Q_n(K^c) \le 1 - (1 - \eta) = \eta.$$

Also given $\varepsilon > 0$ select $\delta > 0$ such that $\sup_{\omega \in K} \nu_\omega(\delta) < \varepsilon$. Then

$$Q_n(\{\omega \in C[0, 1] : \nu_\omega(\delta) \ge \varepsilon\}) \le Q_n(K^c) < \eta \qquad \text{for all } n \ge 1.$$

The converse goes as follows. Given $\eta > 0$, first select B using (i) such that $Q_n(\{\omega : |\omega_0| \le B\}) \ge 1 - \frac{1}{2}\eta$, for $n \ge 1$. Select δ_r using (ii) such that $Q_n(\{\omega : \nu_\omega(\delta_r) < 1/r\}) \ge 1 - 2^{-(r+1)}\eta$ for $n \ge 1$. Now take K to be the closure of

$$\{\omega : |\omega_0| \le B\} \cap \bigcap_{r=1}^{\infty} \left\{ \omega : \nu_\omega(\delta_r) < \frac{1}{r} \right\}.$$

Then $Q_n(K) > 1 - \eta$ for $n \ge 1$, and K is compact by the Arzela–Ascoli theorem. ∎

The next lemma provides a moment inequality that can be useful for checking such Arzela-Ascoli type conditions.

Lemma 3 Let $\Lambda \subset \mathbb{R}^k$ be a bounded rectangle, and let $X^{(n)} = \{X_u^{(n)} : u \in \Lambda\}$, $n \geq 1$, be a sequence of continuous processes with values in a complete metric space (S, ρ) satisfying

$$\mathbb{E}\rho^\alpha(X_u^{(n)}, X_v^{(n)}) \leq c|u - v|^{k+\beta}, \quad \text{for all } u, v \in \Lambda, n \geq 1,$$

for some positive numbers c, α, β. Then, for every given $\varepsilon > 0$ and $0 < \eta < 1$, there is a $\delta > 0$ such that

$$P(\sup\{\rho(X_u^{(n)}, X_v^{(n)}) : u, v \in \Lambda, |u - v| \leq \delta\} > \varepsilon) < \eta, \quad \text{for all } n \geq 1.$$

In particular, for the tightness of the distributions of $X^{(n)} = \{X_t^{(n)} : 0 \leq t \leq 1\}$, $n \geq 1$, it is sufficient that there be positive numbers α, β, M such that

$$\mathbb{E}|X_t^{(n)} - X_s^{(n)}|^\alpha \leq M|t - s|^{1+\beta} \qquad \text{for all } s, t, n. \tag{7.16}$$

Proof Without loss of generality take $\Lambda = [0, 1]^k$. Let $|\cdot|$ denote the *maximum norm* given by $|u| = \max\{|u_i| : 1 \leq i \leq k\}$, $u = (u_1, \dots, u_k)$. For each $N = 1, 2, \dots$, let L_N be the finite lattice $\{j2^{-N} : j = 0, 1, \dots 2^N\}^k$. Write $L = \cup_{N=1}^\infty L_N$. Define $M_N^{(n)} = \max\{\rho(X_u^{(n)}, X_v^{(n)}) : (u, v) \in L_N^2, |u - v| \leq 2^{-N}\}$. Since (i) for a given $u \in L_N$ there are no more than 3^k points in L_N such that $|u - v| \leq 2^{-N}$, (ii) there are $(2^N + 1)^k$ points in L_N, and (iii) for every given pair (u, v), the condition of the theorem holds, one has by Chebyshev's inequality that for $\gamma < \beta/\alpha$

$$P(M_N^{(n)} > 2^{-\gamma N}) \leq c3^k(2^N + 1)^k\left(\frac{2^{-N(k+\beta)}}{2^{-\alpha\gamma N}}\right). \tag{7.17}$$

In particular, since $\gamma < \beta/\alpha$,

$$\sum_{N=1}^\infty P(M_N^{(n)} > 2^{-\gamma N}) < \infty. \tag{7.18}$$

Thus there is a random positive integer $N^* \equiv N^*(\omega)$ and a set Ω^* with $P(\Omega^*) = 1$, such that

$$M_N^{(n)}(\omega) \leq 2^{-\gamma N} \quad \text{for all } N \geq N^*(\omega), \omega \in \Omega^*. \tag{7.19}$$

Fix $\omega \in \Omega^*$ and let $N \geq N^*(\omega)$. We will see by induction that, for all $m \geq N + 1$, one has

$$\rho(X_u^{(n)}, X_v^{(n)}) \leq 2\sum_{j=N}^m 2^{-\gamma j}, \quad \text{for all } u, v \in L_m, |u - v| \leq 2^{-N}. \tag{7.20}$$

For $m = N$ this follows from (7.19). Suppose, as an induction hypothesis, that (7.20) holds for $m = N + 1, \dots, M$. Let $u, v \in L_{M+1}, |u - v| \leq 2^{-N}$. Write $u =$

$(i_1 2^{-M-1}, \ldots, i_k 2^{-M-1})$, $v = (j_1 2^{-M-1}, \ldots, j_k 2^{-M-1})$, where $i_\nu, j_\nu, 1 \le \nu \le k$, belong to $\{0, 1, 2, \ldots, 2^{n+1}\}$. We will find $u^*, v^* \in L_n$ such that $|u - u^*| \le 2^{-n-1}$, $|v - v^*| \le 2^{-M-1}$, and $|u^* - v^*| \le 2^{-N}$. For this let the ν-th coordinate, say $i_\nu^* 2^{-M-1}$ of u^* be the same as that of u if i_ν is even, and $i_\nu^* = i_\nu - 1$ if i_ν is odd and $i_\nu \ge j_\nu$, and $i_\nu^* = i_\nu + 1$ if i_ν is odd and $i_\nu < j_\nu$, $\nu = 1, \ldots, k$. Then $|u^* - u| \le 2^{-M-1}$, and $u^* \in L_M$ (since i_ν^* is even and $i_\nu^* 2^{-M-1} = (i_\nu^*/2) 2^{-M}$). Similarly define v^* with the roles of i_ν and j_ν interchanged, to get $v^* \in L_M$ and $|v - v^*| \le 2^{-M-1}$, with, moreover, $|u^* - v^*| \le |u - v| \le 2^{-N}$. Then by (7.19) and the induction hypothesis,

$$\rho(X_u^{(n)}, X_v^{(n)}) \le \rho(X_u^{(n)}, X_{u^*}^{(n)}) + \rho(X_{u^*}^{(n)}, X_{v^*}^{(n)}) + \rho(X_{u^*}^{(n)}, X_v^{(n)})$$

$$\le 2^{-\gamma(M+1)} + 2 \sum_{\nu=N}^{M} 2^{-\gamma\nu} + 2^{-\gamma(M+1)} = 2 \sum_{\nu=N}^{M+1} 2^{-\gamma\nu},$$

completing the induction argument for (7.20), for all $\omega \in \Omega^*$, $m \ge N + 1$, $N \ge N^*(\omega)$. Since $2 \sum_{\nu=N}^{\infty} 2^{-\gamma\nu} = 2^{-\gamma N+1}(1 - 2^{-\gamma})^{-1}$, and $L = \cup_{m=N+1}^{\infty} L_m$ for all $N \ge N^*(\omega)$, it follows that

$$\sup\{\rho(X_u^{(n)}, X_v^{(n)}) : u, v \in L, |u - v| \le 2^{-N}\}$$
$$= \sup\{\rho(X_u^{(n)}, X_v^{(n)}) : u, v \in \cup_{m=N+1}^{\infty} L_m, |u - v| \le 2^{-N}\}$$
$$\le 2^{-\gamma N+1}(1 - 2^{-\gamma})^{-1}, \quad N \ge N^*(\omega), \omega \in \Omega^*. \tag{7.21}$$

Thus given $\varepsilon > 0$, one has

$$P(\sup\{\rho(X_u^{(n)}, X_v^{(n)}) : u, v \in \Lambda, |u - v| \le \delta\} > \varepsilon)$$
$$\le \theta(N) := c 3^k (2^N + 1)^k \left(\frac{2^{-N(k+\beta)}}{2^{-\alpha\gamma N}}\right). \tag{7.22}$$

Next for $0 < \eta < 1$ find $N(\eta)$ such that, for a given $\gamma \in (0, \beta/\alpha)$, $\sum_{N=N(\eta)}^{\infty} \theta(N) < \eta$, where $\theta(N)$ is determined This provides the asserted probability bound with $(1 - 2^{-\gamma})^{-1} 2^\gamma 2^{-\gamma N(\eta)}$ in place of ε. If this last quantity is larger than ε then find $N(\varepsilon, \eta) \ge N(\eta)$ such that $(1 - 2^{-\gamma}) 2^\gamma 2^{-\gamma N(\varepsilon, \eta)} \le \varepsilon$. Then the asserted bound holds with $\delta = 2^{-N(\varepsilon, \eta)}$. In particular the tightness condition follows from Lemmas 2 and 3   ∎

To complete the extension from $C[0, 1]$ to $C[0, \infty)$ we make use of the following.

Lemma 4 Suppose that $X, X^{(n)}, n \ge 1$, are stochastic processes with values in $C[0, \infty)$ for which one has that $\{X_t^{(n)} : 0 \le t \le T\}$ converges in distribution to $\{X_t : 0 \le t \le T\}$ for each $T = 1, 2, \ldots$. Then $X^{(n)}$ converges in distribution to X as processes in $C[0, \infty)$.

Proof Since $S = C[0, \infty)$ is separable, the proof can be accomplished using the triangle inequality in Prokhorov's metrization of weak convergence as follows:

First define processes X^T and $X^{(n),T}$ with paths in $C[0, \infty)$ by $X^T_t := X_t, 0 \le t \le T$, $X^T_t := X_T, t \ge T$, and $X^{(n),T}_t := X^{(n)}_t, 0 \le t \le T$, and $X^{(n),T}_t := X^{(n)}_T, t \ge K$. Then as $n \to \infty$ one has that $X^{(n),T}$ converges in distribution to X^T in $C[0, \infty)$ by tightness of $\{X^{(n)}_t : 0 \le t \le T\}, n \ge 1$, and convergence of finite-dimensional distributions to $\{X_t : 0 \le t \le T\}$. Next one has $\{X^T_t : 0 \le t < \infty\}$ converges a.s. and hence in distribution to X.   ∎

Proof of Theorem 7.15 under finite fourth moment assumption. By the continuous operation of rescaling if necessary, one may assume $\sigma^2 = 1$ without loss of generality. This proof is given under the further restriction of finite fourth moment $m_4 = \mathbb{E}Z^4_1$. Let us consider the processes restricted to $0 \le t \le 1$. It will suffice to prove tightness of the distributions $Q_n, n \ge 1$, of the polygonal processes $\{B^{(n)}(t) : 0 \le t \le 1\} \in C[0, 1], n = 1, 2, \dots$. To wit, by Prohorov's theorem, one obtains a subsequence Q_{n_k} and a probability Q on $C[0, 1]$ such that $Q_{n_k} \Rightarrow Q$ as $k \to \infty$. But we have already seen that weak convergence of the finite-dimensional distributions of the full sequence is a consequence of classical central limit theory and hence, since finite-dimensional events uniquely determine a probability on $C[0, 1]$, we can conclude $Q_n \Rightarrow Q$ as $n \to \infty$. Moreover Q is a probability on $C[0, 1]$ with the finite-dimensional requirements of the definition of a standard Brownian motion.

We will show that there are positive numbers α, β and M such that

$$\mathbb{E}|B^{(n)}(t) - B^{(n)}(s)|^\alpha \le M|t - s|^{1+\beta} \qquad \text{for } 0 \le s, t \le 1, n = 1, 2, \dots \quad (7.23)$$

By the above lemma this will prove tightness of the distributions of the processes $\{B^{(n)}(t) : t \ge 0\}, n = 1, 2, \dots$.

To establish (7.23), take $\alpha = 4$. First consider the case $s = (j/n) < (k/n) = t$ are at the grid points. Then

$$\mathbb{E}\{B^{(n)}(t) - B^{(n)}(s)\}^4 = n^{-2}\mathbb{E}\{Z_{j+1} + \cdots + Z_k\}^4$$

$$= n^{-2} \sum_{i_1=j+1}^{k} \sum_{i_2=j+1}^{k} \sum_{i_3=j+1}^{k} \sum_{i_4=j+1}^{k} \mathbb{E}\{Z_{i_1} Z_{i_2} Z_{i_3} Z_{i_4}\}$$

$$= n^{-2}\left\{(k - j)\mathbb{E}Z^4_1 + \binom{4}{2}\binom{k-j}{2}(\mathbb{E}Z^2_1)^2\right\}. \quad (7.24)$$

Thus, in this case,

$$\mathbb{E}\{\tilde{X}^{(n)}_t - \tilde{X}^{(n)}_s\}^4 = n^{-2}\{(k - j)m_4 + 3(k - j)(k - j - 1)\}$$

$$\le n^{-2}\{(k - j)m_4 + 3(k - j)^2\} \le (m_4 + 3)\left(\frac{k}{n} - \frac{j}{n}\right)^2$$

$$\le (m_4 + 3)|t - s|^2 = c_1|t - s|^2, \text{ where } c_1 = m_4 + 3.$$

Next, consider the more general case $0 \le s, t \le 1$. Then, for $s < t$,

$$\mathbb{E}\{B^{(n)}(t) - B^{(n)}(s)\}^4 = n^{-2}\mathbb{E}\left\{\sum_{j=[ns]+1}^{[nt]} Z_j + (nt - [nt])Z_{[nt]+1} - ([ns] - ns)Z_{[ns]+1}\right\}^4$$

$$\leq n^{-2}3^4\left\{\mathbb{E}\left(\sum_{j=[ns]+1}^{[nt]} Z_j\right)^4 + (nt - [nt])^4\mathbb{E}Z_{[nt]+1}^4\right. \tag{7.25}$$

$$\left. + (ns - [ns])^4\mathbb{E}Z_{[ns]+1}^4\right\}$$

$$\leq n^{-2}3^4\{c_1([nt] - [ns])^2 + (nt - [nt])^2 m_4 + (ns - [ns])^2 m_4\}$$
$$\leq n^{-2}3^4 c_1\{([nt] - [ns])^2 + (nt - [nt])^2 + (ns - [ns])^2\}$$
$$\leq n^{-2}3^4 c_1\{([nt] - [ns]) + (nt - [nt]) + (ns - [ns])\}^2$$
$$= n^{-2}3^4 c_1\{nt - ns + 2(ns - [ns])\}^2$$
$$\leq n^{-2}3^4 c_1\{nt - ns + 2(nt - ns)\}^2$$
$$= n^{-2}3^6 c_1(nt - ns)^2 = 3^6 c_1(t - s)^2. \tag{7.26}$$

In the above, we used the fact that $(a + b + c)^4 \leq 3^4(a^4 + b^4 + c^4)$ to get the first inequality. The analysis of the first (gridpoint) case (7.24) was then used, along with the fact that for all $t \geq 0, 0 \leq nt - [nt] \leq 1$, to get the second inequality. Take $\beta = 1, M = 3^6(m_4 + 3), \alpha = 4$, in (7.16).

In view of Lemma 4 it now follows that there is a stochastic process $B = \{B(t) : t \geq 0\}$ defined on a probability space $(\Omega, \mathcal{F}, P)$ satisfying the conditions of Definition 7.6 specifying a Brownian motion on $\mathbb{R}$. In particular, there is a probability measure W defined on the Borel σ-field of $C[0, \infty)$ such that for any $0 = t_0 < t_1 < \cdots < t_k, a_1, \ldots, a_k \in \mathbb{R}, k \geq 1$, letting $F := \{\omega \in C[0, \infty) : \omega(0) = 0, \omega(t_j) - \omega(t_{j-1}) \leq a_j, j = 1, \ldots, k\}$,

$$W(F) = \int_{-\infty}^{a_k} \cdots \int_{-\infty}^{a_1} \prod_{j=1}^{k} \frac{1}{\sqrt{2\pi(t_j - t_{j-1})}} e^{\frac{(x_j - x_{j-1})^2}{2(t_j - t_{j-1})}} dx_1 \cdots dx_k.$$

This completes the proof of the functional central limit theorem under the finite fourth moment assumption.   ∎

There are two distinct types of applications of Theorem 7.15. In the first type it is used to calculate probabilities of infinite-dimensional events associated with Brownian motion by directly computing limits of distributions of functionals of the scaled simple random walks. In the second type it (invariance) is used to calculate asymptotic distribution of a large variety of partial sum processes, since the asymptotic probabilities for these are the same as those of simple random walks.

Consider the functional $g(\omega) := \max_{0 \leq t \leq 1} \omega(t), \omega \in C[0, \infty)$. As an application of the FCLT one may obtain the following limit distribution: Let $Z_1, Z_2, \ldots$ be an i.i.d. sequence of real-valued random variables standardized to have mean zero, variance one. Let $S_n := Z_1 + \cdots + Z_n, n \geq 1$. Then $g(\tilde{X}^{(n)})$, and therefore $\max_{0 \leq t \leq 1} X_t^{(n)}$, converges in distribution to $\max_{0 \le t \le 1} B_t$, where $\{B_t : t \ge 0\}$ is standard Brownian motion starting at 0. Thus

$$
\begin{aligned}
\lim_{n \to \infty} P(n^{-\frac{1}{2}} \max_{1 \le k \le n} S_k \ge a) &= \lim_{n \to \infty} P(\max_{0 \le t \le 1} X_t^{(n)} \ge a) \\
&= \lim_{n \to \infty} P(g(\tilde{X}^{(n)}) \ge a) \\
&= P(\max_{0 \le t \le 1} B_t \ge a).
\end{aligned}
\tag{7.27}
$$

For this calculation to be complete one needs to check that the Wiener measure of the boundary of the set $G = \{\omega \in C[0, \infty) : \max_{0 \le t \le 1} \omega(t) \ge a\}$ is zero. This follows from the fact that the extremal random variable $M = \max_{0 \le t \le 1} B_t$ has a density, and consequently, $P(M = a) = 0$; the derivation of the density of M using the strong Markov property (reflection principle) is postponed until Chapter X.

One may note that another point of view is possible in which the FCLT is used to obtain formulae for Brownian motion by making the special choice of simple symmetric random walk for $Z_1, Z_2, \ldots$, do the combinatorics and then pass to the appropriate limit. Both of these perspectives are quite useful.

It is often useful to recognize that it is sufficient that $g : C[0, \infty) \to \mathbb{R}$ be only a.s. continuous with respect to the limiting distribution for the FCLT to apply, i.e., for the convergence of $g(X^{(n)})$ in distribution to $g(X)$. That is, as an immediate consequence of the Mann–Wald Theorem 7.4, one has

Proposition 7.16 If $X^{(n)} := \{X_t^{(n)} : t \ge 0\}$ converges in distribution to $X := \{X_t : t \ge 0\}$ and if $P(X \in D_g) = 0$, where $D_g = \{\mathbf{x} \in C[0, \infty) : g$ is discontinuous at $\mathbf{x}\}$, then $g(X^{(n)})$ converges in distribution to $g(X)$.

Before closing this chapter on weak convergence, let us consider the following alternative metrization of the topology of weak convergence of probabilities on a separable metric space.[5] This particular metric has utility in applications to time asymptotic behavior of Markov chains that is illustrated in Chapter XIII. To state the result it is convenient to have the following notation for the oscillation of a function on a set:

Definition 7.7 (*Oscillation of Function*) Let g be an arbitrary function on a set S. The *oscillation of g on a set* $A \subset S$ is defined by $\omega_g(A) := \sup_{x,y \in A} |g(x) - g(y)|$.

Theorem 7.17 (*Bounded-Lipschitz Metric*) Let (S, ρ) be a separable metric space with Borel sigma field $\mathcal{S} = \mathcal{B}(S)$. Define $d_{BL} : \mathcal{P}(S) \times \mathcal{P}(S) \to [0, \infty)$ by

$$
d_{BL}(\mu, \nu) = \sup\{| \int_S f d\mu - \int_S f d\nu | : \omega_f(S) \le 1, \sup_{x \ne y} \frac{\rho(f(x), f(y))}{\rho(x, y)} \le 1\},
$$

---

[5]While the result here is merely that convergence in the bounded-Lipschitz metric implies weak convergence, the converse is also true. For a proof of this more general result see Bhattacharya and Majumdar (2007), pp. 232–234. This metric was originally studied by Dudley, R.M. (1968): Distances of probability measures and random variables, *Ann. Math.* **39**, 15563–1572.

for $\nu, \mu \in \mathcal{P}(S)$. Then d_{BL} is a metric, referred to as the *bounded-Lipschitz metric*. Moreover, convergence in the bounded-Lipschitz metric implies weak convergence. From here it is straightforward to check that d_{BL} defines a metric on $\mathcal{P}(S)$ as asserted. In particular, $d_{BL}(Q_1, Q_2) = 0$ if and only if $Q_1 = Q_2$ since the above argument shows this class to be measure-determining as well.

Proof Suppose that $d_{BL}(Q_m \Rightarrow Q) \to 0$ as $m \to \infty$. Now, for each integer $n \geq 1$, one may define a convergence-determining sequence $f_n : S \to \mathbb{R}$ as in Proposition 1.6 in Chapter I. But each of these is bounded Lipschitz with Lipschitz constant n. Rescaling by $g_n = f_n / n \vee \omega_{f_n}(S)$. Then, applying the hypothesis to each g_n, one has

$$\lim_{m \to \infty} | \int_S f_n dQ_m - \int_S f_n dQ| = \lim_{m \to \infty} (n \vee \omega_{g_n}(S))| \int_S g_n dQ_m - \int_S g_n dQ| = 0.$$

Since the sequence $\{f_n : n \geq 1\}$ has already been shown to be convergence-determining the proof is complete.   ∎

Remark 7.5 One may note from the proof that the uniformity over $\mathcal{P}(S)$ defining the metric is not actually necessary to obtain weak convergence. On the other hand, the converse (see Footnote) does imply the uniformity contained in the metric.

Exercise Set VII

1. (i) Show that if F is closed, $\delta > 0$, then $\partial\{x : \rho(x, F) \leq \delta\} \subset \{x : \rho(x, F) = \delta\}$. [*Hint*: If y belongs to the set on the left, there is a sequence $y_n \to y$ such that $\rho(y_n, F) \geq \delta$.] (ii) Let $(\Omega, \mathcal{F}, P)$ be an arbitrary probability space. Suppose A_δ, $\delta > 0$, is a collection of disjoint measurable sets. Show that $P(A_\delta) > 0$ for at most countably many δ. [*Hint*: For each positive integer n, the set $\{\delta > 0 : P(A_\delta) > 1/n\}$ must be a finite set.] (iii) Let $h : S_1 \to S_2$ be Borel measurable and P-a.s. continuous. With F_h as in Theorem 7.4, show that $F_h^- \subset F_h \cup D_h$. [*Hint*: If $y \in F_h^- \backslash F_h \subset \partial F_h$, then $h(y) \notin F$, but there is a sequence $y_n \to y$ such that $h(y_n) \in F$ for all $n \geq 1$.]
2. Let $\{X_n\}_{n=1}^\infty$ be a sequence of random maps with values in a metric space S with metric ρ and Borel σ-field $\mathcal{S} = \mathcal{B}(S)$.

    (i) Show that X_n converges in probability to an a.s. constant c if and only if the sequence of probabilities $Q_n := P \circ X_n^{-1}$ converge weakly to δ_c. [Here *convergence in probability* means that given $\varepsilon > 0$ one has $P(\rho(X_n, c) > \varepsilon) \to 0$ as $n \to \infty$.]
    (ii) Show that convergence in probability to a random map X implies $P \circ X_n^{-1} \Rightarrow P \circ X^{-1}$ as $n \to \infty$.

3. Let S be a metric space with Borel σ-field $\mathcal{B}(S)$. (a) Give an example to show that vague convergence does not imply weak convergence, referred to as *escape of probability mass to infinity*. [*Hint*: Consider, for example, $Q_n = \frac{2}{3}\delta_{\{\frac{1}{n}\}} + \frac{1}{3}\delta_{\{n\}}$.] (b) Show that if $\{Q_n\}_{n=1}^\infty$ is tight, then vague convergence and weak convergence are equivalent for $\{Q_n\}_{n=1}^\infty$.

4. Let (S, ρ) be a compact metric space. Give a proof of Proposition 7.6. [*Hint*: Let $\{f_n\}$ be a countable dense sequence in $C(S)$. For a sequence of probabilities Q_n, first consider the bounded sequence of numbers $\int_S f_1 dQ_n, n \geq 1$. Extract a subsequence $Q_{n_{1k}}$ such that $L(f_1) := \lim \int_S f_1 dQ_{n_{1k}}$ exists. Next consider the bounded subsequence $\int_S f_2 dQ_{n_{1k}}$, etc. Use Cantor's diagonalization to obtain a densely defined bounded linear functional L (on the linear span of $\{f_k : k \geq 1\}$) and extend by continuity to $C(S)$. Use the Riesz representation theorem (Appendix A) to obtain the weak limit point.]

5. Let $\{X_n\}_{n=1}^{\infty}$ be a sequence of real-valued random variables on $(\Omega, \mathcal{F}, P)$.

   (i) Suppose that each X_n is in $L^p, n \geq 1$, for some $p \geq 1$, and $\sup_n \mathbb{E}|X_n|^p < \infty$. Show that $\{Q_n = P \circ X_n^{-1}\}_{n=1}^{\infty}$ is a tight sequence. [*Hint*: Use a Chebyshev-type inequality.]

   (ii) Suppose there is a $\delta > 0$ such that for each $-\delta \leq t \leq \delta$, $\mathbb{E}e^{tX_n} < \infty$ for each n, and $\lim_{n\to\infty} \mathbb{E}e^{tX_n} = m(t)$ exists and is finite. Show that $\{Q_n = P \circ X_n^{-1}\}_{n=1}^{\infty}$ is tight. [*Hint*: Apply the Markov inequality to the event $[e^{\delta|X_n|} > e^{\delta a}]$.]

6. Define probabilities on $\mathbb{R}$ absolutely continuous with respect to Lebesgue measure with density $Q_\varepsilon(dx) = \rho_\varepsilon(x)dx$, where $\rho_\varepsilon(x)$ was introduced to obtain C^∞-approximations with compact support in (4.2). Let $\delta_{\{0\}}$ denote the Dirac probability concentrated at 0, and show that $Q_\varepsilon \Rightarrow \delta_{\{0\}}$ as $\varepsilon \downarrow 0$. [*Hint*: Consider probabilities of open sets in Alexandrov's theorem.]

7. Suppose that $Q_n, n \geq 1$, is a sequence of probabilities concentrated on $[a, b]$. Suppose that one may show for each positive integer r that $\int_{[a,b]} x^r Q_n(dx) \to m_r \in \mathbb{R}$ as $n \to \infty$. Show that there is a probability Q such that $Q_n \Rightarrow Q$ as $n \to \infty$ and $\int_{[a,b]} x^r Q(dx) = m_r$ for each $r \geq 1$.

8. Let (S, ρ) be a metric space and B a Borel subset of S given the relative (metric) topology. Let $\{Q_n : n \geq 1\}$ be a sequence of probabilities in $\mathcal{P}(S)$ such that $Q_n(B) = 1$ for all n. If the restrictions of $Q_n, n \geq 1$, to B converge weakly to a probability $Q \in \mathcal{P}(B)$, show that $Q_n \Rightarrow Q$, when considered in $\mathcal{P}(S)$, i.e., extending Q to S by setting $Q(S \backslash B) = 0$.

9. Complete the following steps to prove the equivalence of (a) and (d) of Theorem 4.1 in the case $k \geq 2$.

   (i) Show that F is *continuous from above* at $\mathbf{x}$ in the sense that given $\varepsilon > 0$ there is a $\delta > 0$ such that $|F(\mathbf{x}) - F(\mathbf{y})| < \varepsilon$ whenever $x_i \leq y_i < x_i + \delta, i = 1, \ldots, k$. [*Hint*: Use the continuity of probability measures from above.]

   (ii) Say that F is *continuous from below* at $\mathbf{x}$ if given $\varepsilon > 0$ there is a $\delta > 0$ such that $|F(\mathbf{x}) - F(\mathbf{y})| < \varepsilon$ whenever $x_i - \delta < y_i \leq x_i, i = 1, \ldots, k$. Show that $\mathbf{x}$ is a continuity point of F if and only if continuity holds from below. Moreover, $\mathbf{x}$ is a continuity point of F if and only if $F(\mathbf{x}) = Q(\cap_{i=1}^{k}\{\mathbf{y} \in \mathbb{R}^k : y_i < x_i\})$, where Q is the probability measure whose distribution function is F.

   (iii) Show that $\mathbf{x}$ is a continuity point of F if and only if $\cap_{i=1}^{k}\{\mathbf{y} \in \mathbb{R}^k : y_i \leq x_i\}$ is a Q-continuity set; i.e., its boundary has probability zero. [*Hint*: The boundary of $\cap_{i=1}^{k}\{\mathbf{y} \in \mathbb{R}^{k} : y_i \leq x_i\}$ is the relative complement $\cap_{i=1}^{k}\{\mathbf{y} \in \mathbb{R}^{k} : y_i \leq x_i\}\backslash\cap_{i=1}^{k}\{\mathbf{y} \in \mathbb{R}^{k} : y_i < x_i\}$.]

(iv) Show that if $Q_n \Rightarrow Q$ then $F_n(\mathbf{x}) \to F(\mathbf{x})$ at all continuity points $\mathbf{x}$ of F.

(v) Let $\mathcal{A}$ be a π-system of Borel subsets of $\mathbb{R}^{k}$, i.e., closed under finite intersections. Assume that each open subset of $\mathbb{R}^{k}$ is a finite or countable union of elements of $\mathcal{A}$. Show that if $Q_n(A) \to Q(A)$ for each $A \in \mathcal{A}$ then $Q_n \Rightarrow Q$. [*Hint:* Use the inclusion–exclusion principle to show that $Q_n(\cup_{m=1}^{N} A_m) \to Q(\cup_{m=1}^{N} A_m)$ if $A_m \in \mathcal{A}$ for $m = 1, \ldots, m$. Verify for $\varepsilon > 0$ and open $G = \cup_m A_m, A_m \in \mathcal{A}$, that there is an N such that $Q(G) - \varepsilon \leq Q(\cup_{m=1}^{N} A_m) = \lim_n Q_n(\cup_{m=1}^{N} A_m) \leq \liminf_n Q_n(G)$.]

(vi) Let $\mathcal{A}$ be a π-system of sets such that for each $x \in \mathbb{R}^{k}$ and every $\varepsilon > 0$ there is an $A \in \mathcal{A}$ such that $x \in A^{\circ} \subset A \subset B_{\varepsilon}(x) := \{y \in \mathbb{R}^{k} : |y - x| < \varepsilon\}$, where A° denotes the set of points belonging to the *interior* of A. Show that $Q_n(A) \to Q(A)$ for all $A \in \mathcal{A}$ then $Q_n \Rightarrow Q$. [*Hint:* Check that $\mathcal{A}$ satisfies the conditions required in the previous step.]

(vii) Show that if $F_n(\mathbf{x}) \to F(\mathbf{x})$ at each point $\mathbf{x}$ of continuity of F then $Q_n \Rightarrow Q$. [*Hint:* Take $\mathcal{A}$ to be the collection of sets of the form $A = \{\mathbf{x} : a_i < x_i \leq b_i, i = 1, \ldots, k\}$ for which the $2k$ $(k-1)$-dimensional hyperplanes determining each of its faces has Q-measure zero. The Q, Q_n-probabilities of $A \in \mathcal{A}$ are sums and differences of values of $F(\mathbf{x})$, $F_n(\mathbf{x})$, respectively, as $\mathbf{x}$ varies over the 2^{k} vertices of A. Moreover, vertices of $A \in \mathcal{A}$ are continuity points of F, and at most countably many parallel hyperplanes can have positive Q-measure.]

10. Use Prohorov's theorem to give a simple derivation for Exercise 9. [*Hint:* Suppose that $F_n(x) \to F(x)$ at all points x of continuity of F. Show that $\{Q_n : n \geq 1\}$ is tight, using $Q_n((a, b]) \geq F_n(b) - \sum_{i=1}^{k} F_n(b_1, \ldots, b_{i-1}, a_i, b_{i+1}, \ldots, b_k)$, $1 \leq i \leq k$, for $a_i < b_i, \forall i$, where $a = (a_1, \ldots, a_k), b = (b_1, \ldots, b_k)$.]

Chapter VIII
Random Series of Independent Summands

The convergence of an infinite series $\sum_{n=1}^{\infty} X_n$ is a tail event. Thus, by the Kolmogorov zero–one law (Theorem 5.1), if $X_1, X_2, \ldots$ is a sequence of independent random variables, the convergence takes place with probability one or zero. For a concrete example, consider the so-called random signs question for the divergent series $\sum_{n=1}^{\infty} \frac{1}{n}$. Namely, while $\sum_{n=1}^{\infty} \frac{(-1)^{n+1}}{n}$ is convergent, one might ask what happens if the signs are assigned by i.i.d. tosses of a balanced coin (see Exercise 1)?

To answer questions about almost-sure convergence of a random series, one often proceeds with an effort to show that the sequence $\{S_n = X_1 + \cdots + X_n : n \geq 1\}$ of partial sums is *not* Cauchy with probability zero. A "non-Cauchy with probability zero" statement may be formulated as follows:

$$P\left(\sup_{j,k \geq n} |S_j - S_k| \geq \varepsilon\right) \leq 2P\left(\sup_{m \geq 1} |S_{m+n} - S_n| \geq \frac{\varepsilon}{2}\right)$$

$$= 2 \lim_{N \to \infty} P\left(\max_{1 \leq m \leq N} |S_{n+m} - S_n| \geq \frac{\varepsilon}{2}\right). \quad (8.1)$$

Thus, to prove non-Cauchy with probability zero it is sufficient to show that

$$\lim_{n,N \to \infty} P\left(\max_{1 \leq m \leq N} |S_{n+m} - S_n| \geq \varepsilon\right) = 0. \quad (8.2)$$

This approach is facilitated by the use of maximal inequalities of the type found previously for martingales. At the cost of some redundancy, here is another statement and derivation of Kolmogorov's maximal inequality for sums of independent random variables.

© Springer International Publishing AG 2016
R. Bhattacharya and E.C. Waymire, *A Basic Course in Probability Theory*,
Universitext, DOI 10.1007/978-3-319-47974-3_VIII

Theorem 8.1 (*Kolmogorov's Maximal Inequality*) Let $X_1, \ldots, X_n$ be independent random variables with $\mathbb{E}X_j = 0$, $\operatorname{Var} X_j < \infty$, for $j = 1, \ldots, n$. Let $S_k = \sum_{j=1}^{k} X_j$. For $\delta > 0$ one has $P(\max_{1 \leq k \leq n} |S_k| \geq \delta) \leq \frac{\operatorname{Var} S_n}{\delta^2}$.

Proof Let $\tau = \min\{k \leq n : |S_k| \geq \delta\}$, with $\tau = \infty$ if $|S_k| < \delta$ for all $k \leq n$. Then

$$
\mathbb{E}(S_n^2) \geq \sum_{k=1}^{n} \mathbb{E}(S_n^2 \mathbf{1}_{[\tau=k]})
$$

$$
= \sum_{k=1}^{n} \mathbb{E}(\{S_k + (S_n - S_k)\}^2 \mathbf{1}_{[\tau=k]})
$$

$$
= \sum_{k=1}^{n} \mathbb{E}(\{S_k^2 + 2S_k(S_n - S_k) + (S_n - S_k)^2\} \mathbf{1}_{[\tau=k]})
$$

$$
\geq \sum_{k=1}^{n} \mathbb{E}\{S_k^2 + 2S_k(S_n - S_k)\} \mathbf{1}_{[\tau=k]}. \tag{8.3}
$$

Now observe that $[\tau = k] \in \sigma(X_1, \ldots, X_k)$ and S_k is $\sigma(X_1, \ldots, X_k)$-measurable. Thus $\mathbf{1}_{[\tau=k]}S_k$ and $S_n - S_k$ are independent. Since the latter has mean zero, the expected value of their product is zero, and the above bound reduces to

$$
\mathbb{E}S_n^2 \geq \sum_{k=1}^{n} \mathbb{E}\{S_k^2 \mathbf{1}_{[\tau=k]}\} \geq \sum_{k=1}^{n} \delta^2 P(\tau = k).
$$

Noting that $\sum_{k=1}^{n} P(\tau = k) = P(\max_{1 \leq k \leq n} |S_k| \geq \delta)$ completes the proof.   ∎

A related inequality is given by

Theorem 8.2 (*Skorokhod Maximal Inequality*) Let $X_1, X_2, \ldots, X_n$ be independent random variables. Define $S_m = \sum_{j=1}^{m} X_j$, $1 \leq m \leq n$, $S_0 = 0$. Given $\delta > 0$, let

$$
p = \max_{m \leq n} P(|S_n - S_m| > \delta) < 1.
$$

Then

$$
P(\max_{m \leq n} |S_m| > 2\delta) \leq \frac{1}{q} P(|S_n| > \delta), \quad q = 1 - p.
$$

Proof Let $\tau = \min\{m \leq n : |S_m| > 2\delta\}$, and $\tau = \infty$ if $|S_m| \leq 2\delta, \forall m \leq n$. Then $\sum_{m=1}^{n} P(\tau = m) = P(\max_{m \leq n} |S_m| > 2\delta)$, and

$$P(|S_n| > \delta) \geq \sum_{m=1}^{n} P(|S_n| > \delta, \tau = m)$$

$$\geq \sum_{m=1}^{n} P(|S_n - S_m| \leq \delta, \tau = m)$$

$$= \sum_{m=1}^{n} P(|S_n - S_m| \leq \delta) P(\tau = m)$$

$$\geq \sum_{m=1}^{n} q P(\tau = m) = q P(\max_{m \leq n} |S_m| > 2\delta).$$

∎

Theorem 8.3 (*Mean-Square-Summability Criterion*) Let $X_1, X_2, \ldots$ be independent random variables with mean zero. If $\sum_{n=1}^{\infty} \mathrm{Var}(X_n) < \infty$ then $\sum_{n=1}^{\infty} X_n$ converges a.s.

Proof Applying Kolmogorov's maximal inequality to the sum of $X_{n+1}, \ldots, X_{n+m}$ yields for arbitrary $\varepsilon > 0$,

$$P\left(\max_{1 \leq k \leq m} |S_{n+k} - S_n| > \varepsilon\right) \leq \frac{1}{\varepsilon^2} \sum_{k=1}^{m} \mathrm{Var}(X_{n+k}) \leq \frac{1}{\varepsilon^2} \sum_{k=1}^{\infty} \mathrm{Var}(X_{n+k}).$$

Using continuity of the probability P, it follows that $P(\sup_{k \geq 1} |S_{n+k} - S_n| > \varepsilon) = \lim_{m \to \infty} P\left(\max_{1 \leq k \leq m} |S_{n+k} - S_n| > \varepsilon\right) \leq \frac{1}{\varepsilon^2} \sum_{k=1}^{\infty} \mathrm{Var}(X_{n+k})$. Since the bound is by the tail of a convergent series, one has, letting $n \to \infty$, that

$$\lim_{n \to \infty} P(\sup_{k \geq 1} |S_{n+k} - S_n| > \varepsilon) = 0.$$

It follows by the method leading up to (8.2) that the event $[\{S_n\}_{n=1}^{\infty}$ is not a Cauchy sequence] has probability zero. ∎

As a quick application of the mean-square-summability criterion one may obtain a strong law of large numbers for sums of independent centered random variables whose variances do not grow too rapidly; see Exercise 2

We will see below that it can also be employed in a proof of strong laws for rescaled averages of i.i.d. sequences under suitable moment conditions. This will use truncation arguments stemming from the following further consequence; also see Exercises 4, 5.

Corollary 8.4 (*Kolmogorov's Three-Series Criteria: Sufficiency Part*) Let $X_1, X_2, \ldots$ be independent random variables. Suppose that there is a (truncation level) number $a > 0$ such that the following three-series converge: (i) $\sum_{n=1}^{\infty} P(|X_n| > a)$; (ii) $\sum_{n=1}^{\infty} \mathbb{E}(X_n \mathbf{1}_{[|X_n| \leq a]})$; (iii) $\sum_{n=1}^{\infty} \mathrm{Var}(X_n \mathbf{1}_{[|X_n| \leq a]})$. Then $\sum_{n=1}^{\infty} X_n$ converges with probability one.

Proof Convergence of (i) implies that the truncated and nontruncated series converge and diverge together, since by Borel–Cantelli I, they differ by at most finitely many terms with probability one. In view of the mean-square-summability criterion, part (iii) gives a.s. convergence of the centered truncated sum, and adding (ii) gives the convergence of the uncentered truncated sum.   ∎

Remark 8.1   Note that Corollary 8.4 holds if the truncation level $a = a_n$ depends on n.

As an application of the CLT one may also establish the necessity of Kolmogorov's three-series criteria as follows.

Corollary 8.5   *(Kolmogorov's Three-Series Criteria: Necessary Part)* Let X_1, X_2, $\ldots$ be independent random variables. If $\sum_{n=1}^{\infty} X_n$ converges with probability one, then for any (truncation level) number $a > 0$ the following three-series converge: (i) $\sum_{n=1}^{\infty} P(|X_n| > a)$; (ii) $\sum_{n=1}^{\infty} \mathbb{E}[X_n \mathbf{1}_{[|X_n| \leq a]}]$; (iii) $\sum_{n=1}^{\infty} \mathrm{Var}(X_n \mathbf{1}_{[|X_n| \leq a]})$.

Proof Assume that $\sum_{n=1}^{\infty} X_n$ converges a.s. and let $a > 0$. Necessity of condition (i) follows from Borel–Cantelli II. Let $S_n^{(a)} = \sum_{k=1}^{n} X_k \mathbf{1}_{[|X_k| \leq a]}$, and $\sigma_n^2(a) = \mathrm{Var}(S_n^{(a)})$, $\mu_n(a) = \mathbb{E}S_n^{(a)}$. Suppose for the sake of contradiction of (iii) that $\sigma_n(a) \to \infty$. Then, since $S_n^{(a)}$ converges a.s. to a finite limit, $S_n^{(a)}/\sigma_n(a) \to 0$ a.s. as $n \to \infty$, and hence in probability as well. However, since the terms $X_k \mathbf{1}_{[|X_k| \leq a]} - \mathbb{E}\{X_k \mathbf{1}_{[|X_k| \leq a]}\}$ are uniformly bounded, one may use Lindeberg's central limit theorem to compute for an arbitrary interval $J = (c, d]$, $c < d$, and conclude that

$$P\left(\frac{S_n^{(a)} - \mu_n(a)}{\sigma_n(a)} \in J, \frac{|S_n^{(a)}|}{\sigma_n(a)} < 1\right) \geq P\left(\frac{S_n^{(a)} - \mu_n(a)}{\sigma_n(a)} \in J\right) - P\left(\frac{|S_n^{(a)}|}{\sigma_n(a)} \geq 1\right)$$

is bounded away from zero for all sufficiently large n. This is a contradiction since it implies that for sufficiently large n, the numbers $-\mu_n(a)/\sigma_n(a)$ are between $c - 1$ and $d + 1$ for two distinct choices of intervals J more than 2 units apart. Thus condition (iii) holds. The necessity of condition (ii) now follows by applying the mean-square-summability criterion, Theorem 8.3, to see that $\sum_{n=1}^{\infty}\{X_n \mathbf{1}_{[|X_n| \leq a]} - \mu_n(a)\}$ is a.s. convergent. Thus $\sum_{n=1}^{\infty} \mu_n(a)$ must converge.   ∎

In preparation for an extension[1] of the strong law of large numbers, we record here two very basic facts pertaining to the ordinary "calculus of averages"; their proofs are left as Exercise 3.

Lemma 1   Let $\{c_n\}_{n=1}^{\infty}$ be a sequence of positive real numbers such that $c_n \uparrow \infty$ as $n \to \infty$. Let $\{a_n\}_{n=1}^{\infty}$ be an arbitrary sequence of real numbers. (a) If $a_n \to a$ as $n \to \infty$, then defining $c_0 = 0$,

---

[1] Theorem 8.6 is a stronger statement than Kolmogorov's classical strong law. It is due to Marcinkiewicz and Zygmund (1937): Sur les fonctions indépendentes, *Fund. Math.* **29**, 60–90., but clearly contains the classical law as a special case.

$$[\text{Cesàro}] \qquad \lim_{n \to \infty} \frac{1}{c_n} \sum_{j=1}^{n} (c_j - c_{j-1}) a_j = a.$$

(b) If $\sum_{j=1}^{\infty} \frac{a_j}{c_j}$ converges then

$$[Kronecker] \qquad \lim_{n \to \infty} \frac{1}{c_n} \sum_{j=1}^{n} a_j = 0.$$

Theorem 8.6 *(Strong Law of Large Numbers)* Let $X_1, X_2, \ldots$ be an i.i.d. sequence of random variables, and let $0 < \theta < 2$. Then $n^{-\frac{1}{\theta}} \sum_{j=1}^{n} X_j$ converges a.s. if and only if $\mathbb{E}|X_1|^{\theta} < \infty$ and either (i) $\theta \leq 1$ or (ii) $\theta > 1$ and $\mathbb{E}X_1 = 0$. When the limit exists it is $\mathbb{E}X_1$ in the case $\theta = 1$, and is otherwise zero for all other cases of $\theta \in (0, 2)$, $\theta \neq 1$.

Proof The case $\theta = 1$ was considered in Chapter V (also see Exercise 5). Fix $\theta \in (0, 2)$, $\theta \neq 1$, and assume $\mathbb{E}|X_1|^{\theta} < \infty$. For the cases in which $\theta > 1$, one has $\mathbb{E}|X_1| < \infty$, but assume first that $\mathbb{E}X_1 = 0$ in such cases. We will show that $n^{-\frac{1}{\theta}} \sum_{j=1}^{n} X_j \to 0$ a.s. as $n \to \infty$.

The basic idea for the proof is to use "truncation methods" as follows: Let $Y_n = X_n \mathbf{1}_{\left[|X_n| \leq n^{\frac{1}{\theta}}\right]}$, $n \geq 1$. Then it follows from the identical distribution and moment hypothesis, using Borel–Cantelli lemma I, that $P(Y_n \neq X_n i.o) = 0$ since

$$\sum_{n=1}^{\infty} P(Y_n \neq X_n) = \sum_{n=1}^{\infty} P(|X_1|^{\theta} > n) \leq \int_0^{\infty} P(|X_1|^{\theta} > x) dx = \mathbb{E}|X_1|^{\theta} < \infty.$$

Thus, it is sufficient to show that $n^{-\frac{1}{\theta}} \sum_{k=1}^{n} Y_k$ a.s. converges to zero. In view of Kronecker's lemma, for this one needs only to show that $\sum_{n=1}^{\infty} n^{-\frac{1}{\theta}} Y_n$ is a.s. convergent. If $\theta < 1$, then this follows by the direct calculation that

$$\mathbb{E} \sum_{n=1}^{\infty} n^{-\frac{1}{\theta}} |Y_n| = \sum_{n=1}^{\infty} n^{-\frac{1}{\theta}} \mathbb{E}|X_n| \mathbf{1}_{[|X_n| \leq n^{\frac{1}{\theta}}]}$$

$$\leq \int_0^{\infty} x^{-\frac{1}{\theta}} \mathbb{E}|X_1| \mathbf{1}_{[|X_1| \leq x^{\frac{1}{\theta}}]} dx$$

$$= \mathbb{E} \left\{ |X_1| \int_{|X_1|^{\theta}}^{\infty} x^{-\frac{1}{\theta}} dx \right\} \leq c \mathbb{E}|X_1|^{\theta} < \infty,$$

for a positive constant c. Thus $n^{-\frac{1}{\theta}} \sum_{n=1}^{\infty} Y_n$ is a.s. absolutely convergent for $\theta < 1$. For $\theta > 1$, using the three-series theorem (and Remark 8.1), or Theorem 8.3, it suffices to check that $\sum_{n=1}^{\infty} \mathbb{E} \frac{Y_n}{n^{\frac{1}{\theta}}}$ is convergent, and $\sum_{n=1}^{\infty} n^{-\frac{2}{\theta}} \text{Var}(Y_n) < \infty$. For the first of these, noting that $\mathbb{E}Y_n = -\mathbb{E}X_n \mathbf{1}_{[|X_n| > n^{\frac{1}{\theta}}]}$, one has

$$\sum_{n=1}^{\infty} n^{-\frac{1}{\theta}} |\mathbb{E} Y_n| \le \sum_{n=1}^{\infty} n^{-\frac{1}{\theta}} \mathbb{E} |X_n| \mathbf{1}_{[|X_n| > n^{\frac{1}{\theta}}]}$$

$$\le \int_0^{\infty} x^{-\frac{1}{\theta}} \mathbb{E} |X_1| \mathbf{1}_{[|X_1| > x^{\frac{1}{\theta}}]} dx$$

$$= \mathbb{E} \left\{ |X_1| \int_0^{|X_1|^{\theta}} x^{-\frac{1}{\theta}} dx \right\} \le c' \mathbb{E} |X_1|^{\theta} < \infty,$$

for some constant c'. Similarly, for the second one has

$$\sum_{n=1}^{\infty} n^{-\frac{2}{\theta}} \mathrm{Var}(|Y_n|) \le \sum_{n=1}^{\infty} n^{-\frac{2}{\theta}} \mathbb{E} |Y_n|^2$$

$$= \sum_{n=1}^{\infty} n^{-\frac{2}{\theta}} \mathbb{E} |X_n|^2 \mathbf{1}_{[|X_n| \le n^{\frac{1}{\theta}}]}$$

$$\le \int_0^{\infty} x^{-\frac{2}{\theta}} \mathbb{E} |X_1|^2 \mathbf{1}_{[|X_1|^2 \le x^{\frac{1}{\theta}}]} dx$$

$$= \mathbb{E} \left\{ |X_1|^2 \int_{|X_1|^{\theta}}^{\infty} x^{-\frac{2}{\theta}} dx \right\} \le c' \mathbb{E} |X_1|^{\theta} < \infty.$$

For the converse, suppose that $n^{-\frac{1}{\theta}} \sum_{j=1}^{n} X_j$ is a.s. convergent. Let $S_n := \sum_{j=1}^{n} X_j$. Since a.s.

$$\frac{X_n}{n^{\frac{1}{\theta}}} = \frac{S_n}{n^{\frac{1}{\theta}}} - \left(\frac{n-1}{n} \right)^{\frac{1}{\theta}} \frac{S_{n-1}}{(n-1)^{\frac{1}{\theta}}} \to 0$$

as $n \to \infty$, it follows that

$$\mathbb{E} |X_1|^{\theta} = \int_0^{\infty} P(|X_1|^{\theta} > x) dx \le 1 + \sum_{n=1}^{\infty} P(|X_1|^{\theta} > n) = 1 + \sum_{n=1}^{\infty} P(|n^{-\frac{1}{\theta}} X_n| > 1).$$

The last series converges in view of Borel-Cantelli's Lemma II. In the case that $\theta > 1$, one may further conclude that $\mathbb{E} X_1 = 0$ in view of the strong law of large numbers. ∎

Remark 8.2 For $0 < \theta < 1$, the proof of almost-sure convergence of $n^{-\frac{1}{\theta}} \sum_{j=1}^{n} X_j$ does *not* require independence of the X_j, $j \ge 1$, but only that they have the same distribution. However, the proof of the converse does make use of the independence assumption.

Proposition 8.7 (*Almost-Sure & Convergence in Probability for Series of Independent Terms*) Let $X_1, X_2, \ldots$ be independent random variables. Then $\sum_{n=1}^{\infty} X_n$ converges a.s. if and only if $\sum_{n=1}^{\infty} X_n$ converges in probability.

Proof One part is obvious since almost-sure convergence always implies convergence in probability. For the converse suppose, for contradiction, that $\lim_n \sum_{j=1}^{n} X_j$ exists in probability, but with positive probability is divergent. Then there is an $\varepsilon > 0$ and a $\gamma > 0$ such that for any fixed k, $P(\sup_{n>k} |S_n - S_k| > \varepsilon) > \gamma$. Use Skorokhod's maximal inequality (Theorem 8.2) to bound $P(\max_{k<n\leq m} |S_n - S_k| > \varepsilon)$ for fixed k, m. Then note that $p = p_{k,m} := \max_{k<n\leq m} P(|S_m - S_{n-1}| > \varepsilon/2) \to 0$ as $k, m \to \infty$, while $|S_m - S_k| \to 0$ in probability as $k, m \to \infty$. This indicates a contradiction. ∎

Proposition 8.8 (*Almost-Sure and Convergence in Distribution for Series of Independent Summands*) Let $\{X_n : n \geq 1\}$ be a sequence of independent real-valued random variables. Then $\sum_{k=1}^{n} X_k$ converges a.s. as $n \to \infty$ if and only if it converges in distribution.

Proof One way follows from the dominated convergence theorem using characteristic functions. For the other assume $\sum_{k=1}^{n} X_k$ converges in distribution to Y. Then, letting $\varphi_k(\xi) = \mathbb{E}e^{i\xi X_k}$, $\varphi(\xi) = \mathbb{E}e^{i\xi Y}$, one has $\prod_{k=1}^{n} \varphi_k(\xi) \to \varphi(\xi)$ as $n \to \infty$. Thus $\prod_{k=m}^{n} \varphi_k(\xi) \to 1$ for all ξ as $m, n \to \infty$. For every $\varepsilon > 0$, one has $P(|\sum_{k=m}^{n} X_k| > 2\varepsilon) \leq \varepsilon \int_{[-\frac{1}{\varepsilon}, \frac{1}{\varepsilon}]} (1 - \prod_{k=m}^{n} \varphi_k(\xi))d\xi \to 0$ as $m, n \to \infty$ (See the proof of Theorem 6.11). Now use Proposition 8.7 to complete the proof. ∎

Exercise Set VIII

1. (*Random Signs Problem*) Suppose that $a_1, a_2, \ldots$ is a sequence of real numbers, and $X_1, X_2, \ldots$ an i.i.d. sequence of symmetrically distributed Bernoulli ± 1-valued random variables. Show that $\sum_{n=1}^{\infty} X_n a_n$ converges with probability one if and only if $\sum_{n=1}^{\infty} a_n^2 < \infty$. [*Hint*: Use mean-square-summability in one direction and a Kolmogorov's three-series theorem for the other.]

2. (*A Strong Law of Large Numbers*) Use the mean-square-summability criterion to formulate and prove a strong law of large numbers for a sequence of independent random variables $X_1, X_2, \ldots$ such that $\mathbb{E}X_n = 0$ for each $n \geq 1$, and $\sum_{n=1}^{\infty} \frac{\mathbb{E}X_n^2}{n^2} < \infty$. Assuming independence, for what values of θ does one have this strong law with $\text{Var}(X_n) = n^\theta$?

3. (*Cesàro Limits and Kronecker's Lemma*) Give a proof of the Cesàro and Kronecker lemmas. [*Hint*: For the Cesàro limit, let $\varepsilon > 0$ and choose N sufficiently large that $a + \varepsilon > a_j > a - \varepsilon$ for all $j \geq N$. Consider lim sup and lim inf in the indicated average. For Kronecker's lemma make a "summation by parts" to the indicated sum, and apply the Cesàro limit result.]

4. (*Kolmogorov's Truncation Method*) Let $X_1, X_2, \ldots$ be an i.i.d. sequence of random variables with $\mathbb{E}|X_1| < \infty$. Define $Y_n = X_n \mathbf{1}_{[|X_n|\leq n]}$, for $n \geq 1$. Show that in the limit as $n \to \infty$, (a) $\mathbb{E}Y_n \to \mathbb{E}X_1$; (b) $P(Y_n \neq X_n i.o.) = 0$; and (c) $\sum_{n=1}^{\infty} \frac{\text{Var}(Y_n)}{n^2} < \infty$. [*Hint*: For (a), Lebesgue's dominated convergence; for (b), Borel–Cantelli I; for (c), $\text{Var}(Y_n) \leq \mathbb{E}Y_n^2 = \mathbb{E}\{X_1^2 \mathbf{1}[|X_1| \leq n]\}$, and $\sum_{n=1}^{\infty} \frac{1}{n^2} \mathbb{E}X_1^2 \mathbf{1}_{[|X_1|\leq n]} \leq \mathbb{E}X_1^2 \int_{|X_1|}^{\infty} x^{-2}dx = \mathbb{E}|X_1|.$]

5. (*A strong law of large numbers*) Use Kolmogorov's truncation method from the previous exercise together with Exercise 2 to prove the classic strong law for i.i.d. sequences having finite first moment.

Chapter IX
Kolmogorov's Extension Theorem and Brownian Motion

Suppose a probability measure Q is given on a product space $\Omega = \prod_{t \in \Lambda} S_t$ with the product σ-field $\mathcal{F} = \otimes_{t \in \Lambda} \mathcal{S}_t$. Let $\mathcal{C}$ denote the class of all **finite-dimensional cylinders** C of the form

$$C = \left\{ \omega = (x_t, t \in \Lambda) \in \prod_{t \in \Lambda} S_t : (x_{t_1}, x_{t_2}, \dots, x_{t_n}) \in B \right\}, \qquad (9.1)$$

for $n \geq 1$, $B \in \mathcal{S}_{t_1} \otimes \cdots \otimes \mathcal{S}_{t_n}$, and $(t_1, t_2, \dots, t_n)$ an arbitrary n-tuple of distinct elements of Λ. Since $\otimes_{t \in \Lambda} \mathcal{S}_t$ is the smallest σ-field containing $\mathcal{C}$, it is simple to check from the $\pi - \lambda$ theorem that Q is determined by its values on $\mathcal{C}$. Write $\mu_{t_1, t_2, \dots, t_n}$ for the probability measure on the product space $(S_{t_1} \times S_{t_2} \times \cdots \times S_{t_n}, \mathcal{S}_{t_2} \otimes \mathcal{S}_{t_2} \otimes \cdots \otimes \mathcal{S}_{t_n})$ given by

$$\mu_{t_1, t_2, \dots, t_n}(B) := Q(C) \qquad (B \in \mathcal{S}_{t_1} \otimes \cdots \otimes \mathcal{S}_{t_n}), \qquad (9.2)$$

where $C \in \mathcal{C}$ is of the form (9.1) for a given n-tuple $(t_1, t_2, \dots, t_n)$ of distinct elements in Λ. Note that this collection of **finite-dimensional distributions** $\mathcal{P}_f := \{\mu_{t_1, t_2, \dots, t_n} : t_i \in \Lambda, t_i \neq t_j \text{ for } i \neq j, n \geq 1\}$ satisfies the following so-called **consistency properties**.

(a) For any n-tuple of distinct elements $(t_1, t_2, \dots, t_n)$, $n \geq 1$, and all permutations $(t'_1, t'_2, \dots, t'_n) = (t_{\pi(1)}, \dots, t_{\pi(n)})$ of $(t_1, t_2, \dots, t_n)$, $(n \geq 1)$, one has $\mu_{t'_1, t'_2, \dots, t'_n} = \mu_{t_1, t_2, \dots, t_n} \circ T^{-1}$ under the permutation of coordinates $T : S_{t_1} \times \cdots \times S_{t_n} \to S_{t'_1} \times \cdots \times S_{t'_n}$ given by $T(x_{t_1}, x_{t_2}, \dots, x_{t_n}) := (x_{t'_1}, x_{t'_2}, \dots, x_{t'_n})$, i.e., for finite-dimensional rectangles

$$\mu_{t'_1, t'_2, \dots, t'_n}(B_{t'_1} \times \cdots \times B_{t'_n}) = \mu_{t_1, t_2, \dots, t_n}(B_{t_1} \times \cdots \times B_{t_n}), B_{t_i} \in \mathcal{S}_{t_i} (1 \leq i \leq n). \qquad (9.3)$$

© Springer International Publishing AG 2016
R. Bhattacharya and E.C. Waymire, *A Basic Course in Probability Theory*,
Universitext, DOI 10.1007/978-3-319-47974-3_IX

(b) If $(t_1, t_2, \ldots, t_n, t_{n+1})$ is an $(n + 1)$-tuple of distinct elements of Λ, then $\mu_{t_1, t_2, \ldots, t_n}$ is the image of $\mu_{t_1, t_2, \ldots, t_{n+1}}$ under the projection map $\pi : S_{t_1} \times \cdots \times S_{t_{n+1}} \to S_{t_1} \times \cdots \times S_{t_n}$ given by $\pi(x_{t_1}, x_{t_2}, \ldots, x_{t_n}, x_{t_{n+1}}) := (x_{t_1}, x_{t_2}, \ldots, x_{t_n})$,

i.e., for finite-dimensional cylinders,

$$\mu_{t_1, t_2, \ldots, t_n}(B) = \mu_{t_1, t_2, \ldots, t_n, t_{n+1}}(B \times S_{t_{n+1}}) \forall B \in \mathcal{S}_{t_1} \otimes \cdots \otimes \mathcal{S}_{t_n}. \tag{9.4}$$

The theorem below, variously referred to by other names such as **Kolmogorov's existence theorem**, **Kolmogorov's consistency theorem**, or **Kolmogorov extension theorem** says, conversely, that given a family $\mathcal{P}_f$ of consistent finite-dimensional probabilities, there exists a Q on $(S, \mathcal{S})$ with these as the finite-dimensional distributions.

We present two proofs. The first is a more standard measure theoretical construction based on Caratheodory extension theorem, while the second is based on machinery of functional analysis.

Theorem 9.1 (*Kolmogorov's Extension Theorem*) Suppose $S_t, t \in \Lambda$, are Polish spaces and $\mathcal{S}_t = \mathcal{B}(S_t) \forall t \in \Lambda$. Then given any family $\mathcal{P}_f$ of finite-dimensional probabilities, $\mathcal{P}_f = \{\mu_{t_1, \ldots, t_n} : t_i \in \Lambda, t_i \neq t_j \text{ for } i \neq j, n \geq 1\}$ satisfying the consistency properties (a) and (b), there exists a unique probability Q on the product space $(\Omega = \prod_{t \in \Lambda} S_t, \mathcal{F} = \bigotimes_{t \in \Lambda} \mathcal{S}_t)$ satisfying (9.2) for all $n \geq 1$ and every $(t_1, \ldots, t_n)$ (n-tuple of distinct elements of Λ), $\mu_{t_1, t_2, \ldots, t_n}$. Moreover, the stochastic process $\mathbf{X} = (X_t : t \in \Lambda)$ defined on Ω by the coordinate projections $X_t(\omega) = x_t, \omega = (x_t, t \in \Lambda) \in \Omega$ has distribution Q.

Proof **Measure Theory Version** Let $S_{t_j} (j = 1, 2, \ldots)$ be Polish spaces and let $\mu_{t_1, t_2, \ldots, t_n}$ be a consistent sequence of probability measures on $(S_{t_1} \times \cdots \times S_{t_n}, \mathcal{S}_{t_1} \otimes \cdots \otimes \mathcal{S}_{t_n})$ ($n \geq 1$). Define a sequence of probabilities on $S = \times_{j=1}^{\infty} S_{t_j}$, with the product σ-field $\mathcal{S}$, as follows. Fix $\mathbf{x} = (x_{t_1}, x_{t_2}, \ldots) \in S$. Define $P_n(B) := \mu_{t_1, \ldots, t_n}(B_{\mathbf{x}_n^+})$, where $\mathbf{x}_n^+ = (x_{t_{n+1}}, x_{t_{n+2}}, \ldots) \in S$, and $B_{\mathbf{x}_n^+} = \{\mathbf{y} \in B : y_{t_{n+j}} = x_{t_{n+j}} : j = 1, 2, \ldots\}$ ($B \in \mathcal{S}$). Then $\{P_n : n \geq 1\}$ is tight. To see this, fix $\varepsilon > 0$. Since each individual probability on a Polish space is tight, one can get a compact set $K_{t_n} \subset S_{t_n}$ such that $x_{t_n} \in K_{t_n}$ and $\mu(K_{t_n}) > 1 - \frac{\varepsilon}{2^n}$. Then $P_n(\times_{j=1}^{\infty} K_{t_j}) > 1 - \varepsilon$. Thus, there is a sequence $n'(n \geq 1)$, and a probability measure Q such that $P_{n'} \Rightarrow Q$. This Q is the desired probability for the countably indexed case. For the more general case, assume the hypothesis of the theorem with Λ uncountable. On the field $\mathcal{C}$ of all finite-dimensional cylinders (see (9.1)) define the set function Q as in (9.2). Then, (i) Q is a measure on $\mathcal{C}$. To see this suppose that $\{C_n : n = 0, 1, \ldots\}$ is a disjoint collection in $\mathcal{C}$ whose union $C = \cup_{n=1}^{\infty} C_n \in \mathcal{C}$. Then there exists a countable set $T = \{t_j : j = 1, 2, \ldots\}$ such that $C_n, C(n \geq 1)$ belong to the σ-field $\mathcal{F}_T$ on $\Omega = \times_{t \in \Lambda} S_t$ generated by the coordinate projections $\mathbf{x} \mapsto x_{t_j}, t_j \in T$. In view of the above construction for the countable case, there is a unique extension of Q to $\mathcal{F}_T$ that is countably additive. Now, using the Caratheodory extension theorem, one sees that (ii) Q has a unique extension from $\mathcal{C}$ to the product σ-field $\otimes_{t \in \Lambda} \mathcal{S}_t$.

Functional Analysis Version First, let us consider the case[1] that each image space S_t, $t \in \Lambda$, is assumed to be a compact metric space. This makes $\Omega = \prod_{t \in \Lambda} S_t$ compact for the product topology by Tychonoff's[2] Theorem. On the Banach space $C(\Omega)$ of continuous functions on Ω with the uniform norm $\|f\| := \max_{\mathbf{x} \in \Omega} |f(\mathbf{x})|$, define a bounded linear functional h as follows: For a function $f \in C(S)$ that depends on only finitely many coordinates, say

$$f(\mathbf{x}) = \overline{f}(x_{t_1}, \ldots, x_{t_n}),$$

for some $n \geq 1$, distinct $t_1, \ldots, t_n$, and $\overline{f} : S_{t_1} \times \cdots \times S_{t_n} \to \mathbb{R}$, define

$$h(f) = \int_{S_{t_1} \times \cdots \times S_{t_n}} \overline{f} \, d\mu_{t_1,\ldots,t_n}.$$

From the consistency properties it follows that h is well-defined. By the Stone–Weierstrass theorem from real analysis, see Appendix B, the class of functions in $C(\Omega)$ depending on finitely many coordinates is dense in $C(\Omega)$. Thus h uniquely extends to a bounded (i.e., continuous) linear functional defined on all of $C(\Omega)$. Thus, one may then apply the Riesz representation theorem to obtain the desired probability Q; see Appendix A for a proof of the Riesz representation theorem for compact metric spaces S.[3] In particular, since $C(S_{t_1} \times \cdots \times S_{t_n})$ is a measure-determining class of functions on $S_{t_1} \times \cdots \times S_{t_n}$, it follows that $Q \circ \pi_{t_1,\ldots,t_n}^{-1} = \mu_{t_1\ldots t_n}$, where $\pi_{t_1\ldots t_n}(\omega) = (x_{t_1}, \ldots, x_{t_n})$, for $\omega = (x_t : t \in \Lambda)$. This establishes existence for the compact case. This proof is completed by an embedding into a compact metric space. More specifically, every Polish space S_t ($t \in \Lambda$) has a homeomorphic image $h_t(S_t)$ in a compact metric space K_t; see the Hilbert cube embedding Lemma 1, Chapter VII. So the construction of the probability given above holds on $\times_{t \in \Lambda} \overline{h_t(S_t)}$.
■

Remark 9.1 In the full generality of the specification of finite-dimensional distributions for **Kolmogorov's extension theorem**, sufficient topological assumptions are used to prove countable additivity of Q (via some compactness argument). However, for constructing an **infinite product probability measure**, or even the distribution of a discrete parameter **Markov process** with arbitrary measurable state spaces $(S_t, \mathcal{S}_t)$, $t \in \Lambda$, from specified transition probabilities and initial distribution, consistency is sufficient to prove that Q is a probability. The trade-off is that one is assuming more on the type of dependence structure for the finite-dimensional distributions. The extension theorem is referred to as **Tulcea's extension theorem**. The precise statement is as follows in the case $\Lambda = \{0, 1, \ldots\}$.

---

[1]This proof is due to Edward Nelson (1959), *Regular Probability Measures on Function Spaces,* Ann. of Math. **69**, 630–643.

[2]See Appendix B for a proof of Tychonoff's theorem for the case of countable Λ. For uncountable Λ, see Folland (1984).

[3]For general locally compact Hausdorff spaces see Folland (1984), or Royden (1988).

Theorem 9.2 *(Tulcea's Extension Theorem)* Let $(S_m, \mathcal{S}_m)$, $m = 0, 1, 2, \ldots$, be an arbitrary sequence of measurable spaces, and let $\Omega = \prod_{m=0}^{\infty} S_m$, $\mathcal{F} = \otimes_{m=0}^{\infty} \mathcal{S}_m$ denote the corresponding product space and product σ-field. Let μ_0 be a probability on $\mathcal{S}_0$ and suppose that (a) for each $n \geq 1$ and $(x_0, x_1, \ldots, x_n) \in S_1 \times \cdots \times S_n$, $B \to \mu_n(x_0, x_1, \ldots, x_{n-1}, B)$, $B \in \mathcal{S}_n$, is a probability on $\mathcal{S}_n$ and (b) for each $n \geq 1$, $B \in \mathcal{S}_n$, the map $(x_0, x_1, \ldots, x_{n-1}) \mapsto \mu_n(x_0, x_1, \ldots, x_{n-1}, B)$ is a Borel-measurable map from $\prod_{m=0}^{n-1} S_m$ into $[0, 1]$. Then there is a probability Q on $(\Omega, \mathcal{F})$ such that for each finite-dimensional cylinder set $C = B \times S_n \times S_{n+1} \times \cdots \in \mathcal{F}$, $B \in \otimes_{m=0}^{n-1} \mathcal{S}_m$ $(n \geq 1)$,

$$Q(C) = \int_{S_0} \cdots \int_{S_{n-1}} \mathbf{1}_B(x_0, \ldots, x_{n-1}) \mu_{n-1}(x_0, \ldots, x_{n-2}, dx_{n-1}) \cdots \mu_1(x_0, dx_1) \mu_0(dx_0).$$

In the case that the spaces are Polish spaces this is a consistent specification and the theorem is a special case of Kolmogorov's extension theorem. However, in the absence of topology, it stands alone. The proof[4] is essentially a matter of checking countable additivity so that the Carathéodory extension theorem may be applied.

Remark 9.2 For the case of a product probability measure $\prod_{t \in \Lambda} \mu_t$ on $(\times_{t \in \Lambda} S_t, \otimes_{t \in \Lambda} \mathcal{S}_t)$ the component probability spaces $(S_t, \mathcal{S}_t, \mu_t)$, $t \in \Lambda$, may be arbitrary measure spaces, and Λ may be uncountable. On such a space the coordinate projections $X_s(\omega) = \omega_s$, $\omega = (\omega_t : t \in \Lambda)$, define a family of independent random variables with marginal distributions μ_s $(s \in \Lambda)$.

The following example is a recasting of the content of Tulcea's theorem in the language of Markov processes whose transition probabilities are assumed to have densities.

Example 1 *(Discrete Parameter Markov Process)* Let $(S, \mathcal{S})$ be a measurable space, ν a σ-finite measure on $(S, \mathcal{S})$. Let $p(x, y)$ be a nonnegative measurable function on $(S \times S, \mathcal{S} \otimes \mathcal{S})$ such that $\int_S p(x, y)\nu(dy) = 1 \forall x \in S$. The function $p(x, y)$ is the (one-step) **transition probability density** of a Markov process $\{X_n : n = 0, 1, 2, \ldots\}$ constructed here on the infinite product space $(S^\infty, \mathcal{S}^{\otimes \infty})$ of all sequences $\mathbf{x} := (x_0, x_1, x_2, \ldots)$ in S. Here, as usual, $\mathcal{S}^{\otimes \infty}$ is the product σ-field on S^∞ generated by the class of all finite-dimensional rectangles of the form

$$C = \{x = (x_0, x_1, \ldots) \in S^\infty : x_i \in B_i \text{ for } i = 0, 1, 2, \ldots, m\}, \tag{9.5}$$

for $m \geq 1$, $B_i \in \mathcal{S}$, $i = 0, 1, \ldots, m$. For this construction, fix a probability measure μ_0 on $(S, \mathcal{S})$ and define for $B_i \in \mathcal{S}$, $i = 0, 1, \ldots, n$,

$$\mu_{0,1,2,\ldots,n}(B_0 \times B_1 \times \cdots \times B_n) \tag{9.6}$$

$$= \int_{B_0} \int_{B_1} \cdots \int_{B_n} p(x_0, x_1) p(x_1, x_2) \cdots p(x_{n-1}, x_n) \nu(dx_n) \cdots \nu(dx_1) \mu_0(dx_0).$$

---

[4]For a proof of Tulcea's theorem see Ethier and Kurtz (1986), or Neveu (1965).

More generally, $\mu_{0,1,2,\ldots n}(B)$ is defined $\forall B \in \mathcal{S}^{\otimes(n+1)}$ by integration of the function $p(x_0, x_1) \cdots p(x_{n-1}, x_n)$ over B with respect to the product measure $\mu_0 \times \nu \times \cdots \times \nu$. Since $\int_S p(x_n, x_{n+1})\nu(dx_{n+1}) = 1$, the condition (b) for *consistency* of $\mu_{0,1,\ldots,n}$, $n \geq 0$, required by Theorem 9.1 is easily checked. For integers $0 \leq m_0 < m_1 < \cdots < m_n (n \geq 0)$, the finite-dimensional probability $\mu_{m_0,\ldots,m_n}$ can then be consistently *defined* by $\mu_{m_0,\ldots,m_n} = \mu_{0,1,\ldots,m_n} \circ \pi_{m_0,\ldots,m_n}^{-1}$, where $\pi_{m_0,\ldots,m_n}(x_0, \ldots, x_{m_n})$ $:= (x_{m_0}, \ldots, x_{m_n})$, for $(x_0, \ldots, x_{m_n}) \in S^{m_n+1}$. Define $\mu_{\tau(m_0),\tau(m_1),\ldots,\tau(m_n)}$ for any given permutation τ of $(0, 1, 2, \ldots, n)$ as the induced-image measure on $(S^{m_n+1}, \mathcal{S}^{\otimes(m_n+1)})$ by $(x_0, x_1, \ldots, x_{m_n}) \mapsto (x_{m_{\tau(0)}}, x_{m_{\tau(1)}}, \ldots, x_{m_{\tau(n)}})$ on $(S^{m_n+1}, \mathcal{S}^{\otimes(m_n+1)}, \mu_{0,1,2,\ldots,m_n})$ into $(S^{m_n+1}, \mathcal{S}^{\otimes(m_n+1)})$. Then the family $\mathcal{P}_f$ of Theorem 9.1 is obtained, and it automatically satisfies (a) as well as (b). As noted above, according to Tulcea's proof the conclusion of Theorem 9.1 holds without any topological conditions on $(S, \mathcal{S})$. The coordinate process $\{X_n : n = 0, 1, 2, \ldots\}$ defined by $X_n(\omega) = x_n$ $\forall \omega = (x_0, x_1, \ldots, x_n, \ldots) \in S^\infty$ $n = 0, 1, 2, \ldots$ on $(S^\infty, \mathcal{S}^{\otimes\infty}, Q)$ is a **Markov process** in the sense of Theorem 2.12: The **(regular) conditional distribution of** $X_{m+} := (X_m, X_{m+1}, \ldots)$ **given** $\mathcal{F}_m := \sigma(X_0, X_1, \ldots, X_m)$ is $(Q_y)_{y=X_m} \equiv Q_{X_m}$, where $Q_y = Q$ with the **initial distribution** μ_0 taken to be the Dirac delta measure δ_y (i.e., $\mu_0(\{y\}) = 1$, $\mu_0(S\backslash\{y\}) = 0$) (Exercise 2).

Remark 9.3 As illustrated by this example, in the discrete parameter case in which $\Lambda = \{0, 1, 2, \ldots\}$, it is enough to consistently specify $\mu_{0,1,\ldots,n}$ for $n = 0, 1, 2 \ldots$, subject to condition (b) and then consistently *define* the other finite-dimensional probabilities as being induced by the coordinate projections and permutation maps. More generally, the condition (a) on permutation consistency can always be built into the specification of finite-dimensional probabilities when Λ is *linearly ordered*. This is accomplished by specifying $\mu_{t_1,t_2,\ldots,t_n}$ for $t_1 < t_2 < \cdots < t_n$ and then defining $\mu_{\tau(1),\tau(2),\ldots,\tau(n)}$ as the image (measure) of $\mu_{t_1,t_2,\ldots,t_n}$ under the permutation map $(x_{t_1}, x_{t_2}, \ldots, x_{t_n}) \rightarrow (x_{\tau(1)}, x_{\tau(2)}, \ldots, x_{\tau(n)})$. Thus one needs only to check the consistency property (b) to hold for ordered n-tuples $(t_1, t_2, \ldots, t_n)$ with $t_1 < t_2 < \cdots < t_n$.

Remark 9.4 On an arbitrary measurable space $(S, \mathcal{S})$ one defines a **transition probability** $p(x, B) : S \times \mathcal{S} \rightarrow [0, 1]$ requiring only that (i) $x \mapsto p(x, B)$ be *measurable* for each $B \in \mathcal{S}$, and that (ii) for each $x \in S$, $B \mapsto p(x, B)$ is a *probability* on $\mathcal{S}$. The construction of a **Markov process** with a given transition probability $p(\cdot, \cdot)$ and a given **initial distribution** μ_0 is now defined by the *successive iterated integration*, generalizing (9.6), beginning with the integral of $p(x_{n-1}, B_n)$ with respect to the measure $p(x_{n-2}, dx_{n-1})$ to get $\int_{B_{n-1}} p(x_{n-1}, B_n)p(x_{n-2}, dx_{n-1}) = g_{n-2}(x_{n-2})$, say. Then integrate this with respect to $p(x_{n-3}, dx_{n-2})$ to get $\int_{B_{n-2}} g_{n-2}(x_{n-2})p(x_{n-3}, dx_{n-2}) = g_{n-3}(x_{n-3})$, say, and so on. In this manner one has

$$\mu_{0,1,2,\ldots,n}(B_0 \times B_1 \times \cdots \times B_n) = \int_{B_0} g_0(x_0)\mu_0(dx_0),$$

$$g_0(x_0) = \int_{B_1} g_1(x_1)p(x_0, dx_1),$$

$$g_1(x_1) = \int_{B_2} g_2(x_2) p(x_1, dx_2),$$

$$\ldots, g_{n-2}(x_{n-2}) = \int_{B_{n-1}} g_{n-1}(x_{n-1}) p(x_{n-2}, dx_{n-1}),$$

and

$$g_{n-1}(x_{n-1}) = p(x_{n-1}, B_n) \equiv \int_{B_n} p(x_{n-1}, dx_n),$$

beginning with the last term and moving successively backward.

Example 2 *(Gaussian Process/Random Field)* Here we take $S_t = \mathbb{R}, \mathcal{S}_t = \mathcal{B}(\mathbb{R}), t \in \Lambda$. The Kolmogorov extension theorem may be used to construct a probability space $(\Omega, \mathcal{F}, P)$ on which a family of Gaussian, or normal, random variables $\{X_t : t \in \Lambda\}$ are defined with arbitrarily specified (i) means $m_t = \mathbb{E}(X_t), t \in \Lambda$, and (ii) covariances $\sigma_{t,t'} = \text{Cov}(X_t, X_{t'})$, t and $t' \in \Lambda$, with the property that for every n-tuple $(t_1, t_2, \ldots, t_n)$ of distinct indices $(n \geq 1)$, the matrix $((\sigma_{t_i, t_j}))_{1 \leq i, j \leq n}$ is symmetric and nonnegative-definite. In this case, using the notation above, $\mu_{t_1, t_2, \ldots, t_n}$ is the Gaussian probability distribution parameterized by a (mean) vector $(m_{t_1}, m_{t_2}, \ldots, m_{t_n})^t \in \mathbb{R}^n$ and symmetric, nonnegative-definite (covariance) matrix $((\sigma_{t_i, t_j}))_{1 \leq i, j \leq n}$. More specifically, $\mu_{t_1, t_2, \ldots, t_n}$ is defined as the distribution of $Y = AZ + m$, where $Z = (Z_1, \ldots, Z_n)^t$ is n-dimensional standard normal with pdf $\varphi(z_1, \ldots, z_n) = (2\pi)^{-\frac{k}{2}} \exp\{-\frac{1}{2} \sum_{j=1}^n z_j^2\}$, $m = (m_{t_1}, \ldots, m_{t_n})$ and $A^t A = \Gamma := ((\sigma_{t_i, t_j}))_{1 \leq i, j \leq n}$. Consistency properties (a), (b) are easily checked. Hence there exists a probability measure Q on the product space $(\Omega = \mathbb{R}^\Lambda, \mathcal{F} = \mathcal{B}(\mathbb{R})^{\otimes \Lambda})$ such that the coordinate process $\{X_t : t \in \Lambda\}$ is the desired Gaussian process. Here $\mathbb{R}^\Lambda$ is the space of functions from Λ to $\mathbb{R}$, i.e., the product space, is product space, $X_t(\omega) = x_t$ for $\omega = (x_{t'}, t' \in \Lambda) \in \Omega \equiv \mathbb{R}^\Lambda (t \in \Lambda)$. The indexing set Λ is general and includes examples such as $\Lambda = [0, \infty), [0, 1]$, or in a construction, for example, of **Gaussian random fields** where $\Lambda = \mathbb{R}^k$. One generally assumes, for purposes of sample path regularity, that $t \to m_t$ and $(s, t) \to \sigma_{s,t}$ are continuous.[5]

As a special case, let $\Lambda = [0, \infty)$ (or $\Lambda = [0, 1]$), $m_t = 0 \forall t$, and $\sigma_{t,t'} = \min\{t, t'\}$ $(t, t' \in \Lambda)$. The check that $((\sigma_{t_i, t_j}))_{1 \leq i, j \leq n}$ is nonnegative-definite for all n-tuples of distinct indices is outlined in Exercise 1. The process so constructed on $(\Omega = \mathbb{R}^{[0,\infty)}$ or $\mathbb{R}^{[0,1]})$ defines a **Brownian motion process on the Kolmogorov σ-field** $\mathcal{B}(\mathbb{R})^{\otimes[0,\infty)}$ (or $\mathcal{B}(\mathbb{R})^{\otimes[0,1]}$), i.e., on the product σ-field for Ω generated by finite-dimensional cylinders of the form $C = \{\omega = (x_t, t \in \Lambda) : (x_{t_1}, x_{t_2}, \ldots, x_{t_n}) \in B\}$ for arbitrary $n \geq 1$, $t_1 < t_2 < \cdots < t_n$, $B \in \mathcal{B}(\mathbb{R}^n)$. Unfortunately, the Kolmogorov σ-field does not include the set of (all) continuous functions $C[0, \infty)$ (or $C[0, 1]$). The reason for this is that the product σ-field consists only of sets determined by countably many coordinates, rendering this model mathematically inadequate for computing probabilities of many "events" of interest due to nonmeasurability (Exercise 4). The

---

[5]See e.g., Bhattacharya, R. N. and Waymire E.C. (2016), Stationary Processes and Discrete Parameter Markov Processes, Chapter I, Sec 2, Springer (to appear).

first resolution of this situation was obtained by the seminal construction of Norbert Wiener. This led to the following definition of Brownian motion.

Definition 9.1 A stochastic process $B = \{B_t : t \geq 0\}$, $B_0 = 0$, defined on a probability space $(\Omega, \mathcal{F}, P)$ a.s. having continuous sample paths $t \to B_t, t \geq 0$, and such that for any $0 < t_1 < t_2 < \cdots < t_k$, $k \geq 1$ $(B_{t_1}, \ldots, B_{t_k})$ has a k-dimensional Gaussian distribution with zero mean and variance–covariance matrix $((t_i \wedge t_j))_{1 \leq i, j \leq k}$ is referred to as **one-dimensional standard Brownian motion** started at $B_0 = 0$. The distribution $P \circ B^{-1}$ of the process B is a probability concentrated on the Borel σ-field of $C[0, \infty)$, referred to as **Wiener measure**.

Since Wiener's construction, a number of alternative approaches have become known, one of which follows from the functional central limit theorem applied to simple symmetric random walk, e.g., as obtained for Example 2 in Chapter VII under a finite fourth moment condition. Another resolution of the existence problem is given in the following subsection through a "wavelet construction" in close resemblance to the classic "Fourier construction" of Wiener, but technically much simpler. Specifically, a construction is made of a probability space $(\Omega, \mathcal{F}, P)$ and stochastic process $B = \{B_t : t \in [0, \infty)\}$ such that, as above, for each $0 \leq t_1 < t_2 < \cdots < t_k$ $(k \geq 1)$, $(B_{t_1}, \ldots, B_{t_k})$ is Gaussian with mean $\mathbf{0}$ and covariance matrix $((\min\{t_i, t_j\}))_{1 \leq i, j \leq k}$. Equivalently, the increments $B_{t_j} - B_{t_{j-1}}$, $1 \leq j \leq k$, are independent Gaussian random variables with zero mean and variance $t_j - t_{j-1}$, respectively; cf. Exercise 1. Moreover, for such a model of Brownian motion, the subset $[B \in C[0, \infty)] \in \mathcal{F}$ is a measurable event (and has probability one).

IX.1   A Wavelet Construction of Brownian Motion: The Lévy–Ciesielski Construction

A construction[6] of Brownian motion based on a.s. uniform and absolute convergence on the time interval $[0,1]$ of a random series expansion in terms of the integrated Haar wavelet basis, referred to as the Schauder basis, of $L^2[0, 1]$ may be obtained as a consequence of the following sequence of lemmas. First, though, recursively define **Haar wavelet functions** $H_{0,0}, H_{n,k}$ $n = 0, 1, 2, \ldots, 2^n \leq k < 2^{n+1}$, on $0 \leq t \leq 1$, $H_{0,0}(t) \equiv 1$; $H_{0,1}(t) := \mathbf{1}_{[0,1/2]}(t) - \mathbf{1}_{(1/2,1]}(t)$; and

$$H_{n,k}(t) := 2^{\frac{n}{2}} \mathbf{1}_{[k2^{-n}-1, k2^{-n}-1+2^{-n-1}]}(t)$$
$$- 2^{\frac{n}{2}} \mathbf{1}_{(k2^{-n}-1+2^{-n-1}, k2^{-n}-1+2^{-n}]}(t), \quad 0 \leq t \leq 1.$$

Recall the definition of a complete orthonormal basis in a Hilbert space (see Appendix C).

---

[6]This construction originated in Ciesielski, Z. (1961): Hölder condition for realization of Gaussian processes, *Trans. Amer. Math. Soc.* **99** 403–413, based on a general approach of Lévy, P. (1948), p. 209.

Lemma 1 The collection of Haar wavelet functions $\{H_{n,k}\}$ is a complete orthonormal basis for $L^2[0, 1]$. In particular, $\langle f, g \rangle = \sum_{(n,k)} \langle f, H_{n,k} \rangle \langle g, H_{n,k} \rangle$ holds for $f, g \in L^2[0, 1]$.

Proof Orthonormality follows by a direct calculation. To prove completeness one needs to show that if $f \in L^2[0, 1]$ and $\langle f, H_{n,k} \rangle = 0$ for all n, k, then $f = 0$ almost everywhere with respect to Lebesgue measure on $[0, 1]$. Define

$$I_f(t) = \int_0^t f(s)ds, \qquad 0 \le t \le 1.$$

Then I_f is continuous with $I_f(0) = 0$. Moreover, orthogonality with respect to $H_{0,0}$ implies that $I_f(1) = 0$. Next $I_f(\frac{1}{2}) = 0$ since $l_f(0) = l_f(1) = 0$ and orthogonality of f to $H_{0,1}$. Using the orthogonality of f to $H_{1,2}$, one shows that $I_f(\frac{1}{4}) - (I_f(\frac{1}{2}) - I_f(\frac{1}{4})) = 0$, so that $I_f(\frac{1}{4}) = 0$. Orthogonality with $H_{1,3}$ means that $I_f(\frac{3}{4}) - I_f(\frac{1}{2}) - (I_f(1) - I_f(\frac{3}{4})) = 0$, implying $I_f(\frac{3}{4}) = 0$. Continuing by induction one finds that $I_f(k2^{-n}) = 0$ for all dyadic rationals $k2^{-n} \in [0, 1]$. By continuity it now follows that $I_f(t) = 0$ for all $t \in [0, 1]$ and hence $f = 0$ a.e., as asserted. The last equality is then simply Parseval's relation which holds for any complete orthonormal system (see Appendix C). ∎

Definition 9.2 The functions defined by $S_{n,k}(t) := \int_0^t H_{n,k}(s)ds$, $0 \le t \le 1$, are called the **Schauder functions**.

Lemma 2 The Schauder functions $S_{n,k}$ on $[0, 1]$ are continuous, nonnegative, and attain a maximum value at $2^{-(\frac{n}{2}+1)}$. Moreover, for fixed n, the functions $S_{n,k}$, $k = 2^n, \ldots, 2^{n+1} - 1$, have disjoint supports.

Proof Continuity is obvious. The assertions are also clearly true for $S_{0,0}$ and $S_{0,1}$. Since $H_{n,k}$ is positive, with constant value $2^{\frac{n}{2}}$ on the interval $[k2^{-n} - 1, k2^{-n} - 1 + 2^{-n-1}]$ to the left of $(k2^{-n} - 1, k2^{-n} - 1 + 2^{-n-1}]$, where it takes negative constant value $-2^{\frac{n}{2}}$, and it has the value 0 off these two intervals, $S_{n,k}$ is positive and increasing on the first interval with a maximum value $S_{n,k}(t_M) = 2^{\frac{n}{2}}(k2^{-n} - 1 + 2^{-n-1} - k2^{-n} + 1) = 2^{-(\frac{n}{2}+1)}$ at the endpoint $t_M = k2^{-n} - 1 + 2^{-n-1}$. Moreover, it attains a minimum value $S_{n,k}(t_m) = 0$ at the rightmost endpoint $t_m = k2^{-n} - 1 + 2^{-n-1}$. Thus $S_{n,k}$ is nonnegative with disjoint supports $[k2^{-n} - 1, k2^{-n} - 1 + 2^{-n-1}]$ for $k = 2^n, \ldots, 2^{n+1} - 1$. ∎

Lemma 3 For $0 \le s \le t \le 1$,

$$\sum_{n,k} S_{n,k}(s)S_{n,k}(t) = \min\{s, t\} = s.$$

Proof By definition of the Schauder functions one has $S_{n,k}(t) = \langle 1_{[0,t]}, H_{n,k} \rangle$ for fixed $t \in [0, 1]$. Thus one may apply Parseval's equation to obtain for $s \le t$, $\sum_{n,k} S_{n,k}(s)S_{n,k}(t) = \sum_{n,k} \langle 1_{[0,s]}, H_{n,k} \rangle \langle 1_{[0,t]}, H_{n,k} \rangle = \langle 1_{[0,s]}, 1_{[0,t]} \rangle = s$, since $1_{[0,s]} 1_{[0,t]} = 1_{[0,s]}$. ∎

Since the maximum of the Schauder functions are decaying exponentially, there is some room for growth in the coefficients of a series expansion in these functions as furnished by the next lemma.

Lemma 4 If $\max_{2^n \leq k < 2^{n+1}} |a_{n,k}| = O(2^{n\varepsilon})$, for some $0 < \varepsilon < 1/2$, then $\sum_{n,k} a_{n,k} S_{n,k}$ on $[0, 1]$ converges uniformly and absolutely to a continuous function.

Proof The key is to observe that since for given n, the Schauder functions have disjoint supports for $2^n \leq k < 2^{n+1}$, the maximum value of $|\sum_{k=2^n}^{2^{n+1}-1} a_{n,k} S_{n,k}|$ on $[0, 1]$ is $(\max_{2^n \leq k < 2^{n+1}} |a_{n,k}|)2^{-(\frac{n}{2}+1)}$. Thus for some $c > 0$,

$$\sum_{n \geq m} \left| \sum_{k=2^n}^{2^{n+1}-1} a_{n,k} S_{n,k}(t) \right| \leq \sum_{n \geq m} c 2^{n\varepsilon} 2^{-n/2+1}$$

is the tail of a convergent geometric series. In particular, the partial sums are uniformly Cauchy. The assertion follows since the uniform limit of continuous functions on $[0, 1]$ is continuous. ∎

Lemma 5 There is an i.i.d. sequence $a_{n,k} = Z_{n,k}$ of standard normal random variables on a probability space $(\Omega, \mathcal{F}, P)$. Moreover, $\sum_{n,k} Z_{n,k} S_{n,k}$ is uniformly and absolutely convergent on $[0, 1]$ with probability one.

Proof The existence of the i.i.d. sequence follows from Kolmogorov's extension theorem. From here apply Borel–Cantelli I and the Feller's tail probability estimates to obtain by the preceding lemma that with probability one, $\sum_{n,k} Z_{n,k} S_{n,k}$ is uniformly and absolutely convergent on $[0, 1]$. Specifically, for some $c' > 0$,

$$\sum_{n=1}^{\infty} P(\max_{2^n \leq k < 2^{n+1}} |Z_{n,k}| > 2^{n\varepsilon}) \leq c' \sum_{n=1}^{\infty} 2^n 2^{-\frac{n\varepsilon}{2}} e^{-\frac{1}{2} 2^{2\varepsilon n}} < \infty.$$

Thus $\max_{2^n \leq k < 2^{n+1}} |Z_{n,k}|$ is a.s. $O(2^{n\varepsilon})$ for any choice of $0 < \varepsilon < 1/2$. ∎

Lemma 6 Define $B_t := \sum_{n,k} Z_{n,k} S_{n,k}(t)$, $0 \leq t \leq 1$. Then with probability one, $\{B_t : 0 \leq t \leq 1\}$ has continuous sample paths, $B_0 = 0$, and for any $0 = t_0 < t_1 < \cdots < t_m \leq 1, \xi_1, \ldots, \xi_m \in \mathbb{R}, m \geq 1$, the increments $B_{t_j} - B_{t_{j-1}}, j = 1, \ldots, m$, are distributed as independent normal random variables with zero mean and respective variances $t_j - t_{j-1}, j = 1, \ldots, m$.

Proof Observe that using the Parseval's relation as in Lemma 3,

$$\mathbb{E} e^{i\xi B_t} = \prod_{(n,k)} \mathbb{E} e^{i\xi Z_{(n,k)} S_{(n,k)}(t)}$$

$$= \prod_{(n,k)} e^{-\frac{1}{2}\xi^2 S_{(n,k)}^2(t)}$$

$$= \exp\left\{-\frac{1}{2}\xi^2 \sum_{(n,k)} S_{(n,k)}^2(t)\right\} = e^{-\frac{1}{2}\xi^2 t}. \qquad (9.7)$$

Proceed inductively on m, similarly using Parseval's relation, to check that the increments $B_{t_j} - B_{t_{j-1}}$, $j = 1, \ldots, m$, have the multivariate characteristic function $\mathbb{E} \exp\{i \sum_{j=1}^m \xi_j (B_{t_j} - B_{t_{j-1}})\} = \prod_{j=1}^m \exp(-\frac{1}{2}(t_j - t_{j-1})\xi_j^2)$. ∎

Theorem 9.3 There is a stochastic process $B = \{B_t : t \geq 0\}$ defined on a probability space $(\Omega, \mathcal{F}, P)$ with continuous sample paths and having stationary independent Gaussian increments $B_t - B_s$ with mean zero and variance $t - s$ for each $0 \leq s < t$.

Proof First, use Lemma 6 to construct a Brownian motion on its image space $C[0, 1]$. By the Kolmogorov extension theorem one may construct a sequence $B_t^{(r)}, 0 \leq t \leq 1, r = 1, 2, \ldots$ of independent standard Brownian motions on $[0, 1]$ each starting at 0. Inductively extend $B_t := B_t^{(1)}, 0 \leq t \leq 1$, by $B_t := B_{t-r+1}^{(r)} + B_{r-1}, r - 1 \leq t \leq r, r = 1, 2, \ldots$. Then it is now simple to check that the stochastic process $\{B_t : t \geq 0\}$ satisfies all the properties that define a standard Brownian motion on $[0, \infty)$ starting at 0. ∎

Definition 9.3 A **k-dimensional standard Brownian motion** is a stochastic process $\{\mathbf{B}_t = (B_t^{(1)}, \ldots, B_t^{(k)}) : t \geq 0\}$ such that $\{B_t^{(j)} : t \geq 0\}, j = 1, \ldots, k$, are k independent one-dimensional standard Brownian motions.

In the next chapter some fine-scale properties of Brownian motion paths are presented. In particular, see Exercise 7 of Chapter X for a simple application of the wavelet construction in this connection.

Without being circular, many classical limit theorems for sums of independent random variables arise as consequences of more general theories involving Brownian motion having a.s. continuous paths. This FCLT connection is dramatically highlighted in Chapter XI via the so-called **Skorokhod embedding** of suitably centered and scaled random walks in a Brownian motion.

Exercise Set IX

1. (i) Suppose that $Y_1, Y_2, \ldots$ is a sequence of real-valued random variables in $L^2(\Omega, \mathcal{F}, P)$ and let $\Gamma := ((\text{Cov}(Y_i, Y_j))_{1 \leq i, j \leq n}$. Show that Γ is nonnegative-definite. [*Hint*: Expand $0 \leq \mathbb{E}|\sum_{i=}^n c_i Y_i|^2$ for real numbers $c_1, \ldots, c_n$.]

   (ii) Show that $((\sigma_{t_i, t_j}))_{1 \leq i, j \leq n} := ((\min\{t_i, t_j\}))_{1 \leq i, j \leq n}$ is nonnegative-definite for all n-tuples of distinct indices $t_1, \ldots, t_j$. [*Hint*: Take $0 \leq t_1 < t_2 < \cdots < t_n$, and let $Z_1, Z_2, \ldots, Z_n$ be independent mean-zero Gaussian random variables (for example defined on a finite product space $(\mathbb{R}^n, \mathcal{B}(\mathbb{R}^n))$

such that $\mathrm{Var}(Z_1) = t_1$, $\mathrm{Var}(Z_j) = t_j - t_{j-1}(j = 2, \ldots, n)$. Consider $Y_1 = Z_1$, $Y_2 = Z_1 + Z_2, \ldots, Y_n = Z_1 + Z_2 + \cdots + Z_n$. Compute the covariance matrix of $Y_1, Y_2, \ldots, Y_n$.]

2. Prove the assertion in Example 1 that the (regular) conditional distribution of $X_{m+} := (X_m, X_{m+1}, \ldots)$, given $\mathcal{F}_m := \sigma(X_0, X_1, \ldots, X_m)$, is $(Q_y)_{y=X_m} \equiv Q_{X_m}$, where $Q_y = Q$ with the *initial distribution* μ_0 taken to be the Delta measure δ_y (i.e., $\mu_0(\{y\}) = 1$, $\mu_0(S\backslash\{y\}) = 0$). [*Hint:* First, consider a cylinder set C as in (9.5) and show that $Q_{X_m}(C)$, as given under Remark 9.4 with $\mu_0 = \delta_{X_m}$ equals
$$P([X_{m+} \in C]|\sigma(X_0, X_1, \ldots, X_m)) \equiv P([(X_m, X_{m+1}, \ldots, X_{m+n}) \in B_0 \times B_1 \times \cdots \times B_n]|\sigma(X_0, X_1, \ldots, X_m)) \equiv \mathbb{E}(\mathbf{1}_{[(X_m, X_{m+1}, \ldots, X_{m+n}) \in B_0 \times \cdots \times B_n]}|$$
$\sigma(X_0, X_1, \ldots, X_m))$. For this, first check with $n = 0$ and then $n = 1$. For the latter, let $g(X_0, X_1, \ldots, X_m)$ be nonnegative, bounded measurable and calculate $\mathbb{E}(\mathbf{1}_{[X_n \in B_0, X_{n+1} \in B_1]} g(X_0, \ldots, X_m))$ using the relations in Remark 9.4. Finally, use induction and properties of conditional expectation to calculate $\mathbb{E}(\mathbf{1}_{[X_n \in B_0, \ldots, X_{n+m+1} \in B_{m+1}]} g(X_0, X_1, \ldots, X_m))$.]

3. Let $S_n = \{0, 1\}$, with the power set σ-field $\mathcal{S}_n = 2^{S_n}$, $n = 1, 2, \ldots$. Suppose that $p_n : S_1 \times \cdots \times S_n \to [0, 1]$, $n \geq 1$, are probability mass functions, i.e., $\sum_{(s_1, \ldots, s_n) \in \{0,1\}^n} p_n(s_1, \ldots, s_n) = 1$, for each n. Assume the following consistency condition: $p_n(s_1, \ldots, s_n) = p_{n+1}(s_1, \ldots, s_n, 0) + p_{n+1}(s_1, \ldots, s_n, 1)$, $s_i \in \{0, 1\}$, $1 \leq i \leq n$. Give a direct proof of the existence of a probability space $(\Omega, \mathcal{F}, P)$ and a sequence of random variables $X_1, X_2, \ldots$ such that $P(X_1 = s_1, \ldots, X_n = s_n) = p_n(s_1, \ldots, s_n)$, $s_i \in \{0, 1\}$, $1 \leq i \leq n$, $n \geq 1$. [*Hint:* $\prod_{n \in \mathbb{N}} S_n$ may be viewed as a compact space, with Borel σ-field $\mathcal{B} \equiv \otimes_{n \in \mathbb{N}} \mathcal{S}_n$ and such that the finite-dimensional cylinders are both open and closed. Define a set function on the field of finite-dimensional cylinders and use the Heine–Borel compactness property to prove countable additivity on this field. The rest follows by Carathéodory extension theory.]

4. For $\Omega = \mathbb{R}^{[0,1]}$, we write $\omega = (x_t, 0 \leq t \leq 1) \in \Omega$ to denote a real-valued function on $[0, 1]$. Also $\mathcal{B}^{\otimes[0,1]} = \sigma(\mathcal{C})$, where $C \in \mathcal{C}$ if and only if $C \equiv C(T, A_1, A_2, \ldots) := \{\omega = (x_t, 0 \leq t \leq 1) \in \Omega : x_{t_1} \in B_1, \ldots, x_{t_n} \in B_n, \ldots\}$ for some countable set $T = \{t_1, t_2, \ldots\} \subset [0, 1]$, and Borel sets $B_1, B_2, \ldots$. Let $\mathcal{T}$ denote the collection of countable subsets of $[0, 1]$. For fixed $T \in \mathcal{T}$, let $\mathcal{C}_T$ be the collection of all subsets of Ω of the form $C(T, A_1, A_2, \ldots)$.

   (i) Show that $\mathcal{B}^{\otimes[0,1]} = \cup_{T \in \mathcal{T}} \sigma(\mathcal{C}_T)$.
   (ii) For fixed $T \in \mathcal{T}$, let $\varphi_T : \mathbb{R}^{[0,1]} \to \mathbb{R}^\infty$ by $\varphi_T(\omega) = (x_{t_1}, x_{t_2}, \ldots)$, $\omega = (x_t, 0 \leq t \leq 1)$. Show that $\sigma(\mathcal{C}_T) = \sigma(\varphi_T)$ is the smallest σ-field on $\mathbb{R}^{[0,1]}$ that makes φ_T measurable for the product σ-field $\mathcal{B}^\infty$ on $\mathbb{R}^\infty$.
   (iii) Show that if $A \in \sigma(\mathcal{C}_T)$ and $\omega \in A$, then $\omega' \in \Omega$ with $\varphi_T(\omega) = \varphi_T(\omega')$ implies $\omega' \in A$. [*Hint:* $\sigma(\varphi_T) = \{\varphi_T^{-1}(F) : F \in \mathcal{B}^\infty\}$.]
   (iv) Show that $C[0, 1]$ is not measurable for the Kolmogorov product σ-field $\mathcal{B}^{\otimes[0,1]}$.
   (v) Show that $\{x \in \mathbb{R}^{[0,1]} : \sup_{0 \leq t \leq 1} x(t) \leq 1\}$ is not a measurable set for the Kolmogorov product σ-field.

5. Suppose that S is a Polish space with Borel σ-field $\mathcal{S} = \mathcal{B}(S)$. (i) Let $\mu_n(x_0, \ldots, x_{n-1}, B) = p(x_{n-1}, B)$ in Tulcea's theorem 9.2, with $p(x, dy)$ a probability on $(S, \mathcal{S})$ for each $x \in S$, and such that $x \to p(x, B)$ is Borel measurable for each fixed $B \in \mathcal{S}$. Show that the existence of the probability Q asserted in Tulcea's extension theorem follows from Kolmogorov's extension theorem. [*Hint*: Show that the specification of finite-dimensional distributions by the initial distribution μ and the transition probabilities $p(x, dy)$, $x \in S$, satisfy the Kolmogorov consistency condition (b).] (ii) Following Remark 9.4, extend the specification (9.3) to define $\mu_{0,1,\ldots,n}(B)$ to all $B \in \mathcal{S}^{\otimes(n+1)}$. [*Hint*: Successively define, (i) $g_{n-1}(x_{n-1}; x_0, \ldots, x_{n-2}) = p(x_{n-1}, B_{x_0, x_1, \ldots, x_{n-1}})$, where $B_{x_0, x_1, \ldots, x_{n-1}} = \{y \in S : (x_0, x_1, \ldots, x_n) \in B\}$ is the $(x_0, x_1, \ldots, x_{n-1})$-section of B. Then define $g_{n-2}(x_{n-2}; x_0, \ldots, x_{n-3}) = \int g_{n-1}(x_{n-1}; x_0, \ldots, x_{n-2}) p(x_{n-2}, dx_{n-1})$, and so on.] (iii) Show that the **canonical process** given by coordinate projections $\mathbf{x} \to x_n$, $(\mathbf{x} = (x_0, x_1, \cdots \in S^\infty)$, say $X_n (n \geq 0)$, on $(S^\infty, \mathcal{S}^{\otimes\infty}, Q)$, has the **Markov property**: the conditional distribution of X_{m+1} given $\mathcal{F}_m = \sigma(X_j : 0 \leq j \leq m)$ is $p(X_m, dy)$.

Chapter X
Brownian Motion: The LIL and Some Fine-Scale Properties

In this chapter, we analyze the *growth* of the Brownian paths $t \mapsto B_t$ as $t \to \infty$. We will see by a property of "time inversion" of Brownian motion that this leads to small-scale properties as well. First, however, let us record some basic properties of the Brownian motion that follow somewhat directly from its definition.

Theorem 10.1 Let $B = \{B_t : t \geq 0\}$ be a standard one-dimensional Brownian motion starting at 0. Then

1. *(Symmetry)* $W_t := -B_t, t \geq 0$, is a standard Brownian motion starting at 0.
2. *(Homogeneity and Independent Increments)* $\{B_{t+s} - B_s : t \geq 0\}$ is a standard Brownian motion independent of $\{B_u : 0 \leq u \leq s\}$, for every $s \geq 0$.
3. *(Scale-Change Invariance)*. For every $\lambda > 0$, $\{B_t^{(\lambda)} := \lambda^{-\frac{1}{2}} B_{\lambda t} : t \geq 0\}$ is a standard Brownian motion starting at 0.
4. *(Time-Inversion Invariance)* $W_t := t B_{1/t}, t > 0, W_0 = 0$, is a standard Brownian motion starting at 0.

Proof Each of these is obtained by showing that the conditions defining a Brownian motion are satisfied. In the case of the time-inversion property, one may apply the strong law of large numbers to obtain continuity at $t = 0$. That is, if $0 < t_n \to 0$ then write $s_n = 1/t_n \to \infty$ and $N_n := [s_n]$, where $[\cdot]$ denotes the greatest integer function, so that by the strong law of large numbers, with probability one

$$W_{t_n} = \frac{1}{s_n} B_{s_n} = \frac{N_n}{s_n} \frac{1}{N_n} \sum_{j=1}^{N_n} (B_i - B_{i-1}) + \frac{1}{s_n}(B_{s_n} - B_{N_n}) \to 0,$$

since $B_i - B_{i-1}, i \geq 1$, is an i.i.d. mean-zero sequence, $N_n/s_n \to 1$, and $(B_{s_n} - B_{N_n})/s_n \to 0$ a.s. as $n \to \infty$ (see Exercise 2). ∎

Although the Brownian motion paths cannot be differentiable, it is possible to determine an *order of continuity* using the next general theorem.

© Springer International Publishing AG 2016
R. Bhattacharya and E.C. Waymire, *A Basic Course in Probability Theory*,
Universitext, DOI 10.1007/978-3-319-47974-3_X

Definition 10.1 A stochastic process (or random field) $Y = \{Y_u : u \in \Lambda\}$ is a *version* of $X = \{X_u : u \in \Lambda\}$ taking values in a metric space if Y has the same finite dimensional distributions as X.

Theorem 10.2 *(Kolmogorov-Chentsov Theorem)* Suppose $X = \{X_u : u \in \Lambda\}$ is a stochastic process (or random field) with values in a complete metric space (S, ρ), indexed by a bounded rectangle $\Lambda \subset \mathbb{R}^k$ and satisfying

$$\mathbb{E}\rho^\alpha(X_u, X_v) \le c|u - v|^{k+\beta}, \quad \text{for all } u, v \in \Lambda,$$

where c, α, β are positive numbers. Then there is a version $Y = \{Y_u : u \in \Lambda\}$ of X which is a.s. Hölder continuous of any exponent γ such that $0 < \gamma < \frac{\beta}{\alpha}$.

Proof Without essential loss of generality we take $\Lambda = [0, 1]^k$ and the norm $|\cdot|$ to be the *maximum norm* given by $|u| = \max\{|u_i| : 1 \le i \le k\}$, $u = (u_1, \ldots, u_k)$. For each $N = 1, 2, \ldots$, let L_N be the finite lattice $\{j2^{-N} : j = 0, 1, \ldots 2^N\}^k$. Write $L = \cup_{N=1}^\infty L_N$. Define $M_N = \max\{\rho(X_u, X_v) : (u, v) \in L_N^2, |u - v| \le 2^{-N}\}$. Since (i) for a given $u \in L_N$ there are no more than 3^k points in L_N such that $|u - v| \le 2^{-N}$, (ii) there are $(2^N + 1)^k$ points in L_N, and (iii) for every given pair (u, v), the condition of the theorem holds, one has by Chebyshev's inequality that

$$P(M_N > 2^{-\gamma N}) \le c3^k(2^N + 1)^k \left(\frac{2^{-N(k+\beta)}}{2^{-\alpha\gamma N}}\right). \tag{10.1}$$

In particular, since $\gamma < \beta/\alpha$,

$$\sum_{N=1}^\infty P(M_N > 2^{-\gamma N}) < \infty. \tag{10.2}$$

Thus there is a random positive integer $N^* \equiv N^*(\omega)$ and a set Ω^* with $P(\Omega^*) = 1$, such that

$$M_N(\omega) \le 2^{-\gamma N} \quad \text{for all } N \ge N^*(\omega), \omega \in \Omega^*. \tag{10.3}$$

Fix $\omega \in \Omega^*$ and let $N \ge N^*(\omega)$. By exactly the same induction argument as used for the proof of Lemma 3 in Chapter VII, one has for all $m \ge N + 1$,

$$\rho(X_u, X_v) \le 2\sum_{j=N}^m 2^{-\gamma j}, \quad \text{for all } u, v \in L_m, |u - v| \le 2^{-N}. \tag{10.4}$$

Since $2\sum_{\nu=N}^\infty 2^{-\gamma\nu} = 2^{-\gamma N+1}(1 - 2^{-\gamma})^{-1}$, and $L = \cup_{m=N+1}^\infty L_m$ for all $N \ge N^*(\omega)$, it follows that

$$\sup\{\rho(X_u, X_v) : u, v \in L, |u - v| \leq 2^{-N}\}$$
$$= \sup\{\rho(X_u, X_v) : u, v \in \cup_{m=N+1}^{\infty} L_m, |u - v| \leq 2^{-N}\}$$
$$\leq 2^{-\gamma N + 1}(1 - 2^{-\gamma})^{-1}, \quad N \geq N^*(\omega), \omega \in \Omega^*. \tag{10.5}$$

This proves that on Ω^*, $u \to X_u$ is uniformly continuous (from L into (S, ρ)), and is Hölder continuous with exponent γ. Now define $Y_u := X_u$ if $u \in L$ and otherwise $Y_u := \lim X_{u_N}$, with $u_N \in L$ and $u_N \to u$, if $u \notin L$. Because of uniform continuity of $u \to X_u$ on L (for $\omega \in \Omega^*$), and completeness of (S, ρ), the last limit is well-defined. For all $\omega \notin \Omega^*$, let Y_u be a fixed element of S for all $u \in [0, 1]^k$. Finally, letting $\gamma_j \uparrow \beta/\alpha$, $\gamma_j < \beta/\alpha$, $j \geq 1$, and denoting the exceptional set above as Ω_j^*, one has the Hölder continuity of Y for every $\gamma < \beta/\alpha$ on $\Omega^{**} := \cap_{j=1}^{\infty} \Omega_j^*$ with $P(\Omega^{**}) = 1$.

That Y is a version of X may be seen as follows. For any $r \geq 1$ and r vectors $u_1, \ldots, u_r \in [0, 1]^k$, there exist $u_{jN} \in L$, $u_{jN} \to u_j$ as $N \to \infty$ ($1 \leq j \leq r$). Then $(X_{u_{1N}}, \ldots, X_{u_{rN}}) = (Y_{u_{1N}}, \ldots, Y_{u_{rN}})$ a.s., and $(X_{u_{1N}}, \ldots, X_{u_{rN}}) \to (X_{u_1}, \ldots, X_{u_r})$ in probability, $(Y_{u_{1N}}, \ldots, Y_{u_{rN}}) \to (Y_{u_1}, \ldots, Y_{u_r})$ almost surely. ∎

Corollary 10.3 *(Brownian Motion)* Let $X = \{X_t : t \geq 0\}$ be a real-valued Gaussian process defined on $(\Omega, \mathcal{F}, P)$, with $X_0 = 0$, $\mathbb{E}X_t = 0$, and $\text{Cov}(X_s, X_t) = s \wedge t$, for all $s, t \geq 0$. Then X has a version $B = \{B_t : t \geq 0\}$ with continuous sample paths, which are Hölder continuous on every bounded interval with exponent γ for every $\gamma \in (0, \frac{1}{2})$.

Proof Since $\mathbb{E}|X_t - X_s|^{2m} = c(m)(t - s)^m, 0 \leq s \leq t$, for some constant $c(m)$, for every $m > 0$, the Kolmogorov–Chentsov Theorem 10.2 implies the existence of a version $B^{(0)} = \{B_t^{(0)} : 0 \leq t \leq 1\}$ with the desired properties on $[0, 1]$. Let $B^{(n)}, n \geq 1$, be independent copies of $B^{(0)}$, inddependent of $B^{(0)}$. Define $B_t = B_t^{(0)}, 0 \leq t \leq 1$, and $B_t = B_1^{(0)} + \cdots + B_1^{(n-1)} + B_{t-[t]}^{(n)}$, for $t \in [n, n+1), n = 1, 2, \ldots$. ∎

Corollary 10.4 *(Brownian Sheet)* Let $X = \{X_u : u \in [0, \infty)^2\}$ be a real-valued Guassian random field satisfying $\mathbb{E}X_u = 0$, $\text{Cov}(X_u, X_v) = (u_1 \wedge v_1)(u_2 \wedge v_2)$ for all $u = (u_1, u_2), v = (v_1, v_2)$. Then X has a continuous version on $[0, \infty)^2$, which is Hölder continuous on every bounded rectangle contained in $[0, \infty)^2$ with exponent γ for every $\gamma \in (0, \frac{1}{2})$.

Proof First let us note that on every compact rectangle $[0, M]^2$, $\mathbb{E}|X_u - X_v|^{2m} \leq c(M)|u - v|^m$, for all $m = 1, 2, \ldots$. For this it is enough to check that on each horizontal line $u = (u_1, c)$, $0 \leq u_1 < \infty$, X_u is a one-dimensional Brownian motion with mean zero and variance parameter $\sigma^2 = c$ for $c \geq 0$. The same holds on vertical lines. Hence $\mathbb{E}|X_{(u_1, u_2)} - X_{(v_1, v_2)}|^{2m} \leq 2^{2m-1}(\mathbb{E}|X_{(u_1, u_2)} - X_{(v_1, u_2)}|^{2m} + \mathbb{E}|X_{(v_1, u_2)} - X_{(v_1, v_2)}|^{2m}) \leq 2^{m-1}c(m)(u_2^m|u_1 - v_1|^m + v_1^m|u_2 - v_2|^m) \leq 2^{m-1}c(m) M^m 2|u - v|^m$, where $u = (u_1, u_2), v = (v_1, v_2)$. ∎

Remark 10.1 One may define the *Brownian sheet* on the index set $\Lambda_{\mathcal{R}}$ of all rectangles $R = [u, v)$, with $u = (u_1, u_2), v = (v_1, v_2), 0 \leq u_i \leq v_i < \infty$ ($i = 1, 2$), by setting

$$X_R \equiv X_{[u,v)} := X_{(v_1,v_2)} - X_{(v_1,u_2)} - X_{(u_1,v_2)} + X_{(u_1,u_2)}. \tag{10.6}$$

Then X_R is Gaussian with mean zero and variance $|R|$, the area of R. Moreover, if R_1 and R_2 are nonoverlapping rectangles, then X_{R_1} and X_{R_2} are independent. More generally, $\mathrm{Cov}(X_{R_1}, X_{R_2}) = |R_1 \cap R_2|$. Conversely, given a Gaussian family $\{X_R : R \in \Lambda_{\mathcal{R}}\}$ with these properties, one can restrict it to the class of rectangles $\{R = [0, u) : u = (u_1, u_2) \in [0, \infty)^2\}$ and identify this with the Brownian sheet in Corollary 10.4. It is simple to check that for all n-tuples of rectangles $R_1, R_2, \ldots, R_n \subset [0, \infty)^2$, the matrix $((|R_i - R_j|))_{1 \le i,j \le n}$ is symmetric and nonnegative definite. So the finite dimensional distributions of $\{X_R : R \in \Lambda_{\mathcal{R}}\}$ satisfy Kolmogorov's consistency condition.

In order to prove our main result of this section, we will make use of the following important inequality due to Paul Lévy.

Proposition 10.5 (*Lévy's Inequality*) Let X_j, $j = 1, \ldots, N$, be independent and symmetrically distributed (about zero) random variables. Write $S_j = \sum_{i=1}^{j} X_i$, $1 \le j \le N$. Then, for every $y > 0$,

$$P\left(\max_{1 \le j \le N} S_j \ge y \right) \le 2P(S_N \ge y) - P(S_N = y) \le 2P(S_N \ge y).$$

Proof Write $A_j = [S_1 < y, \ldots, S_{j-1} < y, S_j \ge y]$, for $1 \le j \le N$. The events $[S_N - S_j < 0]$ and $[S_N - S_j > 0]$ have the same probability and are independent of A_j. Therefore

$$
\begin{aligned}
P\left(\max_{1 \le j \le N} S_j \ge y \right) &= P(S_N \ge y) + \sum_{j=1}^{N-1} P(A_j \cap [S_N < y]) \\
&\le P(S_N \ge y) + \sum_{j=1}^{N-1} P(A_j \cap [S_N - S_j < 0]) \\
&= P(S_N \ge y) + \sum_{j=1}^{N-1} P(A_j)P([S_N - S_j < 0]) \\
&= P(S_N \ge y) + \sum_{j=1}^{N-1} P(A_j \cap [S_N - S_j > 0]) \\
&\le P(S_N \ge y) + \sum_{j=1}^{N-1} P(A_j \cap [S_N > y]) \\
&\le P(S_N \ge y) + P(S_N > y) \\
&= 2P(S_N \ge y) - P(S_N = y). \tag{10.7}
\end{aligned}
$$

This establishes the basic inequality. ∎

Corollary 10.6 For every $y > 0$ one has for any $t > 0$,

$$P\left(\max_{0 \leq s \leq t} B_s \geq y\right) \leq 2P(B_t \geq y).$$

Proof Partition $[0, t]$ by equidistant points $0 < u_1 < u_2 < \cdots < u_N = t$, and let $X_1 = B_{u_1}$, $X_{j+1} = B_{u_{j+1}} - B_{u_j}$, $1 \leq j \leq N - 1$, in the proposition. Now let $N \to \infty$, and use the continuity of Brownian motion. ∎

In fact one may use a reflection principle argument (strong Markov property) to see that this inequality is sharp for Brownian motion

$$P(\max_{0 \leq s \leq t} B_s \geq y) = 2P(B_t \geq y). \tag{10.8}$$

Alternatively, the following proposition concerns the **simple symmetric random walk** defined by $S_0 = 0$, $S_j = X_1 + \cdots + X_j$, $j \geq 1$, with $X_1, X_2, \ldots$ i.i.d. ± 1-valued with equal probabilities. It also demonstrates the remarkable strength of the **reflection method**, allowing one in particular to compute the distribution of the maximum of a random walk over a finite time. The above-indicated equality (10.8) then becomes a consequence of the functional central limit theorem proved in Section 1.8, (Theorem 7.15); especially see (9.27).

Proposition 10.7 For the simple symmetric random walk one has for every positive integer y,

$$P\left(\max_{0 \leq j \leq N} S_j \geq y\right) = 2P(S_N \geq y) - P(S_N = y).$$

Proof In the notation of Lévy's inequality given in Proposition 10.5 one has, for the present case of the random walk moving by ± 1 units at a time, that $A_j = [S_1 < y, \ldots, S_{j-1} < y, S_j = y]$, $1 \leq j \leq N$. Then in (10.7) the probability inequalities are all equalities for this special case. ∎

Corollary 10.8 Equation (10.8) holds for every $y > 0, t > 0$.

Theorem 10.9 (*Law of the Iterated Logarithm (LIL) for Brownian Motion*) Each of the following holds with probability one:

$$\overline{\lim}_{t \to \infty} \frac{B_t}{\sqrt{2t \log \log t}} = 1, \qquad \underline{\lim}_{t \to \infty} \frac{B_t}{\sqrt{2t \log \log t}} = -1.$$

Proof Let $\varphi(t) := \sqrt{2t \log \log t}$, $t > 0$. Let us first show that for any $0 < \delta < 1$, one has with probability one that

$$\overline{\lim}_{t \to \infty} \frac{B_t}{\varphi(t)} \leq 1 + \delta. \tag{10.9}$$

For arbitrary $\alpha > 1$, partition the time interval $[0, \infty)$ into subintervals of exponentially growing lengths $t_{n+1} - t_n$, where $t_n = \alpha^n$, and consider the event

$$E_n := \left[\max_{t_n \leq t \leq t_{n+1}} \frac{B_t}{(1+\delta)\varphi(t)} > 1 \right].$$

Since $\varphi(t)$ is a nondecreasing function, one has, using Corollary 10.6, a scaling property, and Lemma 2 from Chapter IV, that

$$P(E_n) \leq P\left(\max_{0 \leq t \leq t_{n+1}} B_t > (1+\delta)\varphi(t_n) \right)$$

$$= 2P\left(B_1 > \frac{(1+\delta)\varphi(t_n)}{\sqrt{t_{n+1}}} \right)$$

$$\leq \sqrt{\frac{2}{\pi}} \frac{\sqrt{t_{n+1}}}{(1+\delta)\varphi(t_n)} e^{-\frac{(1+\delta)^2\varphi^2(t_n)}{2t_{n+1}}} \leq c\frac{1}{n^{(1+\delta)^2/\alpha}} \tag{10.10}$$

for a constant $c > 0$ and all $n > \frac{1}{\log\alpha}$. For a given $\delta > 0$ one may select $1 < \alpha < (1+\delta)^2$ to obtain $P(E_n \ i.o.) = 0$ from the Borel–Cantelli lemma (Part I). Thus we have (10.9). Since $\delta > 0$ is arbitrary we have with probability one that

$$\overline{\lim}_{t\to\infty} \frac{B_t}{\varphi(t)} \leq 1. \tag{10.11}$$

Next let us show that with probability one,

$$\overline{\lim}_{t\to\infty} \frac{B_t}{\varphi(t)} \geq 1. \tag{10.12}$$

For this consider the independent increments $B_{t_{n+1}} - B_{t_n}$, $n \geq 1$. For $\theta = \frac{t_{n+1}-t_n}{t_{n+1}} = \frac{\alpha-1}{\alpha} < 1$, using Feller's tail probability estimate (Lemma 2, Chapter IV) and Brownian scale change,

$$P\left(B_{t_{n+1}} - B_{t_n} > \theta\varphi(t_{n+1}) \right) = P\left(B_1 > \sqrt{\frac{\theta}{t_{n+1}}}\varphi(t_{n+1}) \right)$$

$$\geq \frac{c'}{\sqrt{2\theta\log\log t_{n+1}}} e^{-\theta\log\log t_{n+1}}$$

$$\geq \frac{c}{\sqrt{\log n}} n^{-\theta} \tag{10.13}$$

for suitable positive constants c, c' depending on α and for all $n > \frac{1}{\log\alpha}$. It follows from the Borel–Cantelli Lemma (Part II) that with probability one,

$$B_{t_{n+1}} - B_{t_n} > \theta\varphi(t_{n+1}) \ i.o. \tag{10.14}$$

Also, by (10.11) and replacing $\{B_t : t \geq 0\}$ by the standard Brownian motion $\{-B_t : t \geq 0\}$,

$$\underline{\lim}_{t\to\infty} \frac{B_t}{\varphi(t)} \geq -1, \ a.s. \tag{10.15}$$

Since $t_{n+1} = \alpha t_n > t_n$, we have

$$\frac{B_{t_{n+1}}}{\sqrt{2t_{n+1}\log\log t_{n+1}}} = \frac{B_{t_{n+1}} - B_{t_n}}{\sqrt{2t_{n+1}\log\log t_{n+1}}} + \frac{1}{\sqrt{\alpha}}\frac{B_{t_n}}{\sqrt{2t_n(\log\log t_n + \log\log \alpha)}}. \tag{10.16}$$

Now, using (10.14) and (10.15), it follows that with probability one,

$$\overline{\lim}_{n\to\infty} \frac{B_{t_{n+1}}}{\varphi(t_{n+1})} \geq \theta - \frac{1}{\sqrt{\alpha}} = \frac{\alpha - 1}{\alpha} - \frac{1}{\sqrt{\alpha}}. \tag{10.17}$$

Since $\alpha > 1$ may be selected arbitrarily large, one has with probability one that

$$\overline{\lim}_{t\to\infty} \frac{B_t}{\varphi(t)} \geq \overline{\lim}_{n\to\infty} \frac{B_{t_{n+1}}}{\varphi(t_{n+1})} \geq 1. \tag{10.18}$$

This completes the computation of the limit superior. To get the limit inferior simply replace $\{B_t : t \geq 0\}$ by $\{-B_t : t \geq 0\}$. ∎

The time inversion property for Brownian motion turns the law of the iterated logarithm (LIL) into a statement concerning the degree (or lack) of *local smoothness*. (Also see Exercise 7).

Corollary 10.10 Each of the following holds with probability one:

$$\overline{\lim}_{t\to 0} \frac{B_t}{\sqrt{2t\log\log \frac{1}{t}}} = 1, \qquad \underline{\lim}_{t\to 0} \frac{B_t}{\sqrt{2t\log\log \frac{1}{t}}} = -1.$$

Exercise Set X

1. (*Ornstein–Uhlenbeck Process*) Fix parameters $\gamma > 0, \sigma > 0, x \in \mathbb{R}$. Use the Kolmogorov–Chentsov theorem to obtain the existence of a continuous Gaussian process $X = \{X_t : t \geq 0\}$ starting at $X_0 = x$ with $\mathbb{E}X_t = xe^{-\gamma t}$, and $\mathrm{Cov}(X_s, X_t) = \frac{\sigma^2}{\gamma}e^{-\gamma t}\sinh(\gamma s), 0 < s \leq t$.

2. (i) Use Feller's tail estimate (Lemma 2, Chapter IV). to prove that $\max\{|B_i - B_{i-1}| : i = 1, 2, \ldots, N + 1\}/N \to 0$ a.s. as $N \to \infty$.
   (ii) Without using the law of the iterated logarithm for standard Brownian motion B, show directly that $\limsup_{n\to\infty} \frac{B_n}{\sqrt{2n\log n}} \leq 1$ almost surely.

3. Show that with probability one, standard Brownian motion has arbitrarily large zeros. [*Hint*: Apply the LIL.]

4. Fix $t \geq 0$ and use the law of the iterated logarithm to show that $\lim_{h \to 0} \frac{B_{t+h} - B_t}{h}$ exists with probability zero. [*Hint*: Check that $Y_h := B_{t+h} - B_t, h \geq 0$, is distributed as standard Brownian motion starting at 0. Consider $\frac{1}{h} Y_h = \frac{Y_h}{\sqrt{2h \log \log(1/h)}} \cdot \frac{\sqrt{2h \log \log(1/h)}}{h}$.]

5. For the simple symmetric random walk, find the distributions of the extremes: (a) $M_N = \max\{S_j : j = 0, \dots, N\}$, and (b) $m_N = \min\{S_j : 0 \leq j \leq N\}$.

6. Consider the simple symmetric random walk $S_0 = 0, S_n = X_1 + \cdots + X_n, n \geq 1$, where $X_k, k \geq 1$, are iid symmetric Bernoulli ± 1 valued random variables. Denote the range by $R_n = \max_{m \leq n} S_m - \min_{m \leq n} S_m, n \geq 1$. Show that $\frac{R_n}{\sqrt{n}}$ converges in distribution to a nonnegative random variable as $n \to \infty$.

7. (*Lévy Modulus of Continuity*[1]) Use the wavelet construction $B_t := \sum_{n,k} Z_{n,k} S_{n,k}$ $(t), 0 \leq t \leq 1$, of standard Brownian motion to establish the following fine-scale properties.

   (i) Let $0 < \delta < \frac{1}{2}$. With probability one there is a random constant K such that if $|t - s| \leq \delta$ then $|B_t - B_s| \leq K \sqrt{\delta \log \frac{1}{\delta}}$. [*Hint*: Fix N and write the increment as a sum of three terms: $B_t - B_s = Z_{00}(t - s) + \sum_{n=0}^{N} \sum_{k=2^n}^{2^{n+1}-1} Z_{n,k} \int_s^t H_{n,k}(u) du + \sum_{n=N+1}^{\infty} \sum_{k=2^n}^{2^{n+1}-1} Z_{n,k} \int_s^t H_{n,k}(u) du = a + b + c$. Check that for a suitable (random) constant K' one has $|b| \leq |t - s| K' \sum_{n=0}^{N} n^{\frac{1}{2}} 2^{\frac{n}{2}} \leq |t - s| K' \frac{\sqrt{2}}{\sqrt{2}-1} \sqrt{N} 2^{\frac{N}{2}}$, and $|c| \leq K' \sum_{n=N+1}^{\infty} n^{\frac{1}{2}} 2^{-\frac{n}{2}} \leq K' \frac{\sqrt{2}}{\sqrt{2}-1} \sqrt{N} 2^{-\frac{N}{2}}$. Use these estimates, taking $N = [-\log_2(\delta)]$ such that $\delta 2^N \sim 1$, to obtain the bound $|B_t - B_s| \leq |Z_{00}|\delta + 2K' \sqrt{-\delta \log_2(\delta)}$. This is sufficient since $\delta < \sqrt{\delta}$.]

   (ii) The modulus of continuity is sharp in the sense that with probability one, there is a sequence of intervals $(s_n, t_n), n \geq 1$, of respective lengths $t_n - s_n \to 0$ as $n \to \infty$ such that the ratio $\frac{B_{t_n} - B_{s_n}}{\sqrt{-(t_n - s_n) \log(t_n - s_n)}}$ is bounded below by a positive constant. [*Hint*: Use Borel–Cantelli I together with Feller's tail probability estimate for the Gaussian distribution to show that $P(A_n \ i.o.) = 0$, where $A_n := [|B_{k2^{-n}} - B_{(k-1)2^{-n}}| \leq c\sqrt{n2^{-n}}, k = 1, \dots, 2^n]$ and c is fixed in $(0, \sqrt{2 \log 2})$. Interpret this in terms of the certain occurrence of the complimentary event $[A_n \ i.o.]^c$.]

   (iii) The paths of Brownian motion are a.s. nowhere differentiable.

---

[1] The calculation of the modulus of continuity for Brownian motion is due to Lévy, P. (1937). However this exercise follows Pinsky, M. (1999): Brownian continuity modulus via series expansions, *J. Theor. Probab.* **14** (1), 261–266.

Chapter XI
Strong Markov Property, Skorokhod Embedding, and Donsker's Invariance Principle

This chapter ties together a number of the topics introduced in the text via applications to the further analysis of Brownian motion, a fundamentally important stochastic process whose existence was established in Chapter VII and, independently, in Chapter IX.

The discrete-parameter random walk was introduced in Chapter II, where it was shown to have the Markov property. Markov processes on a general state space S with a given transition probability $p(x, dy)$ were introduced in Chapter IX (see Example 1 and Remark 9.4 in Chapter IX). Generalizing from this example, a sequence of random variables $\{X_n : n \geq 0\}$ defined on a probability space $(\Omega, \mathcal{F}, P)$ with values in a measurable space $(S, \mathcal{S})$ has the **Markov property** if for every $m \geq 0$, the conditional distribution of X_{m+1} given $\mathcal{F}_m := \sigma(X_j, 0 \leq j \leq m)$ is the same as its conditional distribution given $\sigma(X_m)$. In particular, the conditional distribution is a function of X_m, denoted by $p_m(X_m, dy)$, where $p_m(x, dy)$, $x \in S$ is referred to as the (one-step) **transition probability** at time m and satisfies the following:

1. For $x \in S$, $p_m(x, dy)$ is a probability on $(S, \mathcal{S})$.
2. For $B \in \mathcal{S}$, the function $x \to p_m(x, B)$ is a real-valued measurable function on S.

In the special case that $p_m(x, dy) = p(x, dy)$, for every $m \geq 0$, the transition probabilities are said to be **homogeneous** or **stationary**. Unless stated otherwise, Markov processes considered in this book are homogenous.

With the random walk example as background, let us recall some basic definitions. Let P_z denote the **distribution** of a discrete-parameter stochastic process $X = \{X_n : n \geq 0\}$, i.e., a probability on the product space $(S^\infty, \mathcal{S}^{\otimes \infty})$, with transition probability $p(x, dy)$ and initial distribution $P(X_0 = z) = 1$. The notation $\mathbb{E}_z$ is used to denote expectations with respect to the probability P_z.

Definition 11.1 Fix $m \geq 0$. The **after-m** (future) process is defined by $X_m^+ := \{X_{n+m} : n \geq 0\}$.

© Springer International Publishing AG 2016
R. Bhattacharya and E.C. Waymire, *A Basic Course in Probability Theory*,
Universitext, DOI 10.1007/978-3-319-47974-3_XI

It follows from the definition of a Markov process $\{X_n : n = 0, 1, 2, \dots\}$ with a stationary transition probability given above that for every $n \geq 0$ the conditional distribution of $(X_m, X_{m+1}, \dots, X_{m+n})$, given $\sigma(X_0, \dots, X_m)$ is the same as the P_x-distribution of $(X_0, \dots, X_n)$, evaluated at $x = X_m$. To see this, let f be a bounded measurable function on $(S^{n+1}, \mathcal{S}^{\otimes(n+1)})$. Then the claim is that

$$\mathbb{E}\big(f(X_m, X_{m+1}, \dots, X_{m+n}) | \sigma(X_0, \dots, X_m)\big) = g_0(X_m), \qquad (11.1)$$

where given $X_0 = x$,

$$g_0(x) := \mathbb{E}_x f(X_0, X_1, \dots, X_n). \qquad (11.2)$$

For $n = 0$ this is trivial. For $n \geq 1$, first take the conditional expectation of $f(X_m, X_{m+1}, \dots, X_{m+n})$, given $\sigma(X_0, \dots, X_m, \dots, X_{m+n-1})$ to get, by the Markov property, that

$$\begin{aligned}
&\mathbb{E}\big(f(X_m, X_{m+1}, \dots, X_{m+n}) \,|\, \sigma(X_0, \dots, X_m, \dots, X_{m+n-1})\big) \\
&= \mathbb{E}\big(f(x_m, \dots, x_{m+n-1}, X_{m+n}) | \sigma(X_{m+n-1})\big)|_{x_m = X_m, \dots, x_{m+n-1} = X_{m+n-1}} \\
&= \int_S f(X_m, \dots, X_{m+n-1}, x_{m+n}) p(X_{m+n-1}, dx_{m+n}) \\
&= g_{n-1}(X_m, \dots, X_{m+n-1}), \quad \text{say.} \qquad (11.3)
\end{aligned}$$

Next take the conditional expectation of the above with respect to $\sigma(X_0, \dots, X_{m+n-2})$ to get

$$\begin{aligned}
&\mathbb{E}\big(f(X_m, X_{m+1}, \dots, X_{m+n}) \,|\, \sigma(X_0, \dots, X_m, \dots, X_{m+n-2})\big) \\
&= \mathbb{E}\big(g_{n-1}(X_m, \dots, X_{m+n-1}) | \sigma(X_0, \dots, X_{m+n-2})\big) \\
&= \mathbb{E}\big(g_{n-1}(x_m, \dots, x_{m+n-2}, X_{m+n-1}) | \sigma(X_{m+n-2})\big)|_{x_m = X_m, \dots, x_{m+n-2} = X_{m+n-2}} \\
&= \mathbb{E}\int_S g_{n-1}(X_m, \dots, X_{m+n-2}, x_{m+n-1}) p(X_{m+n-2}, dx_{m+n-1}) \\
&= g_{n-2}(X_m, \dots, X_{m+n-2}), \quad \text{say.} \qquad (11.4)
\end{aligned}$$

Continuing in this manner one finally arrives at

$$\begin{aligned}
&\mathbb{E}\big(f(X_m, X_{m+1}, \dots, X_{m+n}) \,|\, \sigma(X_0, \dots, \dots, X_m)\big) \\
&= \mathbb{E}\big(g_1(X_m, X_{m+1}) | \sigma(X_0, \dots, \dots, X_m)\big) \\
&= \int_S g_1(X_m, x_{m+1}) p(X_m, dx_{m+1}) = g_0(X_m), \quad \text{say.} \qquad (11.5)
\end{aligned}$$

Now, on the other hand, let us compute $\mathbb{E}_x f(X_0, X_1, \dots, X_n)$. For this, one follows the same steps as above, but with $m = 0$. That is, first take the conditional expectation of $f(X_0, X_1, \dots, X_n)$, given $\sigma(X_0, X_1, \dots, X_{n-1})$, arriving at $g_{n-1}(X_0, X_1, \ldots, X_{n-1})$. Then take the conditional expectation of this given $\sigma(X_0, X_1, \ldots, X_{n-2})$, arriving at $g_{n-2}(X_0, \ldots, X_{n-2})$, and so on. In this way one again arrives at $g_0(X_0)$, which is (11.1) with $m = 0$, or (11.2) with $x = X_m$.

Since finite-dimensional cylinders $C = B \times S^\infty$, $B \in \mathcal{S}^{\otimes(n+1)}$ $(n = 0, 1, 2, \ldots)$ constitute a π-system, and taking $f = \mathbf{1}_B$ in (11.1), (11.2), one has, for every $A \in \sigma(X_0, \ldots, X_m)$,

$$\mathbb{E}\left(\mathbf{1}_A \mathbf{1}_{[X_m^+ \in C]}\right) = \mathbb{E}\left(\mathbf{1}_A \mathbf{1}_{[(X_m, X_{m+1}, \ldots, X_{m+n}) \in B]}\right) = \mathbb{E}\left(\mathbf{1}_A P_x(C)|_{x=X_m}\right), \qquad (11.6)$$

it follows from the π-λ theorem that

$$\mathbb{E}\left(\mathbf{1}_A \mathbf{1}_{[X_m^+ \in C]}\right) = \mathbb{E}\left(\mathbf{1}_A P_x(C)|_{x=X_m}\right), \qquad (11.7)$$

for all $C \in \mathcal{S}^\infty$; here $P_x(C)|_{x=X_m}$ denotes the (composite) evaluation of the function $x \mapsto P_x(C)$ at $x = X_m$. Thus, we have arrived at the following equivalent, but seemingly stronger, definition of the Markov property.

Definition 11.2 (*Markov Property*) We say that $X = \{X_n : n \geq 0\}$ has the (**homogeneous) Markov Property** if for every $m \geq 0$, the conditional distribution of X_m^+, given the σ-field $\mathcal{F}_m = \sigma(X_0, \ldots, X_m)$, is P_{X_m}, i.e., equals P_y on the set $[X_m = y]$.

This notion may be significantly strengthened by considering the future evolution given its history up to and including a random stopping time. Let us recall that given a stopping time τ, the **pre-τ σ-field** $\mathcal{F}_\tau$ is defined by

$$\mathcal{F}_\tau = \{A \in \mathcal{F} : A \cap [\tau = m] \in \mathcal{F}_m, \forall m \geq 0\}. \qquad (11.8)$$

Definition 11.3 The **after-τ process** $X_\tau^+ = \{X_\tau, X_{\tau+1}, X_{\tau+2}, \ldots\}$ is well defined on the set $[\tau < \infty]$ by $X_\tau^+ = X_m^+$ on $[\tau = m]$.

The following theorem shows that for discrete-parameter Markov processes, this stronger (Markov) property that "conditionally given the past and the present the future starts afresh at the present state" holds more generally for a stopping time τ in place of a constant "present time" m.

Theorem 11.1 (*Strong Markov Property*) Let τ be a stopping time for the process $\{X_n : n \geq 0\}$. If this process has the Markov property of Definition 11.2, then on $[\tau < \infty]$ the conditional distribution of the after $-\tau$ process X_τ^+, given the pre-τ σ-field $\mathcal{F}_\tau$, is P_{X_τ}.

Proof Let f be a real-valued bounded measurable function on $(S^\infty, \mathcal{S}^{\otimes\infty})$, and let $A \in \mathcal{F}_\tau$. Then

$$\mathbb{E}(\mathbf{1}_{[\tau<\infty]}\mathbf{1}_A f(X_\tau^+)) = \sum_{m=0}^{\infty} \mathbb{E}(\mathbf{1}_{[\tau=m]}\mathbf{1}_A f(X_m^+))$$

$$= \sum_{m=0}^{\infty} \mathbb{E}(\mathbf{1}_{[\tau=m]\cap A}\mathbb{E}_{X_m} f)$$

$$= \sum_{m=0}^{\infty} \mathbb{E}(\mathbf{1}_{[\tau=m]\cap A}\mathbb{E}_{X_\tau} f) = \mathbb{E}(\mathbf{1}_{[\tau<\infty]}\mathbf{1}_A \mathbb{E}_{X_\tau} f).$$

The second equality follows from the Markov property in Definition 11.2 since $A \cap [\tau = m] \in \mathcal{F}_m$. ∎

Let us now consider the continuous-parameter Brownian motion process along similar lines. It is technically convenient to consider the canonical model of standard Brownian motion $\{B_t : t \geq 0\}$ started at 0 on $\Omega = C[0, \infty)$ with $\mathcal{B}$ the Borel σ-field on $C[0, \infty)$, P_0, referred to as Wiener measure, and $B_t(\omega) := \omega(t), t \geq 0, \omega \in \Omega$, the coordinate projections. However, for continuous-parameter processes it is often useful to make sure that all events that have probability zero are included in the σ-field for Ω. For example, in the analysis of fine-scale structure of Brownian motion certain sets D may arise that *imply* events $E \in \mathcal{B}, D \subset E$, for which one is able to compute $P(E) = 0$. In particular, then, one would want to conclude that D is measurable (and hence assigned $P(D) = 0$ too). For this it may be necessary to replace $\mathcal{B}$ by its σ-field completion $\mathcal{F} = \overline{\mathcal{B}}$. We have seen that this can always be achieved, and there is no loss in generality in assuming that the underlying probability space $(\Omega, \mathcal{F}, P)$ is **complete** from the outset (see Appendix A).

Although the focus is on Brownian motion, just as for the above discussion of random walk, some of the definitions apply more generally and will be so stated in terms of a generic continuous-parameter stochastic process $\{Z_t : t \geq 0\}$, having continuous sample paths (outside a P-null set).

Definition 11.4 For fixed $s > 0$ the **after-s** process is defined by $Z_s^+ := \{Z_{s+t} : t \geq 0\}$.

Definition 11.5 A continuous-parameter stochastic process $\{Z_t : t \geq 0\}$, with a.s. continuous sample paths, such that for each $s > 0$, the conditional distribution of the after-s process Z_s^+ given $\sigma(Z_t, t \leq s)$ coincides with its conditional distribution given $\sigma(Z_s)$ is said to have the **Markov property**.

As will become evident from the calculations in the proof below, the Markov property of a Brownian motion $\{B_t : t \geq 0\}$ follows from the fact that it has independent increments.

Proposition 11.2 (*Markov Property of Brownian Motion*) Let P_x denote the distribution on $C[0, \infty)$ of standard Brownian motion $B^x = \{B_t^x = x + B_t : t \geq 0\}$ started at x. For every $s \geq 0$, the conditional distribution of $(B_s^x)^+ := \{B_{s+t}^x : t \geq 0\}$ given $\sigma(B_u^x : 0 \leq u \leq s)$ is $P_{B_s^x}$.

Proof Write $\mathcal{G} := \sigma(B_u^x : 0 \le u \le s)$. Let f be a real-valued bounded measurable function on $C[0, \infty)$. Then $\mathbb{E}\big(f((B_s^x)^+)|\mathcal{G}\big) = \mathbb{E}\big(\psi(U, V)|\mathcal{G}\big)$, where $U = B_s^x$, $V = \{B_{s+t}^x - B_s^x : t \ge 0\}$, $\psi(y, \omega) := f(\omega^y)$, $y \in \mathbb{R}$, $\omega \in C[0, \infty)$, and $\omega^y \in C[0, \infty)$ is given by $\omega^y(t) = \omega(t) + y$. By the substitution property for conditional expectation (Theorem 2.10), one has

$$\mathbb{E}\big(\psi(U, V)|\mathcal{G}\big) = h(U) = h(B_s^x),$$

where simplifying notation by writing $B_t = B_t^0$ and, in turn, $\{B_t : t \ge 0\}$ for a standard Brownian motion starting at 0,

$$h(y) = \mathbb{E}\psi(y, V) = \mathbb{E}\psi(y, \{B_t : t \ge 0\}) = \mathbb{E}f(B^y) = \int_{C[0,\infty)} f \, dP_y.$$

∎

It is sometimes useful to extend the definition of standard Brownian motion as follows.

Definition 11.6 Let $(\Omega, \mathcal{F}, P)$ be a probability space and $\mathcal{F}_t, t \ge 0$, a filtration. The **k-dimensional standard Brownian motion with respect to this filtration** is a stochastic process $\{B_t : t \ge 0\}$ on $(\Omega, \mathcal{F}, P)$ having (i) stationary, Gaussian increments $B_{t+s} - B_s$ with mean zero and covariance matrix $t I_k$; (ii) a.s. continuous sample paths $t \mapsto B_t$ on $[0, \infty) \to \mathbb{R}^k$; and (iii) for each $t \ge 0$, B_t is $\mathcal{F}_t$-measurable and $B_t - B_s$ is independent of $\mathcal{F}_s$, $0 \le s < t$. Taking $B_0 = 0$ a.s., then $B^x := \{x + B_t : t \ge 0\}$, is referred to as the **standard Brownian motion started at $x \in \mathbb{R}^k$** (with respect to the given filtration). The stochastic process $X_t = x + \mu t + \sigma B_t, t \ge 0$, where $x, \mu \in \mathbb{R}^k$, and σ is a $k \times k$ matrix defines the k-dimensional Brownian motion started at x and having *drift coefficient* μ and *diffusion coefficient* $D = \sigma^t \sigma$.

For example, one may take the completion $\mathcal{F}_t = \overline{\sigma}(B_s : s \le t), t \ge 0$, of the σ-field generated by the coordinate projections $t \mapsto \omega(t), \omega \in C[0, \infty)$. Alternatively, one may have occasion to use $\mathcal{F}_t = \sigma(B_s, s \le t) \vee \mathcal{G}$, where $\mathcal{G}$ is some σ-field independent of $\mathcal{F}$. The definition of the Markov property can be modified accordingly as follows.

Proposition 11.3 The Markov property of Brownian motions B^x on $\mathbb{R}^k$ defined on $(\Omega, \mathcal{F}, P)$ holds with respect to (i) the right-continuous filtration defined by

$$\mathcal{F}_{t+} := \bigcap_{\varepsilon > 0} \mathcal{F}_{t+\varepsilon} \qquad (t \ge 0), \tag{11.9}$$

where $\mathcal{F}_t = \mathcal{G}_t := \sigma(B_u : 0 \le u \le t)$, or (ii) $\mathcal{F}_t$ is the P-completion of $\mathcal{G}_t$, or (iii) $\mathcal{F}_t = \mathcal{G}_t \vee \mathcal{G}$ $(t \ge 0)$, where $\mathcal{G}$ is independent of $\mathcal{F}$.

Proof It is enough to prove that $B_{t+s} - B_s$ is independent of $\mathcal{F}_{s+}$ for every $t > 0$. Let $G \in \mathcal{F}_{s+}$ and $t > 0$. For each $\varepsilon > 0$ such that $t > \varepsilon$, $G \in \mathcal{F}_{s+\varepsilon}$, so that if $f \in C_b(\mathbb{R}^k)$, one has

$$\mathbb{E}(\mathbf{1}_G f(B_{t+s} - B_{s+\varepsilon})) = P(G) \cdot \mathbb{E}f(B_{t+s} - B_{s+\varepsilon}).$$

Letting $\varepsilon \downarrow 0$ on both sides,

$$\mathbb{E}(\mathbf{1}_G f(B_{t+s} - B_s)) = P(G)\mathbb{E}f(B_{t+s} - B_s).$$

Since the indicator of every closed subset of $\mathbb{R}^k$ is a decreasing limit of continuous functions bounded by 1 (see the proof of Alexandrov's theorem in Chapter VII), the last equality also holds for indicator functions f of closed sets. Since the class of closed sets is a π-system, and the class of Borel sets whose indicator functions f satisfy the equality is a σ-field, one can use the π-λ theorem to obtain the equality for all $B \in \mathcal{B}(\mathbb{R}^k)$. The proofs of (ii) and (iii) are left to Exercise 2.   ∎

One may define the σ-field governing the "past up to time τ" as the σ-field of events $\mathcal{F}_\tau$ given by

$$\mathcal{F}_\tau := \sigma(Z_{t \wedge \tau} : t \ge 0). \tag{11.10}$$

The stochastic process $\{\tilde{Z}_t : t \ge 0\} := \{Z_{t \wedge \tau} : t \ge 0\}$ is referred to as the **process stopped at** τ. Events in $\mathcal{F}_\tau$ depend only on the process stopped at τ. The stopped process contains no further information about the process $\{Z_t : t \ge 0\}$ beyond the time τ. Alternatively, in analogy with the discrete-parameter case, a description of the past up to time τ that is often more useful for checking whether a particular event belongs to it may be formulated as follows.

Definition 11.7 Let τ be a stopping time with respect to a filtration $\mathcal{F}_t, t \ge 0$. The **pre-τ σ-field** is

$$\mathcal{F}_\tau = \{F \in \mathcal{F} : F \cap [\tau \le t] \in \mathcal{F}_t \text{ for all } t \ge 0\}.$$

For example, using this definition it is simple to check that

$$[\tau \le t] \in \mathcal{F}_\tau, \forall t \ge 0, \qquad [\tau < \infty] \in \mathcal{F}_\tau. \tag{11.11}$$

Remark 11.1 We will always use[1] Definition 11.7, and not (11.10). Note, however, that $t \wedge \tau \le t$ for all t, so that $\sigma(X_{t \wedge \tau} : t \ge 0\}$ is contained in $\mathcal{F}_\tau$ (see Exercise 1).

The future relative to τ is the **after-τ process** $Z_\tau^+ = \{(Z_\tau^+)_t : t \ge 0\}$ obtained by viewing $\{Z_t : t \ge 0\}$ from time $t = \tau$ onwards, for $\tau < \infty$. This is

$$(Z_\tau^+)_t(\omega) = Z_{\tau(\omega)+t}(\omega), \qquad t \ge 0, \qquad \text{on } [\tau < \infty]. \tag{11.12}$$

---

[1]The proof of the equivalence of (11.10) and that of Definition 11.7 for processes with continuous sample paths may be found in Stroock and Varadahn (1980), p. 30.

Theorem 11.4 (*Strong Markov Property for Brownian Motion*) Let $\{B_t : t \geq 0\}$ be a k-dimensional Brownian motion with respect to a filtration $\{\mathcal{F}_t : t \geq 0\}$ starting at 0 and let P_0 denote its distribution (Wiener measure) on $C[0, \infty)$. For $x \in \mathbb{R}^k$ let P_x denote the distribution of the Brownian motion process $B_t^x := x + B_t, t \geq 0$, started at x. Let τ be a stopping time. On $[\tau < \infty]$, the conditional distribution of B_τ^+ given $\mathcal{F}_\tau$ is the same as the distribution of $\{B_t^y : t \geq 0\}$ starting at $y = B_\tau$. In other words, this conditional distribution is P_{B_τ} on $[\tau < \infty]$.

Proof First assume that τ has countably many values ordered as $0 \leq s_1 < s_2 < \cdots$. Consider a finite-dimensional function of the after-τ process of the form

$$h(B_{\tau+t_1'}, B_{\tau+t_2'}, \ldots, B_{\tau+t_r'}), \qquad [\tau < \infty], \tag{11.13}$$

where h is a bounded continuous real-valued function on $(\mathbb{R}^k)^r$ and $0 \leq t_1' < t_2' < \cdots < t_r'$. It is enough to prove

$$\mathbb{E}\left[h(B_{\tau+t_1'}, \ldots, B_{\tau+t_r'})\mathbf{1}_{[\tau<\infty]} \mid \mathcal{F}_\tau\right] = [\mathbb{E}h(B_{t_1'}^y, \ldots, B_{t_r'}^y)]_{y=B_\tau}\mathbf{1}_{[\tau<\infty]}. \tag{11.14}$$

That is, for every $A \in \mathcal{F}_\tau$ we need to show that

$$\mathbb{E}(\mathbf{1}_A h(B_{\tau+t_1'}, \ldots, B_{\tau+t_r'})\mathbf{1}_{[\tau<\infty]}) = \mathbb{E}\left(\mathbf{1}_A \left[\mathbb{E}h(B_{t_1'}^y, \ldots, B_{t_r'}^y)\right]_{y=B_\tau} \mathbf{1}_{[\tau<\infty]}\right). \tag{11.15}$$

Now

$$[\tau = s_j] = [\tau \leq s_j] \cap [\tau \leq s_{j-1}]^c \in \mathcal{F}_{s_j},$$

so that $A \cap [\tau = s_j] \in \mathcal{F}_{s_j}$. Express the left side of (11.15) as

$$\sum_{j=1}^{\infty} \mathbb{E}\left(\mathbf{1}_{A \cap [\tau=s_j]} h(B_{s_j+t_1'}, \ldots, B_{s_j+t_r'})\right). \tag{11.16}$$

By the Markov property, the jth summand in (11.16) equals

$$\mathbb{E}(\mathbf{1}_A \mathbf{1}_{[\tau=s_j]}[\mathbb{E}h(B_{t_1'}^y, \ldots, B_{t_r'}^y)]_{y=B_{s_j}}) = \mathbb{E}(\mathbf{1}_A \mathbf{1}_{[\tau=s_j]}[\mathbb{E}h(B_{t_1'}^y, \ldots, B_{t_r'}^y)]_{y=B_\tau}).$$

Summing this over j, one obtains the desired relation (11.15). This completes the proof in the case that τ has countably many values $0 \leq s_1 < s_2 < \cdots$.

The case of more general τ may be dealt with by approximating it by stopping times assuming countably many values. Specifically, for each positive integer n define

$$\tau_n = \begin{cases} \frac{j}{2^n} & \text{if } \frac{j-1}{2^n} < \tau \leq \frac{j}{2^n}, \quad j = 0, 1, 2, \ldots \\ \infty & \text{if } \tau = \infty. \end{cases} \tag{11.17}$$

Since

$$\left[\tau_n = \frac{j}{2^n}\right] = \left[\frac{j-1}{2^n} < \tau \le \frac{j}{2^n}\right] = \left[\tau \le \frac{j}{2^n}\right] \setminus \left[\tau \le \frac{j-1}{2^n}\right] \in \mathcal{F}_{j/2^n},$$

it follows that

$$[\tau_n \le t] = \bigcup_{j:j/2^n \le t} \left[\tau_n = \frac{j}{2^n}\right] \in \mathcal{F}_t \qquad \text{for all } t \ge 0.$$

Therefore, τ_n is a stopping time for each n and $\tau_n(\omega) \downarrow \tau(\omega)$ as $n \uparrow \infty$ for each $\omega \in \Omega$. Also one may easily check that $\mathcal{F}_\tau \subset \mathcal{F}_{\tau_n}$ from the definition (see Exercise 1). Let h be a bounded continuous function on $(\mathbb{R}^k)^r$. Define

$$\varphi(y) \equiv \mathbb{E}h(B_{t_1'}^y, \ldots, B_{t_r'}^y). \tag{11.18}$$

One may also check that φ is continuous using the continuity of $y \to (B_{t_1'}^y, \ldots, B_{t_r'}^y)$. Let $A \in \mathcal{F}_\tau (\subset \mathcal{F}_{\tau_n})$. Applying (11.15) to $\tau = \tau_n$ one has

$$\mathbb{E}(\mathbf{1}_A h(B_{\tau_n+t_1'}, \ldots, B_{\tau_n+t_r'})\mathbf{1}_{[\tau_n<\infty]}) = \mathbb{E}(\mathbf{1}_A \varphi(B_{\tau_n})\mathbf{1}_{[\tau_n<\infty]}). \tag{11.19}$$

Since h, φ are continuous, $\{B_t : t \ge 0\}$ has continuous sample paths, and $\tau_n \downarrow \tau$ as $n \to \infty$, Lebesgue's dominated convergence theorem may be used on both sides of (11.19) to get

$$\mathbb{E}(\mathbf{1}_A h(B_{\tau+t_1'}, \ldots, B_{\tau+t_r'})\mathbf{1}_{[\tau<\infty]}) = \mathbb{E}(\mathbf{1}_A \varphi(B_\tau)\mathbf{1}_{[\tau<\infty]}). \tag{11.20}$$

This establishes (11.15). Since finite-dimensional distributions determine a probability on $C[0, \infty)$, the proof is complete. ∎

Remark 11.2 Note that the proofs of the Markov property (Proposition 11.3) and the strong Markov property (Theorem 11.1) hold for $\mathbb{R}^k$-valued Brownian motions on $\mathbb{R}^k$ with arbitrary drift and positive definite diffusion matrix (Exercise 2).

The examples below illustrate the usefulness of Theorem 11.4 in typical computations. In all these examples $B = \{B_t : t \ge 0\}$ is a one-dimensional standard Brownian motion starting at zero. For $\omega \in C([0, \infty) : \mathbb{R})$ define, for every $a \in \mathbb{R}$,

$$\overline{\tau}_a^{(1)}(\omega) \equiv \overline{\tau}_a(\omega) := \inf\{t \ge 0 : \omega(t) = a\}, \tag{11.21}$$

and, recursively,

$$\overline{\tau}_a^{(r+1)}(\omega) := \inf\{t > \overline{\tau}_a^{(r)} : \omega(t) = a\}, \quad r \ge 1, \tag{11.22}$$

with the usual convention that the infimum of an empty set of numbers is ∞.

Similarly, in the context of the simple random walk, put $\Omega = \mathbb{Z}^\infty = \{\omega = (\omega_0, \omega_1, \dots) : \omega_n \in \mathbb{Z}, \forall n \ge 1\}$, and define

$$\overline{\tau}_a^{(1)}(\omega) \equiv \overline{\tau}_a(\omega) := \inf\{n \ge 0 : \omega_n = a\}, \tag{11.23}$$

and, recursively,

$$\overline{\tau}_a^{(r+1)}(\omega) := \inf\{n > \overline{\tau}_a^{(r)} : \omega_n = a\}, \quad r \ge 1. \tag{11.24}$$

Example 1 *(Recurrence of Simple Symmetric Random Walk)* Consider the simple symmetric random walk $S^x := \{S_n^x = x + S_n^0 : n \ge 0\}$ on $\mathbb{Z}$ started at x. Suppose one wishes to prove that $P_x(\overline{\tau}_y < \infty) = 1$ for $y \in \mathbb{Z}$. This may be obtained from the (ordinary) Markov property applied to $\varphi(x) := P_x(\overline{\tau}_y < \overline{\tau}_a), a \le x \le y$. For $a < x < y$, conditioning on S_1^x, and writing $S_1^{x+} = \{S_{1+n}^x : n \ge 0\}$, we have

$$\begin{aligned}
\varphi(x) &= P_x(\overline{\tau}_y < \overline{\tau}_a) = P(\overline{\tau}_y \circ S^x < \overline{\tau}_a \circ S^x) \\
&= P(\overline{\tau}_y \circ S_1^{x+} < \overline{\tau}_a \circ S_1^{x+}) \\
&= \mathbb{E}_x P_{S_1^x}(\overline{\tau}_y < \overline{\tau}_a) = \mathbb{E}\varphi(S_1^x) \\
&= \mathbb{E}(\mathbf{1}_{[S_1^x = x+1]}\varphi(x+1) + \mathbf{1}_{[S_1^x = x-1]}\varphi(x-1)) \\
&= \frac{1}{2}\varphi(x+1) + \frac{1}{2}\varphi(x-1), a < x < y, \tag{11.25}
\end{aligned}$$

with boundary values $\varphi(y) = 1$, $\varphi(a) = 0$. Solving, one obtains $\varphi(x) = (x - a)/(y - a)$. Thus $P_x(\overline{\tau}_y < \infty) = 1$ follows by letting $a \to -\infty$ using basic "continuity properties" of probability measures. Similarly, letting $y \to \infty$, one gets $P_x(\overline{\tau}_a < \infty) = 1$. Write $\overline{\eta}_a := \inf\{n \ge 1 : \omega_n = a\}$ for the **first return time to** a. Then $\overline{\eta}_a = \overline{\tau}_a$ on $\{\omega : \omega_0 \ne a\}$, and $\overline{\eta}_a > \overline{\tau}_a = 0$ on $\{\omega : \omega_0 = a\}$. By conditioning on S_1^x again, one has $P_x(\overline{\eta}_x < \infty) = \frac{1}{2}P_{x-1}(\overline{\tau}_x < \infty) + \frac{1}{2}P_{x+1}(\overline{\tau}_x < \infty) = \frac{1}{2} \cdot 1 + \frac{1}{2} \cdot 1 = 1$. While this calculation required only the Markov property, next consider the problem of showing that the process will return to y infinitely often. One would like to argue that, conditioning on the process up to its return to y, it merely starts over. This of course is the strong Markov property. So let us examine carefully the calculation to show that under P_x, the rth passage time to y, $\overline{\tau}_y^{(r)}$, is a.s. finite for every $r = 1, 2, \dots$. First note that by the (ordinary) Markov property, $P_x(\overline{\tau}_y < \infty) = 1 \ \forall x$. To simplify notation, write $\tau_y^{(r)} = \overline{\tau}_y^{(r)} \circ S^x$, and $S_{\tau_y^{(r)}}^{x+} = \{S_{\tau_y^{(r)}+n}^x : n \ge 0\}$ is then the after-$\tau_y^{(r)}$ process (for the random walk S^x). Applying the strong Markov property with respect to the stopping time $\tau_y^{(r)}$ one has, remembering that $S_{\tau_y^{(r)}}^x = y$,

$$P(\overline{\tau}_y^{(r+1)} < \infty) = P(\tau_y^{(r)} < \infty, \overline{\eta} \circ S_{\tau_y^{(r)}}^{x+} < \infty)$$

$$= \mathbb{E}\left(\mathbf{1}_{[\tau_y^{(r)} < \infty]} P_y(\overline{\eta} < \infty)\right)$$

$$= \mathbb{E}\left(\mathbf{1}_{[\tau_y^{(r)} < \infty]}\right) \cdot 1$$

$$= P(\overline{\tau}_y^{(r)} < \infty) = 1 \quad (r = 1, 2, \ldots), \qquad (11.26)$$

by induction on r. If $x = y$, then $\overline{\tau}_x^{(1)}$ is replaced by $\overline{\eta}_x$. Otherwise, the proof remains the same. This is equivalent to the **recurrence** of the state y in the sense that

$$P(S_n^x = y \text{ for infinitely many } n) = P(\cap_{r=1}^\infty [\tau_y^{(r)} < \infty]) = 1.$$

Example 2 *(Boundary Value Distribution of Brownian Motion)* Let $B^x = \{B_t^x := x + B_t : t \geq 0\}$ be a one-dimensional standard Brownian motion started at $x \in [c, d]$ for $c < d$, and let $\tau_y = \overline{\tau}_y \circ B^x$. The stopping time $\tau_c \wedge \tau_d$ denotes the first time for B^x to reach the "boundary" states $\{c, d\}$, referred to as a **hitting time** for B^x. Define

$$\psi(x) := P(B_{\tau_c \wedge \tau_d}^x = c) \equiv P(\{B_t^x : t \geq 0\} \text{ reaches } c \text{ before } d), \qquad (c \leq x \leq d). \tag{11.27}$$

Fix $x \in (c, d)$ and $h > 0$ such that $[x - h, x + h] \subset (c, d)$. In contrast to the discrete-parameter case there is no "first step" to consider. It will be convenient to consider $\tau = \tau_{x-h} \wedge \tau_{x+h}$, i.e., τ is the first time $\{B_t^x : t \geq 0\}$ reaches $x - h$ or $x + h$. Then $P(\tau < \infty) = 1$, by the law of the iterated logarithm (see Exercise 6 for an alternative argument). Now, by the strong Markov property (Theorem 11.4), applied with respect to τ,

$$\psi(x) = P(\{B_t^x : t \geq 0\} \text{ reaches } c \text{ before } d)$$

$$= P(\{(B_\tau^{x+})_t : t \geq 0\} \text{ reaches } c \text{ before } d)$$

$$= \mathbb{E}(P(\{(B_\tau^{x+})_t : t \geq 0\} \text{ reaches } c \text{ before } d \mid \mathcal{F}_\tau)). \tag{11.28}$$

The strong Markov property now gives that

$$\psi(x) = \mathbb{E}(\psi(B_\tau^x)), \tag{11.29}$$

so that by symmetry of Brownian motion, i.e., B^0 and $-B^0$ have the same distribution,

$$\psi(x) = \psi(x - h) P(B_\tau^x = x - h) + \psi(x + h) P(B_\tau^x = x + h)$$

$$= \psi(x - h)\frac{1}{2} + \psi(x + h)\frac{1}{2}, \tag{11.30}$$

where, by (11.27), $\psi(x)$ satisfies the boundary conditions $\psi(c) = 1$, $\psi(d) = 0$. Therefore,

$$\psi(x) = \frac{d - x}{d - c}. \tag{11.31}$$

Now, by (11.31) (see also Exercise 6),

$$P(\{B_t^x : t \ge 0\} \text{ reaches } d \text{ before } c) = 1 - \psi(x) = \frac{x - c}{d - c} \tag{11.32}$$

for $c \le x \le d$. It follows, on letting $d \uparrow \infty$ in (11.31), and $c \downarrow -\infty$ in (11.32) that

$$P(\bar{\mathcal{T}}_y < \infty) = 1 \qquad \text{for all } x, y. \tag{11.33}$$

As another illustrative application of the strong Markov property one may derive a Cantor-like structure of the random set of zeros of Brownian motion as follows.

Example 3.

Proposition 11.5 With probability one, the set $\mathcal{Z} := \{t \ge 0 : B_t = 0\}$ of zeros of the sample path of a one dimensional standard Brownian motion, starting at 0, is uncountable, closed, unbounded, and has no isolated point. Moreover, $\mathcal{Z}$ a.s. has Lebesgue measure zero.

Proof The law of iterated logarithm (LIL) may be applied as $t \downarrow 0$ to show that with probability one, $B_t = 0$ for infinitely many t in every interval $[0, \varepsilon]$. Since $t \mapsto B_t(\omega)$ is continuous, $\mathcal{Z}(\omega)$ is closed. Applying the LIL as $t \uparrow \infty$, it follows that $\mathcal{Z}(\omega)$ is unbounded a.s.

We will now show that for $0 < c < d$, the probability is zero of the event $A(c, d)$, say, that B has a single zero in $[c, d]$. For this consider the stopping time $\tau := \inf\{t \ge c : B_t = 0\}$. By the strong Markov property, B_τ^+ is a standard Brownian motion, starting at zero. In particular, τ is a point of accumulation of zeros from the right (a.s.). Also, $P(B_d = 0) = 0$. This implies $P(A(c, d)) = 0$. Considering all pairs of rationals c, d with $c < d$, it follows that $\mathcal{Z}$ has no isolated point outside a set of probability zero (see Exercise 4 for an alternate argument).

Finally, for each $T > 0$ let $H_T = \{(t, \omega) : 0 \le t \le T, B_t(\omega) = 0\} \subset [0, T] \times \Omega$. By the Fubini–Tonelli theorem, denoting the Lebesgue measure on $[0, \infty)$ by m, one has

$$(m \times P)(H_T) = \int_0^T \left\{ \int_\Omega \mathbf{1}_{[B_t = 0]}(\omega) P(d\omega) \right\} dt = \int_0^T P(B_t = 0) dt = 0, \tag{11.34}$$

so that $m(\{t \in [0, T] : B_t(\omega) = 0\}) = 0$ for P-almost all ω.   ∎

The following general consequence of the Markov property can also be useful in the analysis of the (infinitesimal) fine-scale structure of Brownian motion and may be viewed as a corollary to Proposition 11.3. As a consequence, for example, one sees that for any given function $\varphi(t), t > 0$, the event

$$D_\varphi := [B_t < \varphi(t) \text{ for all sufficiently small } t] \tag{11.35}$$

will certainly occur or is certain not to occur. Functions φ for which $P(D_\varphi) = 1$ are said to belong to the **upper class**. Thus $\varphi(t) = \sqrt{2t \log \log t}$ belongs to the upper class by the law of the iterated logarithm for Brownian motion (Theorem 10.9).

Proposition 11.6 *(Blumenthal's Zero–One Law)* With the notation of Proposition 11.3,

$$P(A) = 0 \text{ or } 1 \qquad \forall A \in \mathcal{F}_{0+}. \tag{11.36}$$

Proof It follows from (the proof of) Proposition 11.3 that $\mathcal{F}_{s+}$ is independent of $\sigma\{B_{t+s} - B_s : t \geq 0\} \; \forall s \geq 0$. Set $s = 0$ to conclude that $\mathcal{F}_{0+}$ is independent of $\sigma(B_t : t \geq 0) \supset \mathcal{F}_{0+}$. Thus $\mathcal{F}_{0+}$ is independent of $\mathcal{F}_{0+}$, so that $\forall A \in \mathcal{F}_{0+}$ one has $P(A) \equiv P(A \cap A) = P(A) \cdot P(A)$. ∎

In addition to the strong Markov property, another powerful tool for the analysis of Brownian motion is made available by observing that both the processes $\{B_t : t \geq 0\}$ and $\{B_t^2 - t : t \geq 0\}$ are martingales. Thus one has available the optional sampling theory.

Example 4 *(Hitting by BM of a Two-Point Boundary)* Let $\{B_t^x : t \geq 0\}$ be a one-dimensional standard Brownian motion starting at x, and let $c < x < d$. Let τ denote the stopping time, $\tau = \inf\{t \geq 0 : B_t^x = c \text{ or } d\}$. Then writing $\psi(x) := P(\{B_t^x\}_{t \geq 0}$ reaches d before c), one has (see (11.31))

$$\psi(x) = \frac{x - c}{d - c} \qquad c < x < d. \tag{11.37}$$

Applying the optional sampling theorem to the martingale $X_t := (B_t^x - x)^2 - t$, one gets $\mathbb{E}X_\tau = 0$, or $(d - x)^2 \psi(x) + (x - c)^2 (1 - \psi(x)) = \mathbb{E}\tau$, so that $\mathbb{E}\tau = [(d - x)^2 - (x - c)^2]\psi(x) + (x - c)^2$, or

$$\mathbb{E}\tau = (d - x)(x - c). \tag{11.38}$$

Consider now a Brownian motion $\{Y_t^x : t \geq 0\}$ with nonzero drift coefficient μ and diffusion coefficient $\sigma^2 > 0$, starting at x. Then $\{Y_t^x - t\mu : t \geq 0\}$ is a martingale, so that (see Exercise 6) $\mathbb{E}(Y_\tau^x - \mu\tau) = x$, i.e., $d\psi_1(x) + c(1 - \psi_1(x)) - \mu \mathbb{E}\tau = x$, or

$$(d - c)\psi_1(x) - \mu \mathbb{E}\tau = x - c, \tag{11.39}$$

where $\psi_1(x) = P(Y_\tau^x = d)$, i.e., $\{Y_t^x : t \geq 0\}$ reaches d before c. There are two unknowns, ψ_1 and $\mathbb{E}\tau$ in (11.39), so we need one more relation to solve for them. Consider the exponential martingale $Z_t := \exp\left\{\xi(Y_t^x - t\mu) - \frac{\xi^2\sigma^2}{2}t\right\}$ $(t \geq 1)$. Then $Z_0 = e^{\xi x}$, so that $e^{\xi x} = \mathbb{E}Z_\tau = \mathbb{E}\exp\{\xi(d - \tau\mu) - \xi^2\sigma^2\tau/2\}\mathbf{1}_{[Y_\tau^x = d]} + \mathbb{E}[\exp\{\xi(c - \tau\mu) - \xi^2\sigma^2\tau/2\}\mathbf{1}_{[Y_\tau^x = c]}]$. Take $\xi \neq 0$ such that the coefficient of τ in the exponent is zero, i.e., $\xi\mu + \xi^2\sigma^2/2 = 0$, or $\xi = -2\mu/\sigma^2$. Then optional stopping yields

$$e^{-2\mu x/\sigma^2} = \exp\{\xi d\}\psi_1(x) + \exp\{\xi c\}(1 - \psi_1(x)),$$

$$= \psi_1(x)\left[\exp\left\{-\frac{2\mu d}{\sigma^2}\right\} - \exp\left\{-\frac{2\mu c}{\sigma^2}\right\}\right] + \exp\left\{-\frac{2\mu c}{\sigma^2}\right\},$$

or

$$\psi_1(x) = \frac{\exp\{-2\mu x/\sigma^2\} - \exp\{-2\mu c/\sigma^2\}}{\exp\{-\frac{2\mu d}{\sigma^2}\} - \exp\{-\frac{2\mu c}{\sigma^2}\}}. \tag{11.40}$$

One may use this to compute $\mathbb{E}\tau$:

$$\mathbb{E}\tau = \frac{(d - c)\psi_1(x) - (x - c)}{\mu}. \tag{11.41}$$

Checking the hypothesis of the optional sampling theorem for the validity of the relations (11.37)–(11.41) is left to Exercise 6.

Our main goal for this chapter is to derive a beautiful result of Skorokhod (1965) representing a general random walk (partial sum process) as values of a Brownian motion at a sequence of successive stopping times (with respect to an enlarged filtration). This will be followed by a proof of the functional central limit theorem (invariance principle) based on the Skorokhod embedding representation. Recall that for $c < x < d$,

$$P(\tau_d^x < \tau_c^x) = \frac{x - c}{d - c}, \tag{11.42}$$

where $\tau_a^x := \overline{\tau}_a(B^x) \equiv \inf\{t \geq 0 : B_t^x = a\}$. Also,

$$\mathbb{E}(\tau_c^x \wedge \tau_d^x) = (d - x)(x - c). \tag{11.43}$$

Write $\tau_a = \tau_a^0$, $B^0 = B = \{B_t : t \geq 0\}$. Consider now a two-point distribution $F_{u,v}$ with support $\{u, v\}$, $u < 0 < v$, having mean zero. That is, $F_{u,v}(\{u\}) = v/(v - u)$ and $F_{u,v}(\{v\}) = -u/(v - u)$. It follows from (11.42) that with $\tau_{u,v} = \tau_u \wedge \tau_v$, $B_{\tau_{u,v}}$ has distribution $F_{u,v}$ and, in view of (11.43),

$$\mathbb{E}\tau_{u,v} = -uv = |uv|. \tag{11.44}$$

In particular, the random variable $Z := B_{\tau_{u,v}}$ with distribution $F_{u,v}$ is naturally **embedded** in the Brownian motion. We will see by the theorem below that any given non-degenerate distribution F with mean zero may be similarly embedded by randomizing over such pairs (u, v) to get a random pair (U, V) such that $B_{\tau_{U,V}}$ has distribution F, and $\mathbb{E}\tau_{U,V} = \int_{(-\infty,\infty)} x^2 F(dx)$, the variance of F. Indeed, this is achieved by the distribution γ of (U, V) on $(-\infty, 0) \times (0, \infty)$ given by

$$\gamma(du\,dv) = \theta(v - u)F_-(du)F_+(dv), \tag{11.45}$$

where F_+ and F_- are the restrictions of F to $(0, \infty)$ and $(-\infty, 0)$, respectively. Here θ is the normalizing constant given by

$$1 = \theta \left[\left(\int_{(0,\infty)} v F_+(dv) \right) F_-((-\infty, 0)) + \left(\int_{(-\infty,0)} (-u) F_-(du) \right) F_+(0, \infty) \right],$$

or, noting that the two integrals are each equal to $\frac{1}{2} \int_{-\infty}^{\infty} |x| F(dx)$ since the mean of F is zero, one has

$$1/\theta = \left(\frac{1}{2} \int_{-\infty}^{\infty} |x| F(dx) \right) [1 - F(\{0\})]. \tag{11.46}$$

Let $(\Omega, \mathcal{F}, P)$ be a probability space on which are defined (1) a standard Brownian motion $B \equiv B^0 = \{B_t : t \geq 0\}$, and (2) a sequence of i.i.d. pairs (U_i, V_i) independent of B, with the common distribution γ above. Let $\mathcal{F}_t := \sigma\{B_s : 0 \leq s \leq t\} \vee \sigma\{(U_i, V_i) : i \geq 1\}, t \geq 0$. Define the $\{\mathcal{F}_t : t \geq 0\}$-stopping times (Exercise 13)

$$T_0 \equiv 0, \quad T_1 := \inf\{t \geq 0 : B_t = U_1 \text{ or } V_1\},$$
$$T_{i+1} := \inf\{t > T_i : B_t = B_{T_i} + U_{i+1} \text{ or } B_{T_i} + V_{i+1}\} \, (i \geq 1).$$

Theorem 11.7 (*Skorokhod Embedding*) Assume that F has mean zero and finite variance. Then (a) B_{T_1} has distribution F, and $B_{T_{i+1}} - B_{T_i}$ $(i \geq 0)$ are i.i.d. with common distribution F, and (b) $T_{i+1} - T_i$ $(i \geq 0)$ are i.i.d. with

$$\mathbb{E}(T_{i+1} - T_i) = \int_{(-\infty,\infty)} x^2 F(dx). \tag{11.47}$$

Proof (a) Given (U_1, V_1), the conditional probability that $B_{T_1} = V_1$ is $\frac{-U_1}{V_1 - U_1}$. Therefore, for all $x > 0$,

$$P\left(B_{T_1} > x\right) = \theta \int_{\{v>x\}} \int_{(-\infty,0)} \frac{-u}{v-u} \cdot (v-u) F_-(du) F_+(dv)$$
$$= \theta \int_{\{v>x\}} \left\{ \int_{(-\infty,0)} (-u) F_-(du) \right\} F_+(dv)$$
$$= \int_{\{v>x\}} F_+(dv), \tag{11.48}$$

since $\int_{(-\infty,0)} (-u) F_-(du) = \frac{1}{2} \int |x| F(dx) = 1/\theta$. Thus the restriction of the distribution of B_{T_1} on $(0, \infty)$ is F_+. Similarly, the restriction of the distribution of B_{T_1} on $(-\infty, 0)$ is F_-. It follows that $P(B_{T_1} = 0) = F(\{0\})$. This shows that B_{T_1} has distribution F. Next, by the strong Markov property, the conditional distribution of $B_{T_i}^+ \equiv \{B_{T_i+t} : t \geq 0\}$, given $\mathcal{F}_{T_i}$, is $P_{B_{T_i}}$ (where P_x is the distribution of B^x).

Therefore, the conditional distribution of $B_{T_i}^+ - B_{T_i} \equiv \{B_{T_i+t} - B_{T_i} ; t \geq 0\}$, given $\mathcal{F}_{T_i}$, is P_0. In particular, $Y_i := \{(T_j, B_{T_j}) : 1 \leq j \leq i\}$ and $X^i := B_{T_i}^+ - B_{T_i}$ are independent. Since Y_i and X^i are functions of $B \equiv \{B_t : t \geq 0\}$ and $\{(U_j, V_j); 1 \leq j \leq i\}$, they are both independent of (U_{i+1}, V_{i+1}). Since $\tau^{(i+1)} := T_{i+1} - T_i$ is the first hitting time of $\{U_{i+1}, V_{i+1}\}$ by X^i, it now follows that (1) $(T_{i+1} - T_i \equiv \tau^{(i+1)}, B_{T_{i+1}} - B_{T_i} \equiv X^i_{\tau^{(i+1)}})$ is independent of $\{(T_j, B_{T_j}) : 1 \leq j \leq i\}$, and (2) $(T_{i+1} - T_i, B_{T_{i+1}} - B_{T_i})$ has the same distribution as (T_1, B_{T_1}).

(b) It remains to prove (11.47). But this follows from (11.44):

$$
\begin{aligned}
\mathbb{E}T_1 &= \theta \int_{(0,\infty)} \int_{(-\infty,0)} (-uv)(v-u) F_-(du) F_+(dv) \\
&= \theta \left[\int_{(0,\infty)} v^2 F_+(dv) \cdot \int_{(-\infty,0)} (-u) F_-(du) + \int_{(-\infty,0)} u^2 F_-(du) \cdot \int_{(0,\infty)} v F_+(dv) \right] \\
&= \int_{(0,\infty)} v^2 F_+(dv) + \int_{(-\infty,0)} u^2 F_-(du) = \int_{(-\infty,\infty)} x^2 F(dx).
\end{aligned}
$$

∎

We now present an elegant proof of **Donsker's invariance principle**, or **functional central limit theorem**, using Theorem 11.7. Consider a sequence of i.i.d. random variables Z_i ($i \geq 1$) with common distribution having mean zero and variance 1. Let $S_k = Z_1 + \cdots + Z_k$ ($k \geq 1$), $S_0 = 0$, and define the polygonal random function $S^{(n)}$ on $[0, 1]$ as follows:

$$
S_t^{(n)} := \frac{S_{k-1}}{\sqrt{n}} + n\left(n - \frac{k-1}{n}\right) \frac{S_k - S_{k-1}}{\sqrt{n}}
$$

$$
\text{for } t \in \left[\frac{k-1}{n}, \frac{k}{n}\right], \, 1 \leq k \leq n. \tag{11.49}
$$

That is, $S_t^{(n)} = \frac{S_k}{\sqrt{n}}$ at points $t = \frac{k}{n}$ ($0 \leq k \leq n$), and $t \mapsto S_t^{(n)}$ is linearly interpolated between the endpoints of each interval $\left[\frac{k-1}{n}, \frac{k}{n}\right]$.

Theorem 11.8 (*Invariance Principle*) $S^{(n)}$ converges in distribution to the standard Brownian motion, as $n \to \infty$.

Proof Let T_k, $k \geq 1$, be as in Theorem 11.7, defined with respect to a standard Brownian motion $\{B_t : t \geq 0\}$. Then the random walk $\{S_k : k = 0, 1, 2, \dots\}$ has the same distribution as $\{\widetilde{S}_k := B_{T_k} : k = 0, 1, 2, \dots\}$, and therefore, $S^{(n)}$ has the same distribution as $\widetilde{S}^{(n)}$ defined by $\widetilde{S}_{k/n}^{(n)} := n^{-\frac{1}{2}} B_{T_k}$ ($k = 0, 1, \dots, n$) and with linear interpolation between k/n and $(k+1)/n$ ($k = 0, 1, \dots, n-1$). Also, define, for each $n = 1, 2, \dots$, the standard Brownian motion $\widetilde{B}_t^{(n)} := n^{-\frac{1}{2}} B_{nt}$, $t \geq 0$. We will show that

$$
\max_{0 \leq t \leq 1} \left| \widetilde{S}_t^{(n)} - \widetilde{B}_t^{(n)} \right| \longrightarrow 0 \quad \text{in probability as } n \to \infty, \tag{11.50}
$$

which implies the desired weak convergence. Now

$$\max_{0 \le t \le 1} \left| \widetilde{S}_t^{(n)} - \widetilde{B}_t^{(n)} \right| \le n^{-\frac{1}{2}} \max_{1 \le k \le n} \left| B_{T_k} - B_k \right|$$

$$+ \max_{0 \le k \le n-1} \left\{ \max_{\frac{k}{n} \le t \le \frac{k+1}{n}} \left| \widetilde{S}_t^{(n)} - \widetilde{S}_{k/n}^{(n)} \right| + n^{-\frac{1}{2}} \max_{k \le t \le k+1} \left| B_t - B_k \right| \right\}$$

$$= I_n^{(1)} + I_n^{(2)} + I_n^{(3)}, \quad \text{say.} \tag{11.51}$$

Now, writing $\widetilde{Z}_k = \widetilde{S}_k - \widetilde{S}_{k-1}$, it is simple to check (Exercise 14) that as $n \to \infty$,

$$I_n^{(2)} \le n^{-\frac{1}{2}} \max\{|\widetilde{Z}_k| : 1 \le k \le n\} \to 0 \quad \text{in probability,}$$

$$I_n^{(3)} \le n^{-\frac{1}{2}} \max_{0 \le k \le n-1} \max\{|B_t - B_k| : k \le t \le k+1\} \to 0 \quad \text{in probability.}$$

Hence we need to prove, as $n \to \infty$,

$$I_n^{(1)} := n^{-\frac{1}{2}} \max_{1 \le k \le n} \left| B_{T_k} - B_k \right| \longrightarrow 0 \quad \text{in probability.} \tag{11.52}$$

Since $T_n/n \to 1$ a.s., by SLLN, it follows that (Exercise 14)

$$\varepsilon_n := \max_{1 \le k \le n} \left| \frac{T_k}{n} - \frac{k}{n} \right| \longrightarrow 0 \quad \text{as } n \to \infty \text{ (almost surely).} \tag{11.53}$$

In view of (11.53), there exists for each $\varepsilon > 0$ an integer n_ε such that $P(\varepsilon_n < \varepsilon) > 1 - \varepsilon$ for all $n \ge n_\varepsilon$. Hence with probability greater than $1 - \varepsilon$ one has for all $n \ge n_\varepsilon$ the estimate

$$I_n^{(1)} \le \max_{\substack{|s-t| \le n\varepsilon, \\ 0 \le s,t \le n+n\varepsilon}} n^{-\frac{1}{2}} |B_s - B_t| = \max_{\substack{|s-t| \le n\varepsilon, \\ 0 \le s,t \le n(1+\varepsilon)}} \left| \widetilde{B}_{s/n}^{(n)} - \widetilde{B}_{t/n}^{(n)} \right|$$

$$= \max_{\substack{|s'-t'| \le \varepsilon, \\ 0 \le s',t' \le 1+\varepsilon}} \left| \widetilde{B}_{s'}^{(n)} - \widetilde{B}_{t'}^{(n)} \right| \overset{d}{=} \max_{\substack{|s'-t'| \le \varepsilon, \\ 0 \le s',t' \le 1+\varepsilon}} \left| B_{s'} - B_{t'} \right|$$

$$\longrightarrow 0 \quad \text{as } \varepsilon \downarrow 0,$$

by the continuity of $t \to B_t$. Given $\delta > 0$ one may then choose $\varepsilon = \varepsilon_\delta$ such that for all $n \ge n(\delta) := n_{\varepsilon_\delta}$, $P(I_n^{(1)} > \delta) < \delta$. Hence $I_n^{(1)} \to 0$ in probability. ∎

For another application of Skorokhod embedding let us see how to obtain a **law of the iterated logarithm** (LIL) for sums of i.i.d. random variables using the LIL for Brownian motion.

Theorem 11.9 (*Law of the Iterated Logarithm*) Let $X_1, X_2, \ldots$ be an i.i.d. sequence of random variables with $\mathbb{E}X_1 = 0$, $0 < \sigma^2 := \mathbb{E}X_1^2 < \infty$, and let $S_n = X_1 + \cdots + X_n$, $n \ge 1$. Then with probability one,

$$\limsup_{n \to \infty} \frac{S_n}{\sqrt{2\sigma^2 n \log \log n}} = 1.$$

Proof By rescaling if necessary, one may take $\sigma^2 = 1$ without loss of generality. In view of the Skorokhod embedding one may replace the sequence $\{S_n : n \geq 0\}$ by the embedded random walk $\{\tilde{S}_n = B_{T_n} : n \geq 0\}$. By the SLLN one also has $\frac{T_n}{n} \to 1$ a.s. as $n \to \infty$. In view of the law of the iterated logarithm for Brownian motion, it is then sufficient to check that $\frac{\tilde{S}_{[t]} - B_t}{\sqrt{t \log \log t}} \to 0$ a.s. as $t \to \infty$. From $\frac{T_n}{n} \to 1$ a.s., it follows for given $\varepsilon > 0$ that with probability one, $\frac{1}{1+\varepsilon} < \frac{T_{[t]}}{t} < 1 + \varepsilon$ for all t sufficiently large. Let $t_n = (1 + \varepsilon)^n$, $n = 1, 2, \ldots$. Then one has

$$M_{t_n} := \max\left\{|B_s - B_{t_n}| : \frac{t_n}{1+\varepsilon} \leq s \leq t_n(1+\varepsilon)\right\}$$

$$\leq \max\left\{|B_s - B_{t_n}| : \frac{t_n}{1+\varepsilon} \leq s \leq t_n\right\} + \max\left\{|B_s - B_{t_n}| : t_n \leq s \leq t_n(1+\varepsilon)\right\}$$

$$\leq M_{n,1} + M_{n,2}, say.$$

Since $t_n - \frac{t_n}{1+\varepsilon} = \frac{t_n \varepsilon}{1+\varepsilon} < t_n(1+\varepsilon) - t_n = t_n\varepsilon$, $M_{n,2}$ is stochastically larger than $M_{n,1}$, so that $P(M_{t_n} > 2\sqrt{3\varepsilon t_n \log \log t_n}) \leq 2P(M_{n,2} > \sqrt{3\varepsilon t_n \log \log t_n})$. It follows from the scaling property of Brownian motion, using Lévy's Inequality and Feller's tail probability estimate, that

$$P\left(M_{t_n} > 2\sqrt{3\varepsilon t_n \log \log t_n}\right) \leq 2P\left(\max_{0 \leq u \leq 1} |B_u| > \sqrt{3 \log \log t_n}\right)$$

$$\leq 8P\left(B_1 \geq \sqrt{3 \log \log(t_n)}\right)$$

$$\leq \frac{8}{\sqrt{3 \log \log t_n}} \exp\left(-\frac{3}{2} \log \log t_n\right)$$

$$\leq cn^{-\frac{3}{2}}$$

for a constant $c = (\log(1 + \varepsilon))^{\frac{-3}{2}} > 0$. Summing over n, it follows from the Borel–Cantelli lemma I that with probability one, $M_{t_n} \leq \sqrt{3\varepsilon t_n \log \log t_n}$ for all but finitely many n. Since a.s. $\frac{1}{1+\varepsilon} < \frac{T_{[t]}}{t} < 1 + \varepsilon$ for all t sufficiently large, one has that

$$\limsup_{t \to \infty} \frac{|\tilde{S}_{[t]} - B_t|}{\sqrt{t \log \log t}} \leq \sqrt{3\varepsilon}.$$

Letting $\varepsilon \downarrow 0$ one has the desired result.   ∎

Exercise Set XI

1. (i) If τ_1, τ_2 are stopping times, show that $\tau_1 \vee \tau_2$ and $\tau_1 \wedge \tau_2$ are stopping times.
   (ii) If $\tau_1 \leq \tau_2$ are stopping times, show that $\mathcal{F}_{\tau_1} \subset \mathcal{F}_{\tau_2}$.

2. (i) Extend the Markov property for one-dimensional Brownian motion (Proposition 11.2) to k-dimensional Brownian motion with respect to a given filtration.
   (ii) Prove parts (ii), (iii) of Proposition 11.3.

3. Suppose that X, Y, Z are three random variables with values in arbitrary measurable spaces (S_i, S_i), $i = 1, 2, 3$. Assume that regular conditional distributions exist; see Chapter II for general conditions. Show that $\sigma(Z)$ is conditionally independent of $\sigma(X)$ given $\sigma(Y)$ if and only if the conditional distribution of Z given $\sigma(Y)$ a.s. coincides with the conditional distribution of Z given $\sigma(X, Y)$.

4. Prove that the event $A(c, d)$ introduced in the proof of Proposition 11.5 is measurable, i.e., the event $[\tau < d, B_t > 0 \,\forall \tau < t \leq d]$ is measurable.

5. Consider a Markov chain $X = \{X_n : n = 0, 1, 2 \ldots\}$ on a countable state space. Assume i is (point) recurrent: $P(X_n = i \text{ i.o.} | X_0 = i) = 1$. If j is a state such that $p_{ij}^{(n)} > 0$ for some n, prove that (i) the probability that j is reached starting from i is one, and (ii) j is (point) recurrent. [*Hint*: Consider visiting j between successive returns to i as i.i.d. events.]

6. Check the conditions for the application of the optional sampling theorem (Theorem 3.8(b)) for deriving (11.37)–(11.41). [*Hint*: For Brownian motion $\{Y_t^x : t \geq 0\}$ with a drift μ and diffusion coefficient $\sigma^2 > 0$, let $Z_1 = Y_1^x - x$, $Z_k = Y_k^x - Y_{k-1}^x (k \geq 1)$. Then $Z_1, Z_2, \ldots$ are i.i.d. and Corollary 3.10 applies with $a = c, b = d$. This proves $P(\tau < \infty) = 1$. The uniform integrability of $\{Y_{t \wedge \tau}^x : t \geq 0\}$ is immediate, since $c \leq Y_{t \wedge \tau}^x \leq d$ for all $t \geq 0$.]

7. Let $u' < 0 < v'$. Show that if $F = F_{u',v'}$ is the mean-zero two-point distribution concentrated at $\{u', v'\}$, then $P((U, V) = (u', v')) = 1$ in the Skorokhod embedding of F defined by $\gamma(du\,dv)$.

8. Given any distribution F on $\mathbb{R}$, let $\tau := \inf\{t \geq 0 : B_t = Z\}$, where Z is independent of $B = \{B_t : t \geq 0\}$ and has distribution F. Then $B_\tau = Z$. One can thus embed a random walk with (a nondegenerate) step distribution F (say, with mean zero) in different ways. However, show that $\mathbb{E}\tau = \infty$. [*Hint*: The stable distribution of $\tau_a := \inf\{t \geq 0 : B_t = a\}$ has infinite mean for every $a \neq 0$. To see this, use Corollary 10.6 to obtain $P(\tau_a > t) \geq 1 - 2P(B_t > a) = P(|B_t| \leq a) = P(|B_1| \leq \frac{a}{\sqrt{t}})$, whose integral over $[0, \infty)$ is divergent.]

9. Prove that $\varphi(\lambda) := \mathbb{E} \exp\{\lambda \tau_{u,v}\} \leq \mathbb{E} \exp\{\lambda \tau_{-a,a}\} < \infty$ for $\lambda < \lambda_0(a)$ for some $\lambda_0(a) > 0$, where $a = \max\{-u, v\}$. Here $\tau_{u,v}$ is the first passage time of standard Brownian motion to $\{u, v\}$, $u < 0 < v$. [*Hint*: Use Corollary 3.10 with $X_n := B_n - B_{n-1} (n \geq 1)$.]

10. (i) Show that for every $\lambda \geq 0$, $X_t := \exp\{\sqrt{2\lambda} B_t - \lambda t\}$, $t \geq 0$, is a martingale.
    (ii) Use the optional sampling theorem to prove $\varphi(\lambda) = 2\left(e^{\sqrt{2\lambda} a} + e^{-\sqrt{2\lambda} a}\right)^{-1}$, where $\varphi(\lambda) = \mathbb{E} \exp(\lambda \tau_{-a,a})$, in the notation of the previous exercise.

11. Refer to the notation of Theorem 11.8.

    (i) Prove that $T_i - T_{i-1}$ $(i \geq 1)$ has a finite moment-generating function in a neighborhood of the origin if F has compact support.

(ii) Prove that $\mathbb{E}T_1^2 < \infty$ if $\int |z|^2 F(dz) < \infty$. [*Hint*: $\tau_{u,v} \leq \tau_{-a,a}$ with $a :=$ $\max\{-u, v\} \leq v - u$ and $\mathbb{E}\tau_{U,V}^2 \leq c\theta \int (v - u)^2 F_+(dv) F_-(du)$ for some $c > 0$.]

12. In Theorem 11.7 suppose F is a symmetric distribution. Let X_i ($i \geq 1$) be i.i.d. with common distribution F and independent of $\{B_t : t \geq 0\}$. Let $\widetilde{T}_1 := \inf\{t \geq 0 : B_t \in \{-X_1, X_1\}$, $\widetilde{T}_i := \widetilde{T}_{i-1} + \inf\{t \geq 0 : B_{\widetilde{T}_{i-1}+t} \in \{-X_i, X_i\}\}$ ($i \geq 1$), $\widetilde{T}_0 = 0$.

   (i) Show that $B_{\widetilde{T}_i} - B_{\widetilde{T}_{i-1}}$ ($i \geq 1$) are i.i.d. with common distribution F, and $\widetilde{T}_i - \widetilde{T}_{i-1}$ ($i \geq 1$) are i.i.d.
   (ii) Prove that $\mathbb{E}\widetilde{T}_1 = \mathbb{E}X_1^2$, and $\mathbb{E}\widetilde{T}_1^2 = c\mathbb{E}X_1^4$, where c is a constant to be computed.
   (iii) Compute $\mathbb{E}e^{-\lambda\widetilde{T}_1}$ for $\lambda \geq 0$.

13. Prove that T_i ($i \geq 0$) defined by (11.47) are $\{\mathcal{F}_t\}$–stopping times, where $\mathcal{F}_t$ is as defined there.

14. (i) Let Z_k, $k \geq 1$, be i.i.d. with finite variance. Prove that $n^{-\frac{1}{2}} \max\{|Z_k| : 1 \leq k \leq n\} \to 0$ in probability as $n \to \infty$. [*Hint*: $nP(Z_1 > \sqrt{n}\,\varepsilon) \leq \frac{1}{\varepsilon^2} \mathbb{E}Z_1^2 \mathbf{1}[z : 1 \geq \sqrt{n}\,\varepsilon]$, $\forall \varepsilon > 0$].
   (ii) Derive (11.47). [*Hint*: $\varepsilon_n = \max_{1 \leq k \leq n} |\frac{T_k}{k} - 1| \cdot \frac{k}{n} \leq \{\max_{1 \leq k \leq k_0} |\frac{T_k}{k} - 1|\} \cdot \frac{k_0}{n} + \max_{k \geq k_0} |\frac{T_k}{k} - 1| \, \forall k_0 = 1, 2, \ldots.]$

Chapter XII
A Historical Note on Brownian Motion

Historically, the mathematical roots of Brownian motion lie in the central limit theorem (CLT). The first CLT seems to have been obtained in the early Eighteenth century by DeMoivre for the normal approximation to the binomial distribution (i.e., sum of i.i.d. Bernoulli 0 or 1-valued random variables).[1] In his treatise a century later, Laplace[2] obtained the far reaching generalization to sums of arbitrary independent and identically distributed random variables having finite moments of all orders. Although by the standards of rigor of present day mathematics, Laplace's derivation would not be considered complete, the essential ideas behind this remarkable result may be found in his work. The first rigorous proof[3] of the CLT was given by Lyapounov yet another 100 years later using characteristic functions under the Lyapounov condition for sums of independent, but not necessarily identically distributed, random variables having finite $(2 + \delta)$th moments for some $\delta > 0$. This moment condition was relaxed in 1922 by Lindeberg[4] to prove the more general CLT, and in 1935, Feller[5] showed that the conditions are necessary (as well as sufficient), under uniform asymptotic negligibility of summands. The most popular form of the CLT is that for i.i.d. summands with finite second moments due to Paul Lévy.[6]

There are not many results in mathematics that have had such a profound impact as the CLT, not only on probability and statistics but also on many other branches of mathematics, as well as the natural and physical sciences and engineering

---

[1]DeMoivre (1718).

[2]Laplace, P.-S. (1878–1912).

[3]Lyapunov, A.M. (1901). Nouvelle forme du théorème sur la limite de probabilités. *Mem. Acad. Imp. Sci. St.-Petersberg* **12** (5), 1–24.

[4]Lindeberg, J.W. (1922). Eine neue Herleitung des Exponentialgesetzes in der Wahrscheinlichkeitsrechnung. *Math. Zeitschr.* **15**, 211–225.

[5]Feller, W. (1935). Über den zentralen Grenzwertsatz der Wahrscheinlichkeitsrechnung. *Math. Zeitschr.* **40**, 521–559. Also, ibid (1937), **42**, 301–312.

[6]Lévy, P. (1925).

© Springer International Publishing AG 2016
R. Bhattacharya and E.C. Waymire, *A Basic Course in Probability Theory*,
Universitext, DOI 10.1007/978-3-319-47974-3_XII

as a whole. The idea of a stochastic process $\{B_t : t \geq 0\}$ that has independent Gaussian increments also derives from it. One may consider an infinite i.i.d. sequence $\{X_m : m \geq 1\}$ with finite second moments as in the CLT, and consider sums $S_n, S_{2n} - S_n, S_{3n} - S_{2n}, \ldots$, over consecutive disjoint blocks of n of these random variables X_m having mean μ and variance σ^2. The block sums are independent, each approximately Gaussian with mean $n\mu$ and variance $n\sigma^2$. If one scales the sums as $\frac{S_n - n\mu}{\sigma\sqrt{n}}, \frac{S_{2n} - S_n - n\mu}{\sigma\sqrt{n}} \ldots$, then in the limit one should get a process with independent Gaussian increments. If time is scaled so that one unit of time in the new *macroscopic* scale is equal to n units of time in the old scale, the $B_1, B_2 - B_1, B_3 - B_2, \ldots$ are independent Gaussian $\Phi_{0,1}$. Brownian motion is precisely such a process, but constructed for all times $t \geq 0$ and having continuous sample paths. The conception of such a process was previously introduced in a PhD thesis written in the year 1900 by Bachelier[7] as a model for the movements of stock prices.

Brownian motion is named after the Nineteenth century botanist Robert Brown, who observed under the microscope perpetual irregular motions exhibited by small grains or particles of the size of colloidal molecules immersed in a fluid. Brown[8] himself credited earlier scientists for having made similar observations. After some initial speculation that the movements are those of living organisms was discounted, the movements were attributed to inherent molecular motions. Independently of this debate and unaware of the massive experimental observations that had been made concerning this matter, Einstein[9] published a paper in 1905 in which he derived the *diffusion equation*

$$\frac{\partial C(t, x)}{\partial t} = D \left(\frac{\partial^2 C(t, x)}{\partial x_1^2} + \frac{\partial^2 C(t, x)}{\partial x_2^2} + \frac{\partial^2 C(t, x)}{\partial x_3^2} \right), \quad x = (x_1, x_2, x_3),$$

$$(12.1)$$

for the concentration $C(t, x)$ of large solute molecules of uniform size and spherical shape in a stationary liquid at a point x at time t. The argument (at least implicit in the above article) is that a solute molecule is randomly displaced frequently by collisions with the molecules of the surrounding liquid. Regarding the successive displacements as independent (and identically distributed) with mean vector zero and dispersion matrix $\mathrm{Diag}(d, d, d)$, one deduces a Gaussian distribution of the position

---

[7]Bachelier, L. (1900). Théorie de la spéculation. *Ann. Sci. École Norm. Sup.* **17**, 21–86; also see M. Davis & A. Etheridge (2006) for an English translation with a forward by Paul Samuelson.

[8]Brown, R. (1828). A brief account of microscopical observations made in the months of June, July, and August, 1827, on the particles contained in the pollen of plants; and on the general existence of active molecules in organic and inorganic bodies. *Philos. Magazine N.S.* **14**, 161–173.

[9]Einstein, A. (1905): Uber die von der molekularkinetischen Theorie der Warme geforderte Bewegung von in ruhenden Flussigkeiten suspendierten Teilchen, *Ann. der Physik*, **322** (8), 549560. Similar discoveries of Brownian motion were being made in Poland by the physicist Marian Smoluchoski who published his basic results in the paper von Smoluchowski, M. (1906): Zur kinetischen Theorie der Brownschen Molekularbewegung und der Suspensionen, *Ann. der Physik*, **326** (14), 756–780.

of the solute molecule at time t with mean vector zero and a dispersion matrix[10] $2t \, \text{Diag}(D, D, D)$, where $2D = fd$ with f as the average number of collisions, or displacements, per unit time. The law of large numbers (assuming that the different solute molecules move independently) then provides a Gaussian concentration law that is easily seen to satisfy the Eq. (12.1), away from the boundary. It is not clear that Laplace was aware of the profound fact that the operator $\Delta = \sum_1^3 \partial^2/\partial x_i^2$ in (12.1) bearing his name is intimately related to the central limit theorem he had derived.

Apprised of the experimental evidence concerning the so-called Brownian movement, Einstein titled his next article[11] on the subject, "On the theory of the Brownian movement." In addition to deriving the form of the Eq. (12.1), Einstein used classical thermodynamics, namely the Maxwell–Boltzmann steady-state (Gaussian) velocity distribution and Stokes' law of hydrodynamics (for the frictional force on a spherical particle immersed in a liquid) to express the *diffusion coefficient D* by $D = kT/3\pi\eta a$, where a is the radius of the spherical solute molecule, η is the coefficient of viscosity, T is the temperature, and k is the Boltzmann constant. In particular, the *physical parameters* are embodied in a *statistical parameter*. Based on this derivation, Jean Baptiste Perrin[12] estimated k or, equivalently, Avogadro's number, for which he was awarded the Nobel Prize in 1926. Meanwhile, in 1923, Wiener[13] proved that one may take Brownian paths to be continuous almost surely. That is, he constructed the probability measure Q, the so-called *Wiener measure* on $C[0, \infty)$, extending the normal distribution to infinitely many dimensions in the sense that the coordinate process $X_t(\omega) := \omega(t)$, $\omega \in C[0, \infty)$, $t \geq 0$, has independent Gaussian increments. Specifically, $X_{t+s} - X_t$ has the normal distribution $\Phi_{0,s} \equiv N(0, s)$, $\forall \, 0 \leq t < \infty$, $s > 0$, and $\{X_{t_{i+1}} - X_{t_i} : i = 1, 2, \ldots, m - 1\}$ are independent $\forall \, 0 \leq t_1 < t_2 < \cdots < t_m$ ($\forall \, m > 1$). This was a delicate result, especially since the Brownian paths turned out to have very little smoothness beyond continuity. Indeed, in 1933 it was shown by Paley, Wiener, and Zygmund[14] that with probability one, a Brownian path is continuous but nowhere differentiable. This says that a Brownian particle has no velocity, confirming some remarkable empirical observations in the early physics of Brownian motion. In his monograph "Atoms," Perrin exclaims: "The trajectories are confused and complicated so often and so rapidly that it is impossible to follow them; the trajectory actually measured is very much simpler and shorter than the real one. Similarly, the apparent mean speed of a grain during a given time varies *in the wildest way* in magnitude and direction, and does not tend to a limit as the time taken for an observation decreases, as may be easily shown by noting, in the camera lucida, the positions occupied by a grain from minute to minute, and then every 5 seconds, or,

---

[10]We have generally adapted a convention in which D is referred to as the diffusion coefficient, however this may not be universally held.

[11]Einstein, A. (1906). On the theory of the Brownian movement. *Ann. der Physik* **19**, 371–381. An English translation appears in Fürth (1954).

[12]Jean Perrin (1990), (French original, 1913).

[13]Wiener, N. (1923). Differential space. *J. Math. Phys.* **2**, 131–174.

[14]Paley, R.E.A.C., Wiener, N. and Zygmund, A. (1933). Notes on random functions. *Math. Zietschr.* **37**, 647–668.

better still, by photographing them every twentieth of a second, as has been done by Victor Henri Comandon, and de Broglie when kinematographing the movement. It is impossible to fix a tangent, even approximately, at any point on a trajectory, and we are thus reminded of the continuous underived functions of the mathematicians."

A more dynamical theory of Brownian (particle) motion was given by Ornstein and Uhlenbeck,[15] following the turn-of-the-century work of Langevin.[16]

The so-called Langevin equation used by Ornstein and Uhlenbeck is a *stochastic differential equation* given (in one dimension) by

$$dv(t) = -\beta v(t)dt + \sigma dB(t), \tag{12.2}$$

where $v(t)$ is the velocity of a Brownian molecule of mass m, $-m\beta v$ is the frictional force on it, and $\sigma^2 = 2\beta^2 D$ (D as above). By integrating $v(t)$ one gets a differentiable model of the Brownian molecule. If β and $\sigma^2 \to \infty$ such that $s^2/2\beta^2 = D$ remains a constant, then the position process converges to Einstein's model of Brownian motion (with diffusion coefficient $2D$), providing a scale range for which the models approximately agree.[17] Within the framework of stochastic differential equations one sees that the *steady-state velocity distribution* for the Langevin equation is a Gaussian distribution. On physical grounds this can be equated with the Maxwell–Boltzmann velocity distribution known from statistical mechanics and thermodynamics. In this way one may obtain Einstein's fundamental relationship between the physical parameters and statistical parameters mentioned above.

Brownian motion is a central notion throughout the theoretical development of stochastic processes and its applications. This rich history and its remarkable consequences are brought to life under several different guises in major portions of the theory of stochastic processes.

---

[15]Uhlenbeck, G.E. and Ornstein, L.S. (1930). On the theory of Brownian motion. *Phys. Rev.* **36**, 823–841; reprinted in Wax (1954). Also see Chandrasekhar, S. (1943). Stochastic problems in physics and astronomy. *Rev. Modern Physics* **15**, 2–91; reprinted in Wax (1954).

[16]Langevin, P. (1908). Sur La théorie du movement brownien. *C.R. Acad. Sci. Paris* **146**, 530–533.

[17]For a complete dynamical description see Nelson, E. (1967).

Chapter XIII
Some Elements of the Theory of Markov Processes and Their Convergence to Equilibrium

Special examples of Markov processes, such as random walks in discrete time and Brownian motion in continuous time, have occurred many times in preceding chapters as illustrative examples of martingales and Markov processes. There has also been an emphasis on their recurrence and transience properties. Moreover general discrete parameter Markov processes, also called *Markov chains*, were introduced in Chapter IX, and their important *strong Markov property* is derived in Chapter XI. In the present chapter, we begin afresh and somewhat differently with a focus on the existence of, and convergence to, a unique steady state distribution.

Suppose that $\mathbf{X} = \{X_0, X_1, X_2, \dots\}$ is a (discrete parameter) sequence of random variables on a probability space $(\Omega, \mathcal{F}, P)$ taking values in a measurable space $(S, \mathcal{S})$. The **Markov property** refers to the special type of statistical dependence that arises when the conditional distribution of the *after-n sequence* $\mathbf{X}^{n+} = \{X_n, X_{n+1}, \dots\}$ given $\sigma(X_0, X_1, \dots, X_n)$ coincides with that given $\sigma(X_n)$. If the sequence $\mathbf{X}$ has the Markov property then we refer to it as a **Markov chain** with state space $(S, \mathcal{S})$. The **initial distribution** μ of the initial state X_0 is a probability on $(S, \mathcal{S})$, and the **one-step transition probabilities** are defined by

$$p_n(x, B) = P(X_{n+1} \in B | X_0, \dots, X_n), x \in S, B \in \mathcal{S}, \text{ on } [X_n = x], n \geq 0.$$
(13.1)

The case in which these transition probabilities do not depend explicitly on n is referred to as that of **homogeneous** or **stationary transition probabilities**. Unless otherwise specified, we only consider Markov chains with homogeneous transition probabilities in this chapter.

Suppose $\mathbf{X}$ is a Markov chain having stationary one-step transition probabilities $p(x, B) \equiv p_n(x, B)$. Then, when the initial distribution is μ,

$$P(X_n \in B) = \int_S p^{(n)}(x, B)\mu(dx), \quad B \in \mathcal{S},$$
(13.2)

© Springer International Publishing AG 2016
R. Bhattacharya and E.C. Waymire, *A Basic Course in Probability Theory*,
Universitext, DOI 10.1007/978-3-319-47974-3_XIII

where $p^{(n)}(x, B)$ is the n-**step transition probability** defined recursively as $p^{(1)}(x, B)$
$= p(x, B)$, and

$$p^{(n+1)}(x, B) = \int_S p^{(n)}(y, B) p(x, dy), \quad B \in \mathcal{S}, x \in S, (n = 1, 2, \ldots). \quad (13.3)$$

Given an initial distribution μ and a transition probability $p(x, B)$, $(x \in S, B \in \mathcal{S})$,
a canonical construction of the Markov chain on the sequence space $(S^\infty, \mathcal{S}^{\otimes\infty})$
is discussed in Chapter IX. We denote this distribution by Q^μ, and write Q^x in
place of $Q^{\delta_x}, x \in S$. The Markov property may then be stated as: The conditional
distribution of $\mathbf{X}^{n+}$ given $\sigma(X_0, X_1, \ldots, X_n)$ is Q^{X_n}. That is, on the subset $[X_n = x]$,
this conditional distribution is Q^x, namely, the distribution of the Markov chain
starting at $x \in S$.

Definition 13.1 A probability π on $(S, \mathcal{S})$ is said to be an **invariant distribution** if

$$\int_S p(x, B) \pi(dx) = \pi(B), \quad \forall B \in \mathcal{S}. \quad (13.4)$$

This Definition 13.1 says that if X_0 has distribution π then so does X_1 and, by
iteration, X_n has distribution π for all $n \geq 1$. In fact the initial distribution π makes
the Markov chain a *stationary process* in the sense that the process $\mathbf{X}^{n+}$ has the same
distribution as $\mathbf{X}$ for each $n \geq 1$; Exercise 11.

Two of the most familiar examples of Markov chains are the following:

Example 1 *(Independent Sequence)* Let $X_1, X_2, \ldots$ be an i.i.d. sequence of S-
valued random variables with common distribution π, and let X_0 be an S-valued
random variable, independent of this sequence, and having distribution μ. Then
$\mathbf{X} = \{X_0, X_1, X_2, \ldots\}$ is a Markov chain with initial distribution μ and one-step transition probabilities $p(x, B) = \pi(B)$, $B \in \mathcal{S}, x \in S$. Clearly π is the unique invariant
distribution defined by (13.4).

Example 2 *(General Random Walk on $\mathbb{R}^k$)* Let $\{Y_n : n \geq 1\}$ be an i.i.d. sequence
with common distribution π on $\mathbb{R}^k$, and let Y_0 be an $\mathbb{R}^k$-valued random variable independent of this sequence. These define the *displacements* of the random walk. The
position process for the random walk is defined by $X_0 = Y_0, X_n = Y_0 + Y_1 + \cdots +$
$Y_n, n \geq 1$. Then $\mathbf{X} = \{X_0, X_1, X_2, \ldots\}$ is a Markov chain with initial distribution
that of Y_0 and transition probabilities $p(x, B) = \pi(B - x), x \in S, B \in \mathcal{B}(S)$. This
Markov chain has no invariant probability if $\pi(\{0\}) < 1$.

The following are some basic issues concerning invariant probabilities.

- Existence; not always, $S = \{1, 2, \ldots\}$, $p(x, \{x + 1\}) = 1, x = 1, 2 \ldots$. (also see
  Exercise 1)
- Uniqueness; not always, $S = \{1, 2\}$, $p(x, \{x\}) = 1, x = 1, 2$. (also see Exercise 5).
- Convergence; not always $S = \{1, 2\}$, $p(1, \{2\}) = p(2, \{1\}) = 1$. (also see Exam-
  ple 3 and Exercises 3).

- Rates of convergence; e.g., exponential versus algebraic bounds on an appropriate metric? (see Theorem 13.1 below, Exercise 3(d)).

The following theorem provides a benchmark result that eliminates the obstructions captured by the counterexamples. It covers a broad range of examples but is far from exhaustive.

Theorem 13.1 (*Doeblin Minorization*) Assume that there is a nonzero measure λ on $(S, \mathcal{S})$ and an integer $N \geq 1$ such that

$$p^{(N)}(x, B) \geq \lambda(B), \quad \forall x \in S, B \in \mathcal{S}.$$

Then, there is a unique invariant probability π such that

$$\sup_{x \in S} \sup_{B \in \mathcal{S}} |p^{(n)}(x, B) - \pi(B)| \leq (1 - \delta)^{\lceil \frac{n}{N} \rceil}, \quad n = 1, 2, \ldots, \tag{13.5}$$

where $\delta = \lambda(S)$.

Proof Notice that if $\lambda(S) = 1$ then, considering that the minorization inequality applies to both B and B^c, it follows that $p^{(N)}(x, B) = \lambda(B), x \in S, B \in \mathcal{S}$ is the invariant probability; use (13.3) to see $p^{(n)}(x, B) = \lambda(B)$ does not depend on $x \in S$, and both sides of (13.5) are zero. Now assume $\delta = \lambda(S) < 1$. Let d denote the total variation metric on $\mathcal{P}(S)$. Then recall Proposition 1.9 of Chapter I that $(\mathcal{P}(S), d)$ is a complete metric space and $d_1(\mu, \nu) := \sup\{| \int_S f d\mu - \int_S f d\nu | : f \in \mathbb{B}(S), |f| \leq 1\} = 2d(\mu, \nu)$, for all $\mu, \nu \in \mathcal{P}(S)$. Define $T^* : \mathcal{P}(S) \to \mathcal{P}(S)$ by $T^*\mu(B) = \int_S p(x, B)\mu(dx), B \in \mathcal{B}(S)$. One may use (13.5) to write

$$p^{(N)}(x, B) = \delta\gamma(B) + (1 - \delta)q(x, B), \tag{13.6}$$

where $\gamma(B) := \frac{\lambda(B)}{\delta}$, and $q(x, B) := \frac{p^{(N)}(x,B) - \lambda(B)}{1 - \delta}$ are both probability measures. It follows that for all measurable $f, |f| \leq 1$, and $\mu, \nu \in \mathcal{P}(S)$,

$$\int_S f(y) T^{*N} \mu(dy) - \int_S f(y) T^{*N} \nu(dy)$$

$$= \int_S \int_S f(y) p^{(N)}(x, dy)\mu(dx) - \int_S \int_S f(y) p^{(N)}(x, dy)\nu(dx)$$

$$= (1 - \delta)[\int_S \int_S f(y)q(x, dy)\mu(dx) - \int_S \int_S f(y)q(x, dy)\nu(dx). \tag{13.7}$$

This implies $d_1(T^{*N}\mu, T^{*N}\nu) \leq (1 - \delta)d_1(\mu, \nu)$. Iterating this one obtains (by induction)

$$d_1(T^{*Nk}\mu, T^{*Nk}\nu) \leq (1 - \delta)^k d_1(\mu, \nu), \quad k \geq 1. \tag{13.8}$$

Next observe that $\forall \mu \in \mathcal{P}(S)$, the sequence $\{T^{*Nk}\mu : k \geq 1\}$ is Cauchy for the metric d_1 since $T^{*N(k+r)}\mu = T^{*Nk}(T^{*Nr}\mu)$, and therefore has a limit π which is the unique invariant probability. Take $\mu(\cdot) = p(x, \cdot)$, and $\nu = \pi$ in (13.8) to complete the proof.   ∎

The following is a simple consequence.

Corollary 13.2 Suppose that $S = \{1, 2, \ldots, M\}$ is a finite set and $\{X_0, X_1, \ldots\}$ is a Markov chain on S with one-step transition probabilities $P(X_{n+1} = j | X_n = i)$ given by the transition probability matrix $p = ((p_{ij}))_{i,j \in S}$. If there is an N such that all entries of $p^N = ((p_{ij}^{(N)}))_{1 \leq i, j \leq M}$ are positive, then there is a unique invariant probability π on S, and $p_{i\cdot}^{(n)}$ converges to π exponentially fast and uniformly for all $i \in S$.

Proof Define $\lambda(B) = \sum_{j \in B} \lambda(\{j\})$, where $\lambda(\{j\}) = \min_{i \in S} p_{ij}^{(N)}$, $j \in S$, and the empty sum is defined to be zero. Then for each $B \subset S$,

$$p^{(N)}(i, B) = \sum_{j \in B} p_{ij}^{(N)} \geq \lambda(B).$$

The uniform exponential convergence follows from (13.5).   ∎

Example 3 *(Simple Symmetric Random Walk with Reflection)* Here $S = \{0, 1, \ldots, d - 1\}$ for some $d > 2$, and $p_{i,i+1} \equiv p(i, \{i + 1\}) = \frac{1}{2} = p(i, \{i - 1\}) \equiv p_{i,i-1}$, $1 \leq i \leq d - 2$, and $p_{0,1} = p_{d-1,d-2} = 1$. The unique solution to (13.4) is $\pi(\{0\}) = \pi(\{d - 1\}) = \frac{1}{2(d-1)}$, and $\pi(\{i\}) = 1/(d - 1)$, $1 \leq i \leq d - 2$. However, the hypothesis of Corollary 13.2, (or that of Theorem 13.1), does not hold. Indeed $p_{ij}^{(N)} = 0$ if N and $|i - j|$ have opposite parity; also see Exercise 4 in this regard.

Example 4 *(Fluctuation-Dissipation Effects)* Let $\theta \in (0, 1)$ and let $\varepsilon_1, \varepsilon_2, \ldots$ be an i.i.d. sequence of Gaussian mean zero, variance σ^2 random variables. Define a Markov process on $S = \mathbb{R}$ by $X_{n+1} = \theta X_n + \varepsilon_{n+1}$, $n = 0, 1, 2, \ldots$ for an initial state $X_0 = x \in S$. Then

$$X_n = \sum_{j=0}^{n-1} \theta^j \varepsilon_{n-j} =^{dist} \sum_{j=0}^{n-1} \theta^j \varepsilon_j, n = 1, 2, \ldots.$$

In particular, the limit distribution is Gaussian with mean zero and variance $\frac{1}{1-\theta^2}\sigma^2$.

Remark 13.1 Theorem 13.1, Corollary 13.2, and Example 4 concern so-called *irreducible Markov chains*, in the sense that for each $x \in S$ there is a positive integer $n = n(x)$ such that the n-step transition probability $p^{(n)}(x, B)$ is positive for every $B \in \mathcal{S}$ such that $\lambda(B) > 0$, for some nonzero reference measure λ on $(S, \mathcal{S})$. On the other hand, the Markov chain in Example 3 is not irreducible.

While the time asymptotic theory for irreducible Markov processes is quite well-developed, there are important examples for which irreducibility is too strong an hypothesis. The following example is presented to illustrate some useful theory in cases of non-irreducible Markov processes.

Example 5 *(A Fractional Linear Dynamical System; Products of Random Matrices)*
Let $S = [0, \infty)$ and let $A_n, B_n, n = 1, 2, \ldots$ be an i.i.d. sequence positive random
variables with $\mathbb{E} \log A_1 < 0$. Define a Markov chain on S by $X_0 = x \in S$, and

$$X_{n+1} = \frac{A_{n+1} X_n}{A_{n+1} X_n + B_{n+1}}, n = 0, 1, 2 \ldots.$$

Then $\pi = \delta_0$ is the unique invariant distribution. To see this observe that the compo-
sition of two fractional linear maps $\alpha_1 \circ \alpha_2(x) = \alpha_1(\alpha_2(x))$, $\alpha_j(x) = \frac{a_j x}{a_j x + b_j}$, $x \geq$
0, $j = 1, 2$, may be identified with the result of matrix multiplications of the two
matrices $\begin{pmatrix} a_j & 0 \\ a_j & b_j \end{pmatrix}$, $j = 1, 2$, to compute the composite coefficients.[1] In particular,
X_n may be identified as an n-fold matrix product whose diagonal entries are each dis-
tributed as $\prod_{j=1}^{n} A_j = \exp\{n \frac{\sum_{j=1}^{n} \log A_j}{n}\} \sim \exp\{n \mathbb{E} \log A_1\} \to 0$ almost surely, and
hence in distribution, as $n \to \infty$. The upper off-diagonal entry is zero, and the
lower off-diagonal entry is $\sum_{j=1}^{n} \prod_{i=1}^{j-1} B_i \prod_{i=j}^{n} A_i$. Since $\mathbb{E} \log A_1 < 0$, one has
$\frac{d}{dh} \mathbb{E} A_1^h = \mathbb{E} A_1^h \log A_1 = \mathbb{E} \log A_1 < 0$ at $h = 0$, and $\mathbb{E} A_1^h = 1$ at $h = 0$. One may
choose sufficiently small $h \in (0, 1)$ such that $\mathbb{E} A_1^h < 1$. For such a choice one then has
from sublinearity that $\mathbb{E} \big(\sum_{j=1}^{n} \prod_{i=1}^{j-1} B_i \prod_{i=j}^{n} A_i \big)^h \leq n (\mathbb{E} A_1^h)^n = n e^{n \log \mathbb{E} A_1^h} \to 0$
as $n \to \infty$. In fact by this, Chebyshev's inequality and the Borel–Cantelli argument
$\sum_{j=1}^{n} \prod_{i=1}^{j-1} B_i \prod_{i=j}^{n} A_i \to 0$ a.s. as $n \to \infty$, as well.

The previous two examples are illustrations of Markov processes that arise as
iterations of i.i.d. random maps, or so-called *random dynamical systems*.[2]

Example 6 *(Ehrenfest urn model)* The following model for heat exchange was introduced by P. and T. Ehrenfest in 1907, and later by Smoluchowski in 1916, to explain
an apparent paradox that threatened to destroy the basis of Boltzmann's kinetic theory of matter. In the kinetic theory, heat exchange between two bodies in contact is a
random process involving the exchange of energetic molecules, while in thermodynamics it is an orderly irreversible progression toward an equilibrium state in which
the (macroscopic) temperatures of two bodies in contact become (approximately)
equal. The main objective of kinetic theory was to explain how the larger scale thermodynamic equilibrium could be achieved, while allowing for statistical recurrence
of the random process. In fact, Zermelo argued forcefully that recurrence would
contradict thermodynamic irreversibility. However, Boltzmann was of the view that
the time required by the random process to pass from the equilibrium state to one of

---

[1]The authors thank our colleague Yevgeniy Kovchegov for suggesting this example to illustrate
products of random matrices. Such examples as this, including the positivity constraints, arise
naturally in the context of mathematical biology.

[2]A comprehensive treatment of such Markov processes can be found in Bhattacharya, R., and M.
Majumdar (2007). Limit distributions of products of random matrices has been treated in some
generality by Kaijser, T.(1978): A limit theorem for Markov chains on compact metric spaces
with applications to random matrices, *Duke Math. J.* **45**, 311–349; Kesten, H. and F. Spitzer (1984):
Convergence in distribution of products of random matrices, *Z. Wahrsch. Verw. Gebiete* **67** 363–386.

macroscopic nonequilibrium would be so large that such recurrence would be of no physical significance. Not all physicists were convinced of this reasoning.

So enter the Ehrenfests. Suppose that $2d$ balls labelled $1, 2, \ldots 2d$ are distributed between two boxes A and B at time zero. At each instant of time, a ball label is randomly selected, independently of the number of balls in either box, and that ball is moved from its current box to the other box. Suppose that there are initially Y_0 balls in box A, and let Y_n denote the number of balls in box A at the nth stage of this process. Then one may check that $Y = \{Y_0, Y_1, \ldots\}$ is a Markov chain on the state space $S = \{0, 1, 2, \ldots, 2d\}$ with one-step transition probabilities $p(y, y+1) = \frac{2d-y}{2d}$, $p(y, y-1) = \frac{y}{2d}$, $p(y, z) = 0$ otherwise. Moreover, Y has a unique invariant probability π with mean d, given by the binomial distribution with parameters $\frac{1}{2}, 2d$, i.e.,

$$\pi_j = \binom{2d}{j} 2^{-2d}, \quad j = 0, 1, \ldots, 2d.$$

Viewing the average state d of the invariant distribution π as thermodynamic equilibrium, the paradox is that, as a result of recurrence of the Markov chain, the state $j = 0$ of extreme disequilibrium is certain to eventually occur. The paradox can be resolved by calculating the average length of time to pass from $j = d$ to $j = 0$ in this kinetic theoretical model.[3]

The following proposition provides a general framework for such calculations.

Proposition 13.3 (*Birth–Death Markov Chain with Reflection*) Let $Y = \{Y_n : n = 0, 1, 2, \ldots\}$ be a Markov chain on the state space $S = \{0, 1, \ldots, N\}$ having stationary one-step transition probabilities $p_{i,i+1} = \beta_i, i = 0, 1, \ldots, N-1$, $p_{i,i-1} = \delta_i, i = 1, 2, \ldots, N$, $p_{0,1} = p_{N,N-1} = 1$, and $p_{ij} = 0$ otherwise, where $0 < \beta_i = 1 - \delta_i < 1$. Let

$$T_j = \inf\{n \geq 0 : Y_n = j\}, \quad j \in S,$$

denote the first-passage time to state $j \in S$. Then

$$m_i = \mathbb{E}_i T_0 = \sum_{j=1}^{i} \frac{\beta_j \beta_{j+1} \cdots \beta_{N-1}}{\delta_j \delta_{j+1} \cdots \delta_{N-1}} + \sum_{j=1}^{i} \sum_{k=j}^{N-1} \frac{\beta_j \cdots \beta_{k-1} \beta_k}{\delta_j \delta_{j+1} \cdots \delta_k \beta_k}, \quad 1 \leq i \leq N-1.$$

Proof The idea for the proof involves a scale-change technique that is useful for many Markov chains that do not skip over adjacent states; including one-dimensional diffusions having continuous paths. Specifically, one relabels the states $j \to u_j$ by

---

[3]The original calculations of the Ehrenfests and Smoluchowski were for the mean recurrence times. Such calculations are easily made from the general mean return-time formula $\mathbb{E}_i \tau_i = \frac{1}{\pi_i}$, where $\tau_i = \inf\{n \geq 1 : Y_n = i\}, i \in S$, for irreducible, ergodic Markov chains. In particular, using the formula for π and Stirling's formula, $\mathbb{E}_0 \tau_0 \sim 2^{20,000}$, $\mathbb{E}_d \tau_d \sim 100\sqrt{\pi}$, for the same numerical values for the number of balls and transition rate; e.g., see Kac (1947): Random walk and the theory of Brownian motion, *Am. Math. Monthly*, **54**(7), 369–391. The mean-return time formula and more general theory can be found in standard treatments of discrete parameter Markov processes.

an increasing sequence $0 = u_0 < u_1 < \cdots < u_N = 1$ determined by the requirement that the probabilities of reaching one boundary before another, starting in-between, is proportional to the respective distance to the boundary, as in the examples of simple symmetric random walk on $\mathbb{Z}$, and one-dimensional standard Brownian motion. That is,

$$\psi(i) = P(Y \text{ reaches } 0 \text{ before } N \mid Y_0 = i) = \frac{u_N - u_i}{u_N - u_0}, \quad i \in S.$$

Since

$$\psi(i) = \beta_i \psi(i+1) + \delta_i \psi(i-1), \, 1 \le i \le N-1,$$

and $\psi(0) = 1$, $\psi(N) = 0$, one has

$$u_{i+1} - u_i = \frac{\delta_i}{\beta_i}(u_i - u_{i-1}) = \frac{\delta_1 \cdots \delta_i}{\beta_1 \cdots \beta_i}(u_1 - u_0). \tag{13.9}$$

Thus, one obtains the appropriate *scale function*

$$u_{j+1} = 1 + \sum_{i=1}^{j} \frac{\delta_1 \cdots \delta_i}{\beta_1 \cdots \beta_i}, \quad 1 \le j \le N-1.$$

The transformed Markov chain u_Y is said to be on *natural scale*. Now write $m(u_j) = m_j, \, j \in S$.

$$\{m(u_{j+1}) - m(u_j)\}\beta_j - \{m(u_j) - m(u_{j-1})\}\delta_j = -1, \quad 1 \le j \le N-1, \tag{13.10}$$

with boundary conditions

$$m(u_0) = m(0) = 0, \quad m(u_N) - m(u_{N-1}) = 1.$$

Using (13.9), one has

$$\frac{m(u_{j+1}) - m(u_j)}{u_{j+1} - u_j} - \frac{m(u_j) - m(u_{j-1})}{u_j - u_{j-1}} = -\frac{\beta_0 \beta_1 \cdots \beta_{j-1}}{\delta_1 \delta_2 \cdots \delta_j}, 1 \le j \le N-1.$$

Summing over $j = i, i+1, \ldots, N-1$ and using the boundary conditions, one has

$$(u_N - u_{N-1})^{-1} - \frac{m(u_i) - m(u_{i-1})}{u_i - u_{i-1}} = -\sum_{j=i}^{N-1} \frac{\beta_0 \beta_1 \cdots \beta_{j-1}}{\delta_1 \delta_2 \cdots \delta_j}, \quad 1 \le i \le N-1.$$

This and (13.10) lead to

$$m(u_i) - m(u_{i-1}) = \frac{\beta_i \beta_{i+1} \cdots \beta_{N-1}}{\delta_i \delta_{i+1} \cdots \delta_{N-1}} + \sum_{j=i}^{N-1} \frac{\beta_i \cdots \beta_{j-1} \beta_j}{\delta_i \cdots \delta_j \beta_j}, \quad 1 \le i \le N-1.$$

The factor β_j / β_j was introduced to accommodate the term corresponding to $j = i$. The asserted formula now follows by summing over i, using $m(u_0) = 0$. ∎

In the application to the Ehrenfest model one obtains

$$m_d = \sum_{j=1}^{d} \frac{(2d-j)!(j-1)!}{(2d-1)!} + \sum_{j=1}^{d} \sum_{k=j}^{2d-1} \frac{(2d-j)!(j-1)!}{(2d-k)!k!} = \frac{2^{2d}}{2d}\left(1 + O(\frac{1}{d})\right),$$

in the limit as $d \to \infty$. For $d = 10,000$ balls and an exchange rate of one ball per second, it follows that $m_d = 10^{6000}$ years. The companion calculation of the mean time to thermodynamic equilibrium from a state far away,

$$\tilde{m}_0 = \mathbb{E}_0 T_d \le d + d \log d + O(1), d \to \infty, \tag{13.11}$$

is left as Exercise 6. For the same numerical values one obtains from this that $\tilde{m}_0 \le 29$ h. In particular, it takes about a day on average for the system to reach thermodynamic equilibrium from a state farthest away, but it takes an average time that is inconceivably large for the system to go from a state of thermodynamic equilibrium to the same state far from equilibrium.

We saw that Brownian motion is an example of a continuous parameter Markov process having continuous sample paths. More generally, any right-continuous stochastic process $\mathbf{X} = \{X(t) : t \ge 0\}$ having independent increments has the *Markov property* since for $0 \le s < t$, the conditional distribution of $X(t) = X(s) + X(t) - X(s)$ given $\sigma(X(u) : 0 \le u \le s)$ is the same as that given $\sigma(X(s))$. In view of the independence of $X(t) - X(s)$ and $\sigma(X(u) : 0 \le u \le s)$, the former is the distribution of $x + X(t) - X(s)$ on $[X(s) = x]$. If the Markov process is *homogeneous*, i.e., the conditional distribution of $X(t+s)$ given $\sigma(X(s))$ does not depend on s, then this distribution is the *transition probability* $p(t; x, dy)$ on $[X(s) = x]$, namely the distribution of $X(t)$ when $X(0) = x$. Exercise 12.

The following is another example of a continuous parameter Markov process.

Example 7 (*Ornstein–Uhlenbeck process*) The Ornstein–Uhlenbeck process provides an alternative to the Brownian motion model for the molecular diffusion of a suspended particle in a liquid. It is obtained by considering the particle's *velocity* rather than its position. Considering one coordinate, say $V = \{V(t) : t \ge 0\}$, one assumes that the motion is driven by a combination of inertial drag and the momentum provided by random bombardments by surrounding molecules. Specifically, in a small amount of time $h > 0$,

$$V(t+h) - V(t) \approx -\beta V(t)h + \sigma(B(t+h) - B(t)), \quad t \ge 0,$$

where $\beta > 0$ is a constant drag coefficient, $\sigma^2 > 0$ is the molecular diffusion coefficient, and B denotes standard Brownian motion. The frictional term embodies Stokes law from fluid dynamics which asserts that the frictional force decelerating a spherical particle of radius $r > 0$, mass m, is given by

$$\beta = \frac{6\pi r \eta}{m},$$

where $\eta > 0$ is the coefficient of viscosity of the surrounding fluid. To achieve this modeling hypothesis one may consider the integrated form in which V is specified as a process with continuous sample paths satisfying the so-called *Langevin equation*

$$V(t) = -\beta \int_0^t V(s)ds + \sigma B(t), \quad V(0) = u. \tag{13.12}$$

Theorem 13.4 For each initial state $V(0)$, there is a unique Markov process V with state space $S = \mathbb{R}$ having continuous sample paths defined by (13.12). Moreover, V is Gaussian with transition probability density

$$p(t; u, v) = \frac{1}{\sqrt{2\pi\sigma^2(1 - e^{-2\beta t})}} \exp\{-\frac{1}{2\sigma^2(1 - e^{-2\beta t})}(v - ue^{-\beta t})^2\}, \ u, v \in \mathbb{R}.$$

Proof The proof is by the Picard iteration method. First define a process $V^{(0)}(t) = u$ for all $t \geq 0$. Next recursively define $V^{(n+1)}$ by

$$V^{(n+1)}(t) = \int_0^t V^{(n)}(s)ds + \sigma B(t), \quad t \geq 0, n = 0, 1, 2, \ldots.$$

Iterating this equation for $n = 1, 2, 3$, changing the order of integration as it occurs, one arrives at the following induction hypothesis

$$V^{(n)}(t) = u \sum_{j=0}^{n} \frac{(-\beta t)^j}{j!} + \sum_{j=1}^{n-1}(-\beta)^j \sigma \int_0^t \frac{(t-s)^{j-1}}{(j-1)!} B(s)ds + \sigma B(t), \quad t \geq 0.$$

$$\tag{13.13}$$

Letting $n \to \infty$ one obtains sample pathwise that

$$V(t) := \lim_{n\to\infty} V^{(n)}(t) = e^{-\beta t} u - \beta\sigma \int_0^t e^{-\beta(t-s)} B(s)ds + \sigma B(t), \quad t \geq 0.$$

In particular V is a linear functional of the Brownian motion B. That V has continuous paths and is Gaussian follows immediately from the corresponding properties of Brownian motion. Moreover, this solution is unique. To prove uniqueness, suppose that $Y = \{Y(s) : 0 \leq s \leq T\}$ is another a.s. continuous solution to (13.12) and consider

$$\Delta(t) = \mathbb{E}(\max_{0 \leq s \leq t} |X(s) - Y(s)|^2), 0 \leq t \leq T.$$

Then,

$$\Delta(T) \leq 2\beta^2 \mathbb{E}(\int_0^T |V(s) - Y(s)|ds)^2 \leq 2\beta^2 \int_0^T \Delta(s)ds. \qquad (13.14)$$

Since $t \to \Delta(t)$ is nondecreasing on $0 \leq t \leq T$, applying this inequality to the integrand $\Delta(s)$ and reversing the order of integration yields $\Delta(T) \leq (2\beta^2)^2 \int_0^T (T - s)\Delta(s)ds \leq \frac{(2\beta^2 T)^2}{2} \Delta(T)$. Iterating, one sees by induction that

$$\Delta(T) \leq \frac{(2\beta^2 T)^n}{n!} \Delta(T), n = 2, 3, \ldots.$$

Thus $\Delta(T) = 0$ and $Y = V$ a.s. on $[0, T]$. Since T is arbitrary this establishes the uniqueness. From uniqueness one may prove the Markov property holds for V as follows. First, let us note that the solution starting at u at time s, i.e.,

$$V^{(s,u)}(t) = u - \beta \int_s^t V^{(s,u)}(s)ds + \sigma(B(t) - B(s)), t \geq s, \qquad (13.15)$$

can be obtained by Picard iteration as a unique measurable function $\theta(s, t; u, B(t) - B(s)), t \geq s$. Since $V(t), t \geq s$ is a solution starting at $u = V(s)$, i.e.,

$$V(t) = V(s) - \beta \int_s^t V(r)dr + \sigma(B(t) - B(s)), \quad 0 \leq s < t,$$

it follows from uniqueness that $V(t) = \theta(s, t; V(s), B(t) - B(s)), t \geq s$. Thus, the conditional distribution of $V(t)$ given $\mathcal{F}_s = \sigma(B(r) : r \leq s)$ is the distribution of $\theta(x, t; u, B(t) - B(s))$ evaluated at $u = V(s)$. Since $\sigma(V(r) : r \leq s) \subset \mathcal{F}_s, s \geq 0$, this proves the Markov property.

Let us now compute the transition probabilities, from which we will also see that they are homogenous in time. In view of the linearity of the functional θ of Brownian motion it is clear that the conditional distribution is Gaussian. Thus, it is sufficient to compute the conditional mean and variance of $V(t)$ started at $u = V(s), s < t$. In particular, one obtains

$$p(t; u, v) = \sqrt{\frac{\beta}{\pi\sigma^2(1 - e^{-2\beta t})}} \exp\left\{-\frac{\beta(v - ue^{-\beta t})^2}{\sigma^2(1 - e^{-2\beta t})}\right\}$$

is Gaussian with mean $ue^{-\beta t}$ and variance $\frac{\sigma^2}{2\beta}(1 - e^{-2\beta t})$.   ∎

Remark 13.2 A simpler construction of the Ornstein–Uhlenbeck process is given in Exercise 8 which expresses it as a functional of Brownian motion. The Markov property is also immediate from this representation. However, the above derivation is significant because of its historic relation to physics, in particular, significant in its role as a precursor to the development of the mathematical theory of stochastic differential equations. In this regard, the Ornstein–Uhlenbeck example provides an example of a stochastic differential equation

$$dV(t) = -\beta V(t)dt + \sigma dB(t), \quad V(0) = u,$$

which, because σ is a constant, requires no special calculus to interpret. In fact, the definition is provided for (13.12) using ordinary Riemann integrals $\int_0^t V(s)ds$ of the (continuous) paths of V. The extension to more general equations of the form

$$dV(t) = \mu(V(t), t)dt + \sigma(V(t), t)dB(t), \quad V(0) = 0,$$

in one and higher dimensions is the subject of stochastic differential equations and Itô calculus to define integrals of the form $\int_0^t \sigma(V(s), s)dB(s)$ for nonconstant integrands $\sigma(V(s), s)$. K. Itô's development of a useful calculus in this regard provides a striking illustration of the power of martingale theory.

Exercise Set XIII

1. (*Unrestricted Simple Symmetric Random Walk on* $\mathbb{Z}$) Define a transition probability on $S = \mathbb{Z}$ by $p_{i,i+1} = \frac{1}{2} = p_{i,i-1}, i \in \mathbb{Z}$. Show that there is not an invariant probability for this Markov chain.
2. (*Uniqueness of an Invariant Probability*) (a) Suppose $\frac{1}{N} \sum_{n=1}^{N} p^{(n)}(x, dy)$ converges, for each $x \in S$, to a probability $\pi(dy)$ in total variation norm as $N \to \infty$. Show that π is the unique invariant probability. (b) Suppose that the convergence in (a) to $\pi(dy)$ is weak convergence of the probabilities $\frac{1}{N} \sum_{n=1}^{N} p^{(n)}(x, dy)$ on a metric space $(S, \mathcal{B}(S))$. Show the same conclusion as in (a) holds if the transition probability $p(x, dy)$ has the *Feller property*: Namely, for each bounded, continuous function f on S the function $x \to \int_S f(y)p(x, dy), x \in S$ is continuous.
3. (*Asymmetric Simple Random Walk with Reflection*) Let $S = \{0, 1, \ldots, d-1\}$ for some $d > 2$, and for some $0 < p < 1$, define $p_{i,i+1} = p, p_{i,i-1} = 1 - p, 1 \leq i \leq d - 2$, and $p_{0,1} = 1 = p_{d-1,d-2}$. (a) Show that there is a unique invariant probability and compute it. (b) Show that $p_{ij}^{(n)} = 0$ if n and $|i - j|$ have opposite parity. (c) Show that $\tilde{p}_{i,j} := p_{i,j}^{(2)}$ defines a transition probability on each of the state spaces $S_0 = \{i \in S : i \text{ is even}\}$, and $S_1 = \{i \in S : i \text{ is odd}\}$, and that the hypothesis of Corollary 13.2 holds for each of these Markov chains. (d) Show that $\frac{1}{N} \sum_{n=1}^{N} p_{ij}^{(n)}$ converges to the unique invariant probability π on S. Moreover, show that the convergence is exponentially fast as $N \to \infty$, and uniform over all $i, j \in S$.
4. (*Lazy Random Walk*) Suppose the transition probabilities in Exercise 3 are modified to assign positive probability $p_{ii} = \varepsilon > 0$ to each state in S while keeping

$p_{i,i+1} = p_{i,i-1} = (1 - \varepsilon)/2, 1 \le i \le d - 2$, and $p_{0,1} = p_{d-1,d-2} = 1 - \varepsilon$, and $p_{i,j} = 0$ if $|i - j| > 1$. Show that Doeblin's Theorem 13.1 applies to this Markov chain.

5. (*Simple Random Walk with Absorption*) Suppose that the transition probabilities in Exercise 3 are modified so that $p_{0,0} = p_{1,1} = 1$. Show that there are two invariant probabilities $\delta_{\{0\}}$ and $\delta_{\{1\}}$, and hence infinitely many.

6. (*Ehrenfest model continued*) Calculate $\tilde{m}_0$ in (13.11) for the Ehrenfest model by the following steps:

   (i) Write $\tilde{m}(u_i) = \tilde{m}_i, 1 \le i \le d - 1$, and show that the same equations as for $m(u_i)$ apply with boundary conditions $\tilde{m}(u_0) = 1 + \tilde{m}(u_1), \tilde{m}(u_d) = 0$.

   (ii) Summing over $j = 1, 3, \ldots, d - 1$, show that $\tilde{m}_0 = 1 + \sum_{j=1}^{d-1} \frac{j!}{(2d-1)\cdots(2d-j)}$ $+ \sum_{j=1}^{d-1} \sum_{k=1}^{j} \frac{(j+1)j\cdots(k+2)(k+1)}{(2d-k)\cdots(2d-j)(j+1)}$

   (iii) Verify that $\tilde{m}_0 \le d + d\log d + O(1)$ as $d \to \infty$.

7. (*Stationary Ornstein–Uhlenbeck/Maxwell-Boltzmann Steady State*) (a) Show that the time-asymptotic distribution of the Ornstein–Uhlenbeck process is Gaussian with mean zero and variance $\frac{\sigma^2}{2\beta}$ regardless of the initial distribution. (b) Show that this is the unique invariant distribution of V.[4] (c) What general features do the Erhenfest model and Ornstein–Uhlenbeck diffusion have in common ? [*Hint*: Consider the conditional mean and variance of displacements of the process $v_n = Y_n - d, n = 0, 1, 2, \ldots$. Namely, $\mathbb{E}(v_{n+1} - v_n | v_0, \ldots, v_n)$ and $\mathbb{E}((v_{n+1} - v_n)^2 | v_0, \ldots, v_n)$.]

8. (*Ornstein–Uhlenbeck process; Time change of Brownian Motion*) Assume that $V(0)$ has the stationary distribution for the Ornstein–Uhlenbeck process. Then V can be expressed as a time-change of Brownian motion as follows: $V(t) = e^{-\beta t} B(\frac{\sigma^2}{2\beta} e^{2\beta t})$, $t \ge 0$. [*Hint*: Compute the mean and variance of the Gaussian transition probability densities.]

9. (*Poisson Process*) Let $T_1, T_2, \ldots$ be an i.i.d. sequence of exponentially distributed random variables with intensity $\lambda > 0$, i.e., $P(T_1 > t) = e^{-\lambda t}, t \ge 0$. Define a counting process $N = \{N(t) : t \ge 0\}$ by $N(t) = \max\{n : T_1 + \cdots + T_n \le t\}, t \ge 0$. The random variables $T_1, T_2, \ldots$ are referred to as *interarrival times* of N. Show that N is a continuous parameter Markov process on the state space $S = \{0, 1, 2, \ldots\}$ with transition probabilities $p(t; x, y) = \frac{(\lambda t)^{y-x}}{(y-x)!} e^{-\lambda t}, y = x, x + 1, \ldots, x = 0, 1, 2, \ldots, t \ge 0$. [*Hint*: N has independent increments.]

10. (*Dilogarithmic Random Walk*) The *dilogarithmic random walk*[5] is the multiplicative random walk on the multiplicative group $S = (0, \infty)$ defined by

---

[4]The invariant distribution of the Ornstein–Uhlenbeck process is referred to as the Maxwell–Boltzmann distribution. The physics of fluids requires that the variance be given by the physical parameter $\frac{\kappa T}{m}$ where κ is Boltzmann constant, T is absolute temperature, and m is the mass of the particle.

[5]This random walk plays a role in probabilistic analysis of the incompressible Navier–Stokes equations introduced by Y. LeJan, A. S. Sznitman (1997): Stochastic cascades and three-dimensional Navier–Stokes equations. *Probab. Theory Related Fields* 109, no. 3, 343–366. This particular structure was exploited in Dascaliuc, R., N. Michalowski, E. Thomann, E. Waymire (2015): Symmetry breaking and uniqueness for the incompressible Navier-Stokes equations, *Chaos*, American Physical

$M_n = R_0 \prod_{j=1}^{n} R_j$, $n = 1, 2, \ldots$ where R_0 is a positive random variable independent of the i.i.d. sequence $\{R_n : n \geq 1\}$ having marginal distribution given by
$P(R_1 \in dr) = \frac{2}{\pi^2} \ln \left| \frac{1+r}{1-r} \right| \frac{dr}{r}$, $r > 0$. Show that (a) $\mathbb{E} R_1 = \infty$. (b) $\mathbb{E}|\ln R_1|^m <$
∞ for $m = 1, 2, \ldots$. (c) The distribution of M_n is symmetric about 1, the iden-
tity element of the multiplicative group S, and $\{M_n : n \geq 0\}$ is 1-neighborhood
recurrent. [*Hint*: Show that the additive random walk $S_n = \ln M_n$, $n \geq 0$, is 0-
neighborhood recurrent.]

11. Suppose that X_0 has an invariant distribution π in the sense of (13.4). Show that
the Markov chain $\mathbf{X}$ is stationary (or translation invariant) in the sense that $\mathbf{X}^{n+}$
and $\mathbf{X}$ have the same distribution for each $n \geq 1$.

12. For a homogeneous continuous parameter Markov process show that the conditional distribution of $X(t+s)$ given $\sigma(X(s))$ on $[X(s) = x]$ is the same as the
conditional distribution of of $X(t)$ given $X(0)$ on $[X(0) = x]$.

(Footnote 5 continued)
Society, 25 (7). The dilogarithmic functions are well-studied and arise in a variety of unrelated
contexts.

Appendix A
Measure and Integration

A.1 Measures and the Carathéodory Extension

Let S be a nonempty set. A class $\mathcal{F}$ of subsets of S is a **field**, or an **algebra** if (i) $\emptyset \in \mathcal{F}$, $S \in \mathcal{F}$, (ii) $A \in \mathcal{F}$ implies $A^c \in \mathcal{F}$, (iii) $A, B \in \mathcal{F}$ implies $A \cup B \in \mathcal{F}$. Note that (ii) and (iii) imply that $\mathcal{F}$ is closed under finite unions and finite intersections. If (iii) is replaced by (iii)$'$: $A_n \in \mathcal{F}$ $(n = 1, 2, \ldots)$ implies $\cup_{n=1}^{\infty} A_n \in \mathcal{F}$, then $\mathcal{F}$ is said to be a σ-**field**, or a σ-**algebra**. Note that (iii)$'$ implies (iii), and that a σ-field is closed under countable intersections.

A function $\mu : \mathcal{F} \to [0, \infty]$ is said to be a **measure on a field** $\mathcal{F}$ if $\mu(\emptyset) = 0$ and $\mu(\cup_{n=1}^{\infty} A_n) = \sum_{n=1}^{\infty} \mu(A_n)$ for every sequence of pairwise disjoint sets $A_n \in \mathcal{F}$ $(n = 1, 2, \ldots)$ such that $\cup_{n=1}^{\infty} A_n \in \mathcal{F}$. Note that this property, known as **countable additivity**, implies **finite additivity** (by letting $A_n = \emptyset$ for $n \geq m$ for some m, say). A measure μ on a field $\mathcal{F}$ is σ-**finite** if there exists a sequence $A_n \in \mathcal{F}$ $(n = 1, 2, \ldots)$ such that $\cup_{n=1}^{\infty} A_n = S$ and $\mu(A_n) < \infty$ for every n.

If μ is a measure on a field $\mathcal{F}$, and $A_n \in \mathcal{F}$ $(n \geq 1)$, $A \subset \cup_n A_n$, $A \in \mathcal{F}$, then $\mu(A) \leq \sum_{n=1}^{\infty} \mu(A_n)$ (**subadditivity**). To see this write $B_1 = A_1$, $B_n = A_1^c \cap \cdots \cap A_{n-1}^c \cap A_n (n \geq 2)$. Then $B_n(n \geq 1)$ are disjoint, $\cup_{n=1}^{\infty} A_n = \cup_{n=1}^{\infty} B_n$, so that $\mu(A) = \mu(A \cap (\cup_{n=1}^{\infty} B_n)) = \sum_{n=1}^{\infty} \mu(A \cap B_n) \leq \sum_{n=1}^{\infty} \mu(A_n)$ (since $B_n \subset A_n$ for all n).

Let μ be a measure on a σ-field $\mathcal{F}$ on S. Then $\mathcal{F}$ is said to be μ-**complete** if all subsets of μ-null sets in $\mathcal{F}$ belong to $\mathcal{F}$: $N \in \mathcal{F}$, $\mu(N) = 0$, $B \subset N$ implies $B \in \mathcal{F}$. In this case the measure μ is also said to be **complete**. Given any measure μ on a σ-field $\mathcal{F}$, it is simple to check that the class of subsets

$$\overline{\mathcal{F}} = \{C = A \cup B : A \in \mathcal{F}, B \subset N \text{ for some } N \in \mathcal{F} \text{ such that } \mu(N) = 0\}$$

is a μ-complete σ-field, $\widetilde{\mu}(A \cup B) := \mu(A)$ $(A \in \mathcal{F}, B \subset N, \mu(N) = 0)$ is well defined, and $\widetilde{\mu}$ is a measure on $\overline{\mathcal{F}}$ extending μ. This extension of μ is called the **completion of** μ.

© Springer International Publishing AG 2016
R. Bhattacharya and E.C. Waymire, *A Basic Course in Probability Theory*,
Universitext, DOI 10.1007/978-3-319-47974-3

We now derive one of the most basic results in measure theory, due to Carathéodory, which provides an extension of a measure μ on a field $\mathcal{A}$ to a measure on the σ-field $\mathcal{F} = \sigma(\mathcal{A})$, the smallest σ-field containing $\mathcal{A}$. First, on the set 2^S of all subsets of S, call a set function $\mu^* : 2^S \to [0, \infty]$ an **outer measure** on S if (1) $\mu^*(\emptyset) = 0$, (2) *(monotonicity)* $A \subset B$ implies $\mu^*(A) \leq \mu^*(B)$, and (3) *(subadditivity)* $\mu^*(\cup_{n=1}^{\infty} A_n) \leq \sum_{n=1}^{\infty} \mu^*(A_n)$ for every sequence A_n ($n = 1, 2, \dots$).

Proposition 1.1 Let $\mathcal{A}$ be a class of subsets such that $\emptyset \in \mathcal{A}$, $S \in \mathcal{A}$, and let $\mu : \mathcal{A} \to [0, \infty]$ be a function such that $\mu(\emptyset) = 0$. For every set $A \subset S$, define

$$\mu^*(A) = \inf \left\{ \sum_n \mu(A_n) : A_n \in \mathcal{A} \forall n A \subset \cup_n A_n \right\}. \tag{1.1}$$

Then μ^* is an outer measure on S.

Proof (1) Since $\emptyset \subset \emptyset \in \mathcal{A}$, $\mu^*(\emptyset) = 0$. (2) Let $A \subset B$. Then every countable collection $\{A_n : n = 1, 2, \dots\} \subset \mathcal{A}$ that covers B (i.e., $B \subset \cup_n A_n$) also covers A. Hence $\mu^*(A) \leq \mu^*(B)$. (3) Let $A_n \subset S$ ($n = 1, 2, \dots$), and $A = \cup_n A_n$. If $\mu^*(A_n) = \infty$ for some n, then by (2), $\mu^*(A) = \infty$. Assume now that $\mu^*(A_n) < \infty$ $\forall n$. Fix $\varepsilon > 0$ arbitrarily. For each n there exists a sequence $\{A_{n,k} : k = 1, 2, \cdots\} \subset \mathcal{A}$ such that $A_n \subset \cup_k A_{n,k}$ and $\sum_k \mu(A_{n,k}) < \mu^*(A_n) + \varepsilon/2^n$ ($n = 1, 2, \dots$). Then $A \subset \cup_n \cup_k A_{n,k}$, and therefore $\mu^*(A) \leq \sum_{n,k} \mu(A_{n,k}) < \sum_n \mu^*(A_n) + \varepsilon$. ∎

The technically simplest, but rather unintuitive, proof of Carathéodory's theorem given below is based on the following notion. Let μ^* be an outer measure on S. A set $A \subset S$ is said to be μ^*-**measurable** if the following "balance conditions" are met:

$$\mu^*(E) = \mu^*(E \cap A) + \mu^*(E \cap A^c) \quad \forall E \subset S. \tag{1.2}$$

Theorem 1.2 *(Carathéodory Extension Theorem)* (a) Let μ^* be an outer measure on S. The class $\mathcal{M}$ of all μ^*-measurable sets is a σ-field, and the restriction of μ^* to $\mathcal{M}$ is a complete measure. (b) Let μ^* be defined by (1.1), where $\mathcal{A}$ is a field and μ is a measure on $\mathcal{F}$. Then $\sigma(\mathcal{A}) \subset \mathcal{M}$ and $\mu^* = \mu$ on $\mathcal{A}$. (c) If a measure μ on a field $\mathcal{A}$ is σ-finite, then it has a unique extension to a measure on $\sigma(\mathcal{A})$, this extension being given by μ^* in (1.1) restricted to $\sigma(\mathcal{A})$.

Proof (a) To show that $\mathcal{M}$ is a *field*, first note that $A = \emptyset$ trivially satisfies (1.2) and that if A satisfies (1.2), so does A^c. Now, in view of the subadditivity property of μ^*, (1.2) is equivalent to the inequality

$$\mu^*(E) \geq \mu^*(E \cap A) + \mu^*(E \cap A^c) \quad \forall E \subset S. \tag{1.3}$$

To prove that $\mathcal{M}$ is closed under finite intersections, let $A, B \in \mathcal{M}$. Then $\forall E \subset S$,

$$\mu^*(E) = \mu^*(E \cap B) + \mu^*(E \cap B^c) \quad \text{(since } B \in \mathcal{M})$$
$$= \mu^*(E \cap B \cap A) + \mu^*(E \cap B \cap A^c) + \mu^*(E \cap B^c \cap A)$$
$$+ \mu^*(E \cap B^c \cap A^c) \quad \text{(since } A \in \mathcal{M})$$
$$\geq \mu^*(E \cap (B \cap A)) + \mu^*(E \cap (B \cap A)^c).$$

For the last inequality, use $(B \cap A)^c = B^c \cup A^c = (B^c \cap A) \cup (B^c \cap A^c) \cup (B \cap A^c)$, and subadditivity of μ^*. By the criterion (1.3), $B \cap A \in \mathcal{M}$. Thus $\mathcal{M}$ is a field.

Next, we show that $\mathcal{M}$ is a σ-field and μ^* is countably additive on $\mathcal{M}$. Let $B_n \in \mathcal{M}$ $(n = 1, 2, \ldots)$ be a pairwise disjoint sequence in $\mathcal{M}$, and write $C_m = \cup_{n=1}^m B_n$ $(m \geq 1)$. We will first show, by induction on m, that

$$\mu^*(E \cap C_m) = \sum_{n=1}^m \mu^*(E \cap B_n) \quad \forall \, E \subset S. \tag{1.4}$$

This is true for $m = 1$, since $C_1 = B_1$. Suppose (1.4) holds for some m. Since $B_{m+1} \in \mathcal{M}$, one has for all $E \subset S$,

$$\mu^*(E \cap C_{m+1}) = \mu^*((E \cap C_{m+1}) \cap B_{m+1}) + \mu^*((E \cap C_{m+1}) \cap B_{m+1}^c)$$
$$= \mu^*(E \cap B_{m+1}) + \mu^*(E \cap C_m)$$
$$= \mu^*(E \cap B_{m+1}) + \sum_{n=1}^m \mu^*(E \cap B_m),$$

using the induction hypothesis for the last equality. Thus (1.4) holds for $m + 1$ in place of m, and the induction is complete. Next, writing $A = \cup_{n=1}^\infty B_n$ one has, for all $E \subset S$,

$$\mu^*(E) = \mu^*(E \cap C_m) + \mu^*(E \cap C_m^c) \quad \text{(since } C_m \in \mathcal{M})$$
$$= \sum_{n=1}^m \mu^*(E \cap B_n) + \mu^*(E \cap C_m^c) \geq \sum_{n=1}^m \mu^*(E \cap B_n) + \mu^*(E \cap A^c),$$

since $C_m^c \supset A^c$. Letting $m \to \infty$, one gets

$$\mu^*(E) \geq \sum_{n=1}^\infty \mu^*(E \cap B_n) + \mu^*(E \cap A^c) \geq \mu^*(E \cap A) + \mu^*(E \cap A^c), \tag{1.5}$$

using the subadditivity property for the last inequality. This shows that $A \equiv \cup_{n=1}^\infty B_n \in \mathcal{M}$, i.e., $\mathcal{M}$ is closed under countable disjoint unions. If $\{A_n : n = 1, \ldots\}$ is an arbitrary sequence in $\mathcal{M}$, one may express $A \equiv \cup_{n=1}^\infty A_n$ as $A = \cup_{n=1}^\infty B_n$, where $B_1 = A_1$, $B_2 = A_1^c \cap A_2$, $B_n = A_1^c \cap \cdots \cap A_{n-1}^c \cap A_n$ $(n > 2)$, are pairwise disjoint sets in $\mathcal{M}$. Hence $A \in \mathcal{M}$, proving that $\mathcal{M}$ is a σ-field. To prove countable additivity of μ^* on $\mathcal{M}$, let B_n $(n \geq 1)$ be a pairwise disjoint sequence in $\mathcal{M}$ as before, and take

$E = A \equiv \cup_{n=1}^{\infty} B_n$ in the first inequality in (1.5) to get $\mu^*(\cup_{n=1}^{\infty} B_n) \geq \sum_{n=1}^{\infty} \mu^*(B_n)$.
By the subadditive property of μ^*, it follows that $\mu^*(\cup_{n=1}^{\infty} B_n) = \sum_{n=1}^{\infty} \mu^*(B_n)$.

We have proved that μ^* is a measure on the σ-field $\mathcal{M}$. Finally, if $A \subset N \in \mathcal{M}$, $\mu^*(N) = 0$, then $\mu^*(E \cap A) \leq \mu^*(A) \leq \mu^*(N) = 0$, and $\mu^*(E \cap A^c) \leq \mu^*(E)$, so that (1.3) holds, proving $A \in \mathcal{M}$. Hence $\mathcal{M}$ is μ^*-complete.

(b) Consider now the case in which $\mathcal{A}$ is a field, μ is a measure on $\mathcal{A}$, and μ^* is the outer measure (1.1). To prove $\mathcal{A} \subset \mathcal{M}$, let $A \in \mathcal{A}$. Fix $E \subset S$ and $\varepsilon > 0$ arbitrarily. There exists $A_n \in \mathcal{A}$ $(n = 1, 2, \ldots)$ such that $E \subset \cup_{n=1}^{\infty} A_n$ and $\mu^*(E) \geq \sum_{n=1}^{\infty} \mu(A_n) - \varepsilon$. Also,

$$\mu^*(E \cap A) \leq \mu^* \left(A \cap \bigcup_{n=1}^{\infty} A_n \right) \leq \sum_{n=1}^{\infty} \mu(A \cap A_n),$$

$$\mu^*(E \cap A^c) \leq \mu^* \left(A^c \cap \bigcup_{n=1}^{\infty} A_n \right) \leq \sum_{n=1}^{\infty} \mu(A^c \cap A_n),$$

$$\mu^*(E \cap A) + \mu^*(E \cap A^c) \leq \sum_{n=1}^{\infty} \left\{ \mu(A \cap A_n) + \mu(A^c \cap A_n) \right\}$$

$$= \sum_{n=1}^{\infty} \mu(A_n) \leq \mu^*(E) + \varepsilon.$$

Hence (1.3) holds, proving that $A \in \mathcal{M}$. To prove $\mu = \mu^*$ on $\mathcal{A}$, let $A \in \mathcal{A}$. By definition (1.1), $\mu^*(A) \leq \mu(A)$ (letting $A_1 = A$ and $A_n = \emptyset$ for $n \geq 2$, be a cover of A). On the other hand, $\mu(A) \leq \sum_{n=1}^{\infty} \mu(A_n)$ for every sequence $A_n \in \mathcal{A}$ $(n \geq 1)$ such that $A \subset \cup_{n=1}^{\infty} A_n$, so that $\mu^*(A) \geq \mu(A)$ (by subadditivity of μ on $\mathcal{A}$). Hence $\mu^*(A) = \mu(A)$.

(c) Suppose μ is a σ-finite measure on the field $\mathcal{A}$, and μ^* its extension to the σ-field $\sigma(\mathcal{A})$ $(\subset \mathcal{M})$ as derived in (b). Let ν be another extension of μ to $\sigma(\mathcal{A})$. Since one may express $S = \cup_{n=1}^{\infty} A_n$ with $A_n \in \mathcal{A}$ pairwise disjoint and $\mu(A_n) < \infty$ $\forall n$, it is enough to consider the restrictions of μ and ν to $A_n \cap \sigma(\mathcal{A}) \equiv \{A_n \cap A : A \in \sigma(\mathcal{A})\}$ for each n separately. In other words, it is enough to prove that $\mu = \nu$ on $\sigma(\mathcal{A})$ in the case $\mu(S) < \infty$. But for this case, the class $\mathcal{C} = \{A \in \mathcal{F} : \mu(A) = \nu(A)\}$ is a λ-system, and it contains the π-system $\mathcal{A}$. Hence, by the π-λ theorem, $\mathcal{C} = \sigma(\mathcal{A})$. ∎

Example 1 (*Lebesgue–Stieltjes Measures*) Let $S = \mathbb{R}$. The σ-field generated by the collection of open subsets of a topological space is referred to as the **Borel σ-field**. Let $\mathcal{B}(\mathbb{R})$ denote the Borel σ-field. A measure μ on $\mathcal{B}(\mathbb{R})$ is said to be a **Lebesgue–Stieltjes** (or **L–S**) **measure** if $\mu((a, b]) < \infty$ $\forall -\infty < a < b < \infty$. Given such a measure one may define its **distribution function** $F_\mu : \mathbb{R} \to \mathbb{R}$ by

$$F_\mu(x) = \begin{cases} -\mu((x, 0]) + c & \text{if } x < 0 \\ \mu((0, x]) + c & \text{if } x > 0, \end{cases} \tag{1.6}$$

where c is an arbitrary constant. Note that

$$\mu((a,b]) = F_\mu(b) - F_\mu(a) \qquad (-\infty < a < b < \infty). \qquad (1.7)$$

Moreover, F_μ is nondecreasing and right-continuous. Conversely, *given a function F which is nondecreasing and right-continuous on* $\mathbb{R}$, *there exists a unique Lebesgue–Stieltjes measure* μ *whose distribution function is* F. To prove this, first, fix an interval $S = (c,d]$, $-\infty < c < d < \infty$. The class of all finite unions $\cup_{j=1}^m (a_j, b_j]$ of pairwise disjoint intervals $(a_j, b_j]$ $(c \le a_j < b_j \le d)$ is a field $\mathcal{A}$ on S. Define the set function μ on $\mathcal{A}$ first by (1.7) on intervals $(a,b]$, and then on disjoint unions above as $\sum_{j=1}^m \mu((a_j, b_j])$. It is simple to check that this is *well defined*. That is, if $(c_i, d_i]$, $1 \le i \le n$, is another representation of $\cup_{j=1}^m (a_j, b_j]$ as a union of disjoint intervals, then $\sum_{i=1}^n [F(d_i) - F(c_i)] = \sum_{j=1}^m [F(b_j) - F(a_j)]$ (Show this by splitting each $(a_j, b_j]$ by $(c_i, d_i], 1 \le i \le n$). *Finite additivity of* μ on $\mathcal{A}$ is then a consequence of the definition of μ. In view of this, to prove *countable additivity* of μ on $\mathcal{A}$, it is enough to show that if $I_j = (a_j, b_j]$ $(j = 1,2,\dots)$ is a sequence of pairwise disjoint intervals whose union is $(a,b]$, then $\mu((a,b]) \equiv F(b) - F(a) = \sum_{j=1}^\infty \mu(I_j)$. Clearly, $\sum_{j=1}^n \mu(I_j) = \mu(\cup_{j=1}^n I_j) \le \mu((a,b])$ for all n, so that $\sum_{j=1}^\infty \mu(I_j) \le \mu((a,b])$. For the opposite inequality, fix $\varepsilon > 0$ and find $\delta > 0$ such that $F(a+\delta) - F(a) < \varepsilon$ (by right-continuity of F). Also, find $\delta_j > 0$ such that $F(b_j+\delta_j)-F(b_j) < \varepsilon/2^j$ $(j = 1,2,\dots)$. Then $\{(a_j, b_j + \delta_j) : j \ge 1\}$ is an open cover of the compact interval $[a+\delta, b]$, so that there exists a finite subcover: $[a+\delta, b] \subset \cup_{j=1}^m (a_j, b_j+\delta_j)$, say. Then $\mu((a,b]) = F(b) - F(a) \le F(b) - F(a+\delta) + \varepsilon \le \sum_{j=1}^m [F(b_j + \delta_j) - F(a_j)] \le \sum_{j=1}^m [F(b_j)-F(a_j)]+\varepsilon \le \sum_{j=1}^\infty [F(b_j)-F(a_j)]+\varepsilon \le \sum_{j=1}^\infty \mu(I_j)+\varepsilon$. This proves that μ is a measure on $\mathcal{A}$. Now use Carathéodory's extension theorem to extend uniquely μ to $\sigma(\mathcal{A}) = \mathcal{B}((c,d])$. Since $\mathbb{R} = \cup_{n=-\infty}^\infty (n, n+1]$, one may construct μ on each of $(n, n+1]$ and then piece (or add) them together to construct the unique L-S measure on $\mathcal{B}(\mathbb{R})$ with the given distribution function F.

(a) As a very special L-S measure, one constructs **Lebesgue measure** m on $(\mathbb{R}, \mathcal{B}(\mathbb{R}))$ specified by

$$m((a,b]) = b - a,$$

with distribution function $F(x) = x$. Lebesgue measure is variously denoted by m, λ, or dx.

(b) For an example of a L-S measure with a continuous distribution, but that does not have a density with respect to Lebesgue measure, consider the representation of a real x in $(0,1]$ by its *ternary* representation $x = \sum_{n=1}^\infty a_n 3^{-n}$, where $a_n \in \{0,1,2\}$. By requiring that there be an infinite number of 2's among $\{a_n\}$ one gets a one-to-one correspondence $x \mapsto \{a_n\}$. The **Cantor set** C is defined to be the set of all x in whose ternary expansion the digit 1 does not occur. That is, C is obtained by first omitting the middle third $(1/3, 2/3]$ of $(0,1]$, then omitting the middle thirds of the two remaining intervals $(0, 1/3], (2/3, 1]$, then omitting the middle thirds of the four remaining intervals, and so on. The Lebesgue measure of the omitted set is $\sum_{n=1}^\infty 2^{n-1}3^{-n} = 1$. Hence the remaining set C has Lebesgue measure zero. Define the **Cantor function**

$$F(x) = \sum_{n=1}^{\infty} \frac{a_n}{2} 2^{-n} \quad \text{for } x = \sum_{n=1}^{\infty} a_n 3^{-n} \in C,$$

and extend F to $[0, 1]$ by letting $F(0) = 0$, and F constant between the endpoints of every omitted interval. Then F is continuous and nondecreasing, and the corresponding L–S probability measure (distribution) μ is the **Cantor measure** on $(0, 1]$, which is **nonatomic** (i.e., $\mu(\{x\}) = 0 \, \forall x$) and **singular** in the sense that $\mu(C) = 1$ and $m(C) = 0$, where m is Lebesgue measure.

A.2 Integration and Basic Convergence Theorems

Let $\mathcal{S}$ be a σ-field on S. We say that $(S, \mathcal{S})$ is a **measurable space**. Denote by L the *class of* all (extended) real-valued *measurable functions* $f : S \to \overline{\mathbb{R}} = [-\infty, \infty]$, i.e., $f^{-1}(B) \in \mathcal{S} \, \forall B \in \mathcal{B}(\mathbb{R})$ and $f^{-1}(\{-\infty\}) \in \mathcal{S}$, $f^{-1}(\{+\infty\}) \in \mathcal{S}$. The subclass of nonnegative *measurable functions* is denoted by L^+. A **simple function** is of the form $f = \sum_1^m a_j \mathbf{1}_{A_j}$, where $m \geq 1$, $a_j \in \mathbb{R} \, \forall j$, A_j's are pairwise disjoint sets in $\mathcal{S}$. The class of all simple functions is denoted by L_s, and the subclass of nonnegative simple functions by L_s^+.

In general, if $(S_i, \mathcal{S}_i)$ $(i = 1, 2)$ are measurable spaces, a *map*, or function, $f : S_1 \to S_2$ *is* said to be **measurable** if $f^{-1}(B) \in \mathcal{S}_1 \, \forall B \in \mathcal{S}_2$. In particular, if S_i is a metric space with Borel σ-field $\mathcal{S}_i$ $(i = 1, 2)$, then a *continuous map* $f : S_1 \to S_2$ is measurable, since $f^{-1}(B)$ is an open subset of S_1 if B is an open subset of S_2, and since $\mathcal{F} \equiv \{B \in \mathcal{S}_2 : f^{-1}(B) \in \mathcal{S}_1\}$ is a σ-field (containing the class of all open subsets of S_2). It is simple to check that *compositions* of measurable maps are measurable. As an example, let $(S, \mathcal{S})$ be a measurable space, and let f, g be measurable maps on S into $\mathbb{R}^k$. Then $\alpha f + \beta g$ is measurable for all constants $\alpha, \beta \in \mathbb{R}$. To see this, consider the map $h(x, y) \mapsto \alpha x + \beta y$ on $\mathbb{R}^k \times \mathbb{R}^k$ into $\mathbb{R}^k$. Since h is continuous, it is measurable. Also, $\varphi(s) := (f(s), g(s))$ is a measurable map on S into $S \times S$, with the **product σ-field** $\mathcal{S} \otimes \mathcal{S}$ (i.e., the smallest σ-field on $S \times S$ containing the class of all **measurable rectangles** $A \times B$, $A \in \mathcal{S}$, $B \in \mathcal{S}$). Now $\alpha f + \beta g$ equals the composition $h \circ \varphi$ and is therefore measurable.

Example 1 (a) Let $f_1 f_2, \ldots, f_k$ be (extended) real-valued measurable functions on a measurable space $(S, \mathcal{S})$. Then $M \equiv \max\{f_1, \ldots, f_k\}$ is measurable, since $[M \leq x] = \cap_{j=1}^k [f_j \leq x] \, \forall x \in \mathbb{R}^1$. Similarly, $\min\{f_1, \ldots, f_k\}$ is measurable. (b) Let f_n $(n \geq 1)$ be a sequence of (extended) real-valued measurable functions on a measurable space $(S, \mathcal{S})$. Then $h \equiv \liminf f_n$ is measurable. For $g_n := \inf\{f_j : j \geq n\}$ is measurable $(n \geq 1)$, since $[g_n \geq x] = \cap_{j=n}^{\infty} [f_j \geq x]$. Also, $g_n \uparrow h$, so that $[h \leq x] = \cap_{n=1}^{\infty} [g_n \leq x] \, \forall x \in \mathbb{R}$. Similarly, $\limsup f_n$ is measurable.

Let μ be a σ-finite measure on $(S, \mathcal{S})$ (i.e., on $\mathcal{S}$). For $f \in L_s^+$ define the **integral** $\int f d\mu$, $\int_S f(x)\mu(dx)$, or simply $\int f$ when there is no ambiguity about the underlying measure μ, by

$$\int f \equiv \int_S f \, d\mu := \sum_{j=1}^m a_j \mu(A_j), \tag{2.1}$$

with the convention $0 \cdot \infty = 0$. If $f = \sum_1^n b_i \mathbf{1}_{B_i}$ is another representation of f, then $a_j = b_i$ on $A_j \cap B_i$, and using the finite additivity of μ, one has $\sum_1^m a_j \mu(A_j) = \sum_j \sum_i a_j \mu(A_j \cap B_i) = \sum_j \sum_i b_i \mu(A_j \cap B_i) = \sum_i b_i \mu(B_i)$. Thus $\int f$ is well defined for $f \in L_s^+$. Using a similar splitting where necessary, one can prove the following properties of the integral on L_s^+:

(i) $\int cf = c \int f$ $\forall c \geq 0$,
(ii) $\int f \leq \int g$ if $f \leq g$,   (2.2)
(iii) $\int (f + g) = \int f + \int g$

for an arbitrary $f \in L^+$ (set of all extended nonnegative measurable functions on S) define

$$\int f := \lim \int f_n, \tag{2.3}$$

where $f_n \in L_s^+$ and $f_n \uparrow f$. To show that $\int f$ is *well defined* let us first observe that there does exist $f_n \in L_s^+$, $f_n \uparrow f$. For example, let f_n be the so-called **standard approximation**,

$$f_n = \sum_{k=1}^{n2^n} (k-1)2^{-n} \mathbf{1}_{[\frac{k-1}{2^n} \leq f < \frac{k}{2^n}]} + n\mathbf{1}_{[f \geq n]}. \tag{2.4}$$

Secondly, suppose $f_n, g_n \in L_s^+$, $f_n \uparrow f$, $g_n \uparrow f$. We will show that

$$\lim_n \int g_n = \lim_n \int f_n. \tag{2.5}$$

For this, fix $c \in (0, 1)$ and $m \geq 1$. One may write $g_m = \sum_1^k a_j \mathbf{1}_{A_j}$, $\int c g_m = c \sum_1^k a_j \mu(A_j) = c \int g_m$. Let $B_n = \{x \in S : f_n(x) \geq c g_m(x)\}$. Then $f_n = f_n \mathbf{1}_{B_n} + f_n \mathbf{1}_{B_n^c}$, so that (by (2.2)), and using $B_n \uparrow S$,

$$\int f_n = \int f_n \mathbf{1}_{B_n} + \int f_n \mathbf{1}_{B_n^c} \geq \int f_n \mathbf{1}_{B_n} \geq \int c g_m \mathbf{1}_{B_n}$$

$$= c \sum_{j=1}^k a_j \mu(A_j \cap B_n) \uparrow c \sum_{j=1}^k a_j \mu(A_j) = c \int g_m \quad \text{as } n \uparrow \infty.$$

Hence $\lim_n \int f_n \geq c \int g_m \; \forall c \in (0,1)$, which implies $\lim_n \int f_n \geq \int g_m$. Letting $m \uparrow \infty$, we obtain $\lim_n \int f_n \geq \lim_m \int g_m$. Reversing the roles of f_n and g_n, we then get (2.5), and $\int f$ is well defined.

As simple consequences of the definition (2.3) and the order property (2.2)(ii), one obtains the following results.

Proposition 2.1 (a) Let $f \in L^+$. Then

$$\int f = \sup \left\{ \int g : g \in L_s^+, g \leq f \right\}, \qquad \forall f \in L^+. \tag{2.6}$$

(b) Let $f, g \in L^+, c \geq 0$. Then (2.2)(i)—(iii) hold.

Proof (a). Clearly, $\int f$ is dominated by the right hand side of (2.6), by the definition (2.3). For the reverse inequality, let $g \in L_s^+, g \leq f$. Then $g_n := \max\{g, f_n\} \uparrow f$ with f_n as in (2.3). Since $g_n \in L_s^+$, it follows that $\int g \leq \int g_n \to \int f$. Hence $\int g \leq \int f$.

(b). (i) and (iii) in (2.2) follow from the definition (2.3), while (ii) follows from (2.6). ∎

A useful convergence result for functions in L^+ is the following.

Proposition 2.2 Let $f_k, f \in L^+, f_n \uparrow f$. Then $\int f_k \uparrow \int f$.

Proof By Proposition 2.1(b), $\lim_k \int f_k \leq \int f$ (order property of integrals). Next, let $g_{k,n} \in L_s^+, g_{k,n} \uparrow f_k$ as $n \uparrow \infty$. Define $g_n = \max\{g_{k,n} : k = 1, \ldots, n\} \in L_s^+$, $g_n \uparrow g$, say. But $g_n \geq g_{k,n} \; \forall k \leq n$, so that $g \geq f_k \; \forall k$, implying $g \geq f$. On the other hand, $g_n \leq f \; \forall n$. Thus $g = f$ and, therefore, $\lim_n \int g_n = \int f$. But $g_n \leq f_n \; \forall n$ which implies $\lim_n \int f_n \geq \lim_n \int g_n = \int f$. ∎

Let $f \in L$, and set $f^+ = \max\{f, 0\}, f^- = -\min\{f, 0\}$. Then $f^+, f^- \in L^+$ and $f = f^+ - f^-$. If at least one of $\int f^+, \int f^-$ is finite, we say that the *integral of f exists* and define

$$\int f = \int f^+ - \int f^-. \tag{2.7}$$

If $\int f^+$ and $\int f^-$ are both finite, then $0 \leq \int |f| = \int f^+ + \int f^- < \infty$ (since $|f| = f^+ + f^-$, Proposition 2.1(b) applies), and f is said to be **integrable** (with respect to μ). The following result is now simple to prove.

Proposition 2.3 Let $f, g \in L$ be integrable and $\alpha, \beta \in \mathbb{R}^1$. Then (i) $\alpha f, \beta g, \alpha f + \beta g$ are integrable and $\int(\alpha f + \beta g) = \alpha \int f + \beta \int g$ *(linearity)*, and (ii) $f \leq g$ implies $\int f \leq \int g$ *(order)*.

Proof (i) First, let $\alpha \geq 0$. Then $(\alpha f)^+ = \alpha f^+, (\alpha f)^- = \alpha f^-$, so that $\int \alpha f = \int \alpha f^+ - \int \alpha f^- = \alpha \int f^+ - \alpha \int f^- = \alpha \int f$, by Proposition 2.1(b). Now let $\alpha < 0$. Then $(\alpha f)^+ = -\alpha f^-, (\alpha f)^- = -\alpha f^+$. Hence $\int \alpha f = \int -\alpha f^- - \int -\alpha f^+ = -\alpha \int f^- - (-\alpha) \int f^+ = \alpha(\int f^+ - \int f^-) = \alpha \int f$. Next if f, g are integrable, then writing $h = f + g$, we have $|h| \leq |f| + |g|$, so that $\int |h| \leq \int |f| + \int |g| < \infty$. Since $h = f + g = f^+ + g^+ - f^- - g^- = h^+ - h^-$, one has $h^+ + f^- + g^- = h^- + f^+ + g^+$ and, by Proposition 2.1(b), $\int h^+ + \int f^- + \int g^- = \int h^- + \int f^+ + \int g^+$. Therefore, $\int h \equiv \int h^+ - \int h^- = \int f^+ - \int f^- + \int g^+ - \int g^- = \int f + \int g$. This proves (i). To prove (ii) note that $f \leq g$ implies $f^+ \leq g^+$, $f^- \geq g^-$. Hence $\int f \equiv \int f^+ - \int f^- \leq \int g^+ - \int g^- \equiv \int g$. ∎

Our next task is to show that the integral of a function f remains unaffected if it is modified arbitrarily (but measurably) on a μ-null set. First, note that if $f \in L^+$, then

$$\int f = 0 \quad \text{iff} \quad f = 0 \text{ a.e. } (\mu) \quad (f \in L^+), \tag{2.8}$$

where **a.e.** (μ) is short-hand for *almost everywhere with respect to* μ, or *outside a μ-null set*. To prove (2.8), let $N = \{x : f(x) > 0\}$. Then one has $f = f\mathbf{1}_N + f\mathbf{1}_{N^c} = f \cdot \mathbf{1}_N$ ($f = 0$ on N^c). If $\mu(N) = 0$, then for all $g \in L_s^+$, $g \leq f$, one has $g = 0$ on N^c, so that $\int g = \int g\mathbf{1}_N + \int g\mathbf{1}_{N_c} = \int g\mathbf{1}_N = 0$, implying $\int f = 0$. Conversely, if $\int f = 0$, then $\mu(N) = 0$. For otherwise there exists $\varepsilon > 0$ such that writing $N_\varepsilon := \{x : f(x) > \varepsilon\}$, one has $\mu(N_\varepsilon) > 0$. In that case, $g := \varepsilon \mathbf{1}_{N_\varepsilon} \leq f$ and $\int f \geq \int g = \varepsilon \mu(N_\varepsilon) > 0$, a contradiction.

As a consequence of (2.8), one has the result that *if $f = g$ a.e., and f, g are integrable, then $\int f = \int g$.* To see this note that $|\int f - \int g| = |\int (f - g)| \leq \int |f - g| = 0$, since $|f - g| = 0$ a.e.

From here on, all functions f, g, h, f_n, g_n, h_n, etc., are assumed to be measurable, unless specified otherwise.

An important notion in measure theory is that of convergence in measure. Let f_n ($n \geq 1$), f be measurable functions on a measure space $(S, \mathcal{S}, \mu)$. The sequence $\{f_n\}_{n \geq 1}$ **converges in measure** to f if

$$\mu ([|f_n - f| > \varepsilon]) \longrightarrow 0 \quad \text{as } n \to \infty, \forall \varepsilon > 0. \tag{2.9}$$

Proposition 2.4 (a) If $f_n \to f$ in measure then there exists a subsequence $\{f_{n_k}\}_{k \geq 1}$ that converges a.e. to f. (b) If $\mu(S) < \infty$, then the convergence $f_n \to f$ a.e. implies $f_n \to f$ in measure.

Proof (a) Assume $f_n \to f$ in measure. For each k one can find n_k such that $\mu([|f_{n_k} - f| > 1/2^k]) < 1/2^k$. Now, for any given $\varepsilon > 0$, $[\limsup_k |f_{n_k} - f| > \varepsilon] \subset \cap_{m=1}^\infty \cup_{k=m}^\infty [|f_{n_k} - f| > 1/2^k] = N$, say. But $\mu(N) \leq \sum_{k=m}^\infty 2^{-k} = 2^{-m+1} \to 0$ as $m \to \infty$, i.e., $\mu(N) = 0$, proving $f_{n_k} \to f$ a.e., as $k \to \infty$. (b) Suppose $\mu(S) < \infty$, and $f_n \to f$ a.e. If f_n does not converge in measure to f, there exist $\varepsilon > 0, \delta > 0$, and a subsequence $\{f_{n_k}\}_{k \geq 1}$ such that $\mu([|f_{n_k} - f| > \varepsilon]) > \delta \; \forall k = 1, 2, \ldots$. But writing $A_k = [|f_{n_k} - f| > \varepsilon]$, one then has $B_m \equiv \cup_{k=m}^\infty A_k \downarrow B = [|f_{n_k} - f| > \varepsilon$ for infinitely many $k]$. Since $B_m^c \uparrow B^c$, it follows from countable additivity of μ that $\mu(B_m^c) \uparrow \mu(B^c)$, so that $\mu(B_m) = \mu(S) - \mu(B_m^c) \to \mu(S) - \mu(B^c) = \mu(B)$. Since $\mu(B_m) > \delta \; \forall m$, one obtains $\mu(B) \geq \delta$, which contradicts the fact that $f_n \to f$ a.e. ∎

Theorem 2.5 *(Basic Convergence Theorems for Integrals)*

(a) *(Monotone Convergence Theorem).* Suppose f_n ($n \geq 1$), f are nonnegative a.e. and $f_n \uparrow f$ a.e., then $\int f_n \uparrow \int f$.

(b) *(Fatou's Lemma).* If $g_n \geq 0$ a.e., then $\int \liminf g_n \leq \liminf \int g_n$.

(c) *(Lebesgue's Dominated Convergence Theorem).* If $f_n \to f$ in μ-measure and $|f_n| \leq h$ a.e., where h is integrable, then $\lim_n \int |f_n - f| = 0$. In particular, $\int f_n \to \int f$.

Proof (a) Since a countable union of μ-null sets is μ-null, there exists N such that $\mu(N) = 0$ and $f_n \geq 0$, $f \geq 0$, $f_n \uparrow f$ on N^c. Setting $f_n = 0$ ($n \geq 1$) and $f = 0$ on N does not change the integrals $\int f_n$, $\int f$. Hence one may apply Proposition 2.2.

(b) As in (a), one may assume $g_n \geq 0$ on S ($\forall n \geq 1$). Let $f_n = \inf\{g_k : k \geq n\}$. Then $0 \leq f_n \uparrow \liminf g_n = f$, say, and $\int f_n \uparrow \int f$ (by (a)). Also $f_n \leq g_n \, \forall n$, so that $\int g_n \geq \int f_n \, \forall n$, implying, in particular, $\liminf \int g_n \geq \liminf \int f_n = \lim \int f_n = \int f$.

(c) First assume $f_n \to f$ a.e. Apply Fatou's lemma to $g_n := 2h - |f_n - f|$, $0 \leq g_n \to 2h$ a.e., to get $\int 2h \leq \liminf \int g_n = \int 2h - \limsup \int |f_n - f|$, proving $\lim \int |f_n - f| = 0$. Now assume $f_n \to f$ in μ-measure. If $\int |f_n - f|$ does not converge to zero, there exist $\delta > 0$ and a subsequence $1 < n_1 < n_2 < \cdots$ such that $\int |f_{n_k} - f| > \delta \, \forall k$. Then there exists, by Proposition 2.4(a), a further subsequence of $\{n_k : k \geq 1\}$, say $\{n_k' : k \geq 1\}$, such that $f_{n_k'} \to f$ a.e. as $k \to \infty$, to which the above result applies to yield $\int |f_{n_k'} - f| \to 0$ as $k \to \infty$, contradicting $\int |f_{n_k'} - f| > \delta$ $\forall k$. ∎

The next result provides useful approximations to functions in the complex Banach space $L^p = L^p(\mathbb{R}^k, \mathcal{B}(\mathbb{R}^k), \mu)$, with norm $\|f\| := \left(\int_{\mathbb{R}^k} |f(x)|^p \mu(dx) \right)^{\frac{1}{p}}$, where $1 \leq p < \infty$. The result can of course be specialized to the real Banach space L^p.

Proposition 2.6 Let μ be a measure on $(\mathbb{R}^k, \mathcal{B}(\mathbb{R}^k))$ that is finite on compact subsets of $\mathbb{R}^k$. Then the set of infinitely differentiable functions with compact support is dense in L^p.

Proof For simplicity of notation we will take $k = 1$. The general case is similar. It is easy to see by considering real and imaginary parts separately, and then splitting a real-valued function f as $f = f^+ - f^-$, that it is enough to consider real-valued, nonnegative $f \in L^p$. Given $\varepsilon > 0$, find $N > 0$ such that $\int_{\{x:|x|\geq N\}} f^p d\mu < \frac{\varepsilon}{5}$. Set $f_N = f \mathbf{1}_{(-N,N]}$. Since $f_{N,M} := f_N \wedge M \equiv \min\{f_N, M\} \uparrow f$ as $M \uparrow \infty$, and $|f_{N,M} - f_N|^p \leq 2^p |f_N|^p \leq 2^p |f|^p$, there exists M such that $\|f_N - f_{N,M}\| < \frac{\varepsilon}{5}$. Because $f_{N,M}$ is bounded, there exists a simple function $g = \sum_{j=1}^{m} x_j \mathbf{1}_{B_j}$, where $x_j > 0$, B_j Borel, $B_j \subset (-N, N]$, $\mu(B_j) < \infty$, $1 \leq j \leq m$, such that $\sup\{|f_{N,M}(x) - g(x)| : x \in \mathbb{R}\} < \frac{\varepsilon}{5}$ (use the standard approximation (2.4)). Then $\|f_{N,M} - g\| < \frac{\varepsilon}{5}$.

We will now approximate g by a μ-a.e. continuous step function. For this, first note that the set of all finite unions of disjoint intervals of the form $(a, b]$, $-N \leq a < b \leq N$, is a field $\mathcal{F}_0$ on $(-N, N]$ such that $\sigma(\mathcal{F}_0) = \mathcal{B}((-N, N])$. Hence by Carathéodory's extension theorem, one can find a sequence of such disjoint intervals whose union contains B_j and approximates it as closely as desired. Since $\mu(B_j) < \infty$, one may take a finite subset of these intervals, say $(a_{ij}, b_{ij}], 1 \leq i \leq n_j$, such that $A_j = \cup_{i=1}^{n_j}(a_{ij}, b_{ij}]$ satisfies $\mu(B_j \Delta A_j) < m^{-1}(\frac{\varepsilon}{5c})^p$, for $c := \max\{x_1, \ldots, x_m\}$ $(1 \leq j \leq m)$. Since the set $\{x : \mu(\{x\}) > 0\}$ is countable, one may use the approximation of $(a_{ij}, b_{ij}]$ from above and below, if necessary, to ensure that $\mu(\{a_{ij}\}) = 0 = \mu(\{b_{ij}\}), 1 \leq i \leq n_j, j = 1, \ldots m$. Note that, with $h = \sum_{j=1}^{m} x_j \mathbf{1}_{A_j}$ and $g = \sum_{j=1}^{m} x_j \mathbf{1}_{B_j}$, as above, one has $\|h - g\|^p \leq mc^p[m^{-1}(\frac{\varepsilon}{5})^p] = (\frac{\varepsilon}{5})^p$, so that $\|h - g\| < \varepsilon/5$. Finally, let ψ be an infinitely differentiable probability density on $\mathbb{R}$ with compact support (e.g. see (4.2) in Chapter V). Define $\psi_n(x) = n\psi(nx)(n = 1, 2, \ldots)$. Then the probabilities $\psi_n(x)dx$ converge weakly to δ_0 as $n \to \infty$. Hence the functions

$$h_n(x) := \int_{\mathbb{R}} h(x - y)\psi_n(y)dy = \int_{\mathbb{R}} h(y)\psi_n(x + y)dy, \quad n \geq 1, \qquad (2.10)$$

are infinitely differentiable with compact support, and $h_n(x) \to h(x)$ at all points x of continuity of h. Since the set of possible discontinuities of h, namely $\{a_{ij} : 1 \leq i \leq n_j, 1 \leq j \leq m\} \cup \{b_{ij} : 1 \leq i \leq n_j, 1 \leq j \leq m\}$ has μ-measure zero, $h_n \to h$ μ-almost everywhere. Also h_n, h have compact support and are uniformly bounded by $c = \max\{x_1, \ldots, x_m\}$. Hence $h_n \to h$ in L^p, and there exists n_0 such that $\|h_{n_0} - h\| < \frac{\varepsilon}{5}$. Therefore,

$$\|h_{n_0} - f\| \leq \|h_{n_0} - f\| + \|h - g\| + \|g - f_{N,M}\| + \|f_{N,M} - f_N\| + \|f_N - f\|$$
$$< 5(\frac{\varepsilon}{5}) = \varepsilon.$$

Since $\varepsilon > 0$ is arbitrary the proof is complete. ∎

Remark 2.1 Note that the proof shows that if μ is finite on compacts sets, then step functions are dense in $L^p(\mathbb{R}, \mathcal{B}(\mathbb{R}), \mu)$. Indeed rational-valued step functions with supporting intervals whose endpoints are dyadic rationals are dense in L^p. In particular, it follows that L^p is **separable**. The argument extends to $\mathbb{R}^k$ for $k > 1$ as well.

A.3 Product Measures

Let $(S_i, \mathcal{S}_i)$ $(i = 1, 2)$ be measurable spaces. The **product σ-field** $\mathcal{S} = \mathcal{S}_1 \otimes \mathcal{S}_2$ on the Cartesian product space $S = S_1 \times S_2$ is the smallest σ-field containing all sets of the form $A \times B$, with $A \in \mathcal{S}_1$ and $B \in \mathcal{S}_2$, called **measurable rectangles**. Let μ_i be a σ-finite measure on $(S_i, \mathcal{S}_i)$ $(i = 1, 2)$. Define the set function μ on the class $\mathcal{R}$ of all measurable rectangles

$$\mu(A \times B) := \mu_1(A)\mu_2(B) \qquad (A \in \mathcal{S}_1, B \in \mathcal{S}_2). \qquad (3.1)$$

Theorem 3.1 There exists a unique extension of μ from $\mathcal{R}$ to a σ-finite measure on the product σ-field $\mathcal{S} = \mathcal{S}_1 \otimes \mathcal{S}_2$.

Proof For the proof we need first the fact that if $C \in \mathcal{S}$, then the x-**section** $C_x := \{y \in S_2 : (x, y) \in C\}$ belongs to $\mathcal{S}_2$, $\forall x \in S_1$. The class $\mathcal{C}$ of all sets C for which this is true contains $\mathcal{R}$, since $(A \times B)_x = B$ (if $x \in A$), or $\emptyset$ (if $x \notin A$). Since it is easy to check that $\mathcal{C}$ is a λ-system containing the π-system $\mathcal{R}$, it follows by the $\pi - \lambda$ Theorem that $\mathcal{C} \supset \mathcal{S}$.

Similarly, if f is an extended real-valued measurable function on the product space $(S, \mathcal{S})$ then every x-**section of** f defined by $f_x(y) = f(x, y)$, $y \in S_2$, is a measurable function on $(S_2, \mathcal{S}_2)$, $\forall x \in S$. For if $D \in \mathcal{B}(\mathbb{R})$ and $x \in S_1$, then $f_x^{-1}(D) \equiv [y : f(x, y) \in D] = (f^{-1}(D))_x \in \mathcal{S}_2$.

Next, for $C \in \mathcal{S}$ the function $x \mapsto \mu_2(C_x)$ is measurable on $(S_1, \mathcal{S}_1)$. This is clearly true for $C \in \mathcal{R}$, and the general assertion again follows from the π-λ theorem. Now define μ on $\mathcal{C}$ by

$$\mu(C) := \int_{S_1} \mu_2(C_x) \mu_1 dx. \tag{3.2}$$

If $C = \cup_n C_n$, where $C_n \in \mathcal{S}$ $(n = 1, 2, \dots)$ are pairwise disjoint, then $C_x = \cup_n (C_n)_x$ and, by countable additivity of μ_2, $\mu_2(C_x) = \sum_n \mu_2((C_n)_x)$, so that $\mu(C) = \int_{S_1} \sum_n \mu_2((C_n)_x) \mu_1(dx) = \sum_n \int_{S_1} \mu_2((C_n)_x) \mu_1(dx) = \sum_n \mu(C_n)$. Here the interchange of the order of summation and integration is valid, by the monotone convergence theorem. Thus (3.2) defines a measure on $\mathcal{S}$, extending (3.1). The measure μ is clearly σ-finite. If ν is another σ-finite measure on $(S, \mathcal{S})$ such that $\nu(A \times B) = \mu_1(A)\mu_2(B) \ \forall A \times B \in \mathcal{R}$, then the class of sets $C \in \mathcal{S}$ such that $\mu(C) = \nu(C)$ is easily seen to be a σ-field and therefore contains $\sigma(\mathcal{R}) = \mathcal{S}$. ∎

The measure μ in Theorem 3.1 is called the **product measure** and denoted by $\mu_1 \times \mu_2$. The measure space $(S, \mathcal{S}, \mu)$ with $S = S_1 \times S_2$, $\mathcal{S} = \mathcal{S}_1 \otimes \mathcal{S}_2$, $\mu = \mu_2 \times \mu_2$, is called the **product measure space**.

Next note that instead of (3.2), one can define the measure $\widetilde{\mu}$ by

$$\widetilde{\mu}(C) = \int_{S_2} \mu_1(C^y) \mu_2(dy), \qquad C \in \mathcal{S}, \tag{3.3}$$

where $C^y = \{x \in S_1 : (x, y) \in C|$ is the y-**section** of C, for $y \in S_2$. But $\widetilde{\mu} = \mu$ on $\mathcal{R}$ and therefore, by the uniqueness of the extension, $\widetilde{\mu} = \mu$ on $\mathcal{S}$. It follows that if $f = 1_C$ for some $C \in \mathcal{S}$ and $f^y(x) := f(x, y)$, then

$$\int_S f d\mu = \int_{S_1} \left\{ \int_{S_2} f_x(y)\mu_2(dy) \right\} \mu_1(dx) = \int_{S_2} \left\{ \int_{S_1} f^y(x)\mu_1(dx) \right\} \mu_2(dy). \tag{3.4}$$

This equality of the iterated integrals with different orders of integration immediately extends to nonnegative simple functions. For arbitrary nonnegative $\mathcal{S}$-measurable f

one uses an approximation $f_n \uparrow f$ by simple functions f_n and applies the monotone convergence theorem to arrive at the following important result.

Theorem 3.2 *(Fubini–Tonelli Theorem)* (a) Let f be a nonnegative measurable function on the product measure space $(S, \mathcal{S}, \mu)$, where $S = S_1 \times S_2$, $\mathcal{S} = \mathcal{S}_1 \otimes \mathcal{S}_2$, $\mu = \mu_1 \times \mu_2$. Then (3.4) holds. (b) If f is μ-integrable, then (3.4) holds.

Proof We have outlined above a proof of (a). For (b), use $f = f^+ - f^-$, linearity of the integral (with respect to μ, μ_1, μ_2), and (a). ∎

Given $k \ (\geq 2)$ σ-finite measure spaces $(S_i, \mathcal{S}_i, \mu_i)$, $1 \leq i \leq k$, the above definitions and results can be extended to define the product measure space $(S, \mathcal{S}, \mu)$ with (1) $S = S_1 \times \cdots \times S_k$ the **Cartesian product** of $S_1, \ldots, S_k$, and (2) $\mathcal{S} = \mathcal{S}_1 \otimes \cdots \otimes \mathcal{S}_k$, the **product σ-field**, i.e., the smallest σ-field on S containing the class $\mathcal{R}$ of all measurable rectangles $A_1 \times A_2 \times \cdots \times A_k$ ($A_i \in \mathcal{S}_i, 1 \leq i \leq k$), and (3) $\mu = \mu_1 \times \mu_2 \times \cdots \times \mu_k$, the σ-finite *product measure* on $\mathcal{S}$ satisfying

$$\mu(A_1 \times A_2 \times \cdots \times A_k) = \mu_1(A_1)\mu_2(A_2) \cdots \mu_k(A_k) \qquad (A_i \in \mathcal{S}_i, a \leq i \leq k). \quad (3.5)$$

Example 1 The **Lebesgue measure on** $\mathbb{R}^k$ is the product measure $\mathbf{m} = m_1 \times m_2 \times \cdots \times m_k$ defined by taking $S_i = \mathbb{R}$, $\mathcal{S}_i = \mathcal{B}(\mathbb{R})$, $m_i =$ Lebesgue measure on $\mathbb{R}$, $1 \leq i \leq k$.

A.4   Riesz Representation on $C(S)$

Suppose that S is a compact metric space with Borel σ-field $\mathcal{B}$, and $C(S)$ the space of continuous real-valued functions on S. If μ is a finite measure on $(S, \mathcal{B})$, then the linear functional $\ell_\mu(f) = \int_S f \, d\mu$, $f \in C(S)$, is clearly a linear functional on $C(S)$. Moreover ℓ_μ is a **positive linear functional** in the sense that $\ell_\mu(f) \geq 0$ for all $f \in C(S)$ such that $f(x) \geq 0$ for all $x \in S$. Additionally, giving $C(S)$ the uniform norm $\|f\| = \sup\{|f(x)| : x \in S\}$, one has that ℓ_μ is a **bounded linear functional** in the sense that $\sup_{\|f\| \leq 1, f \in C(S)} |\ell_\mu(f)| < \infty$. In view of linearity this boundedness is easily checked to be equivalent to continuity of $\ell_\mu : C(S) \to \mathbb{R}$. The **Riesz representation theorem** for $C(S)$ asserts that these are the only bounded linear functionals on $C(S)$.

Theorem 4.1 *(Riesz Representation Theorem on $C(S)$)* Let S be a compact metric space. If ℓ is a bounded positive linear functional on $C(S)$, then there is a unique finite measure μ on $(S, \mathcal{B})$ such that for all $f \in C(S)$,

$$\ell(f) = \int_S f \, d\mu.$$

Moreover,

$$\mu(A) = \inf\{\mu(G) : G \supset A, G \text{ open}\} = \sup\{\mu(F) : F \subset A, F \text{ closed}\}, \quad A \in \mathcal{B}.$$

Definition 4.1 A measure μ defined on the Borel σ-field $\mathcal{B}$ of a topological space is said to be **regular** if

$$\mu(A) = \inf\{\mu(G) : G \supset A, G \text{ open}\} = \sup\{\mu(F) : F \subset A, F \text{ closed}\}, \quad A \in \mathcal{B}.$$

Observe that the uniqueness assertion follows trivially from the fact that $C(S)$ is a measure-determining class of functions for finite measures. The proof will follow from a sequence of lemmas, the first of which already appeared implicitly in the proof of Alexandrov's theorem 7.1.

Lemma 1 *(Urysohn)* For a closed set $F \subset S$ there is a (pointwise) nonincreasing sequence $h_n \in C(S)$, $n \geq 1$, such that $h_n \downarrow \mathbf{1}_F$, and for an open set G there is a (pointwise) nondecreasing sequence $g_n \in C(S)$ such that $g_n \uparrow \mathbf{1}_G$.

For a function $f \in C(S)$, the smallest closed set outside of which f is zero is called the **support** of f and is denoted by $\mathrm{supp}(f)$. Note that if $f \in C(S)$ satisfies $0 \leq f \leq \mathbf{1}_A$, then $\mathrm{supp}(f) \subset \overline{A}$. For open sets $G \subset S$ it is convenient to introduce notation $g \prec G$ to denote a function $g \in C(S)$ **subordinate** to G in the sense that $0 \leq g \leq 1$ and $\mathrm{supp}(g) \subset G$. With this notation we will see that the desired measure may be expressed explicitly for open $G \subset S$ as

$$\mu(G) = \sup\{\ell(g) : g \prec G\}. \tag{4.1}$$

Note that since S is open (and closed), one has $\mu(S) < \infty$ from the boundedness of ℓ. With μ defined for open sets by (4.1), for arbitrary $A \subset S$ let

$$\mu^*(A) = \inf\{\mu(G) : G \supset A, G \text{ open}\}. \tag{4.2}$$

Lemma 2 μ^* is an outer measure and each Borel-measurable subset of S is μ^*-measurable.

Proof For the first part we will in fact show that

$$\mu^*(A) = \inf\left\{\sum_{n=1}^{\infty} \mu(G_n) : \cup_{n=1}^{\infty} G_n \supset A, G_n \text{ open}\right\}.$$

from which it follows by Proposition 1.1 that μ^* is an outer measure. For this formula it suffices to check that for any given sequence G_n, $n \geq 1$, of open sets one has $\mu(\cup_{n=1}^{\infty} G_n) \leq \sum_{n=1}^{\infty} \mu(G_n)$. Let $G = \cup_{n=1}^{\infty} G_n$ and $g \prec G$, $g \in C(S)$. The support $\mathrm{supp}(g) \subset S$ is compact, and hence $\mathrm{supp}(g) \subset \cup_{n=1}^{N} G_n$, for some N. By Urysohn's lemma, there are functions $g_n \in C(S)$, $1 \leq n \leq N$, such that $g_n \prec G_n$, and $\sum_{n=1}^{N} g_n = 1$ on $\mathrm{supp}(g)$. Now $g = \sum_{n=1}^{N} g_n g$ and $g_n g \prec G_n$, so that

$$\ell(g) = \sum_{n=1}^{N} l(g_n g) \le \sum_{n=1}^{N} \mu(G_n) \le \sum_{n=1}^{\infty} \mu(G_n).$$

Since $g \prec G$ is arbitrary, it follows that $\mu(G) \le \sum_{n=1}^{\infty} \mu(G_n)$ as desired. For the second part of the lemma it suffices to check that each open set is μ^*-measurable. That is, if G is an open set, then for any $E \subset S$ one must check $\mu^*(E) \ge \mu^*(E \cap G) + \mu^*(E \cap G^c)$. If E is also open then given $\varepsilon > 0$ there is a $g \in C(S)$, $g \prec E \cap G$, such that $\ell(g) > \mu(E \cap G) - \varepsilon$. Similarly, $E \cap \mathrm{supp}(g)^c$ is open and there is a $\tilde{g} \in C(S)$, $\tilde{g} \prec E \cap \mathrm{supp}(g)^c$, such that $\ell(\tilde{g}) > \mu(E \cap \mathrm{supp}(g)^c) - \varepsilon$. But now $g + \tilde{g} \prec E$ and $\mu(E) > \ell(g) + \ell(\tilde{g}) > \mu(E \cap G) + \mu(E \cap \mathrm{supp}(g)^c) - 2\varepsilon \ge \mu^*(E \cap G) + \mu^*(E \cap G^c) - 2\varepsilon$. Since ε is arbitrary, the desired Carathéodory balance condition (1.2) holds for open E. For arbitrary $E \subset S$ let $\varepsilon > 0$ and select an open set $U \supset E$ such that $\mu(U) < \mu^*(E) + \varepsilon$. Then $\mu^*(E) + \varepsilon \ge \mu(U) \ge \mu^*(U \cap G) + \mu^*(U \cap G^c) > \mu^*(E \cap G) + \mu^*(E \cap G^c)$. ∎

From here one readily obtains a measure space $(S, \mathcal{B}, \mu)$ by restricting μ^* to $\mathcal{B}$. The proof of the theorem is completed with the following lemma.

Lemma 3 For closed $F \subset S$,

$$\mu(F) = \inf\{\ell(h) : h \ge \mathbf{1}_F\}.$$

Moreover,

$$\ell(f) = \int_S f \, d\mu \quad \forall f \in C(S).$$

Proof For closed $F \subset S$ and an arbitrary $h \in C(S)$ with $h \ge \mathbf{1}_F$ consider, for $\varepsilon > 0$, the open set $G_\varepsilon = \{x \in S : h(x) > 1 - \varepsilon\}$. Let $g \in C(S)$, $g \prec G_\varepsilon$. Then $\ell(g) \le (1 - \varepsilon)^{-1}\ell(h)$. It now follows that $\mu(F) \le \mu(G_\varepsilon) \le (1 - \varepsilon)^{-1}\ell(h)$, and hence, since $\varepsilon > 0$ is arbitrary, $\mu(F) \le \ell(h)$. To see that $\mu(F)$ is the greatest lower bound, let $\varepsilon > 0$ and let $G \supset F$ be an open set with $\mu(G) - \mu(F) < \varepsilon$. By Urysohn's lemma there is an $h \in C(S)$, $h \prec G$, with $h \ge \mathbf{1}_F$. Thus, using the definition of μ, $\ell(h) \le \mu(G) \le \mu(F) + \varepsilon$. To establish that μ furnishes the desired representation of ℓ, let $f \in C(S)$. In view of the linearity of ℓ, it suffices to check that $\ell(f) \le \int_S f \, d\mu$; since the same inequality would then be true with f replaced by $-f$, and hence the reverse inequality follows. Let $m = \min\{f(x) : x \in S\}$, $M = \max\{f(x) : x \in S\}$. For $\varepsilon > 0$, partition $[m, M]$ as $y_0 < m < y_1 < \cdots < y_n = M$ such that $y_j - y_{j-1} < \varepsilon$. Let $A_j = f^{-1}(y_{j-1}, y_j] \cap \mathrm{supp}(f)$, $1 \le j \le n$. Then $A_1, \ldots, A_n$ is a partition of $\mathrm{supp}(f)$ into disjoint Borel-measurable subsets. Let $G_j \supset A_j$ be an open set with $\mu(G_j) < \mu(A_j) + \frac{\varepsilon}{n}$, $j = 1, \ldots, n$, with $f(x) < y_j + \varepsilon$, $x \in G_j$. Apply Urysohn's lemma to obtain $g_j \prec G_j$ with $\sum_{j=1}^{n} g_j = 1$ on $\mathrm{supp}(f)$. Then $f = \sum_{j=1}^{n} g_j f$, and since $g_j f \le (y_j + \varepsilon)g_j$, and $y_j - \varepsilon < f(x)$ on A_j, one has

$$\ell(f) = \sum_{j=1}^{n} \ell(g_j f) \le \sum_{j=1}^{n} (y_j + \varepsilon)\ell(g_j) \le \sum_{j=1}^{n} (y_j + \varepsilon)\mu(G_j)$$

$$\le \sum_{j=1}^{n} (y_j + \varepsilon)\mu(A_j) + \sum_{j=1}^{n} (y_j + \varepsilon)\frac{\varepsilon}{n}$$

$$\le \sum_{j=1}^{n} (y_j - \varepsilon)\mu(A_j) + 2\varepsilon\mu(\mathrm{supp}(f)) + (M + \varepsilon)\varepsilon$$

$$\le \sum_{j=1}^{n} \int_{A_j} f d\mu + \{2\mu(\mathrm{supp}(f)) + M + \varepsilon\}\varepsilon$$

$$= \int_{S} f d\mu + \{2\mu(\mathrm{supp}(f)) + M + \varepsilon\}\varepsilon.$$

Since $\varepsilon > 0$ is arbitrary, the desired inequality is established. ■

Example 2 To associate the Riemann integral of continuous functions f on the k-dimensional unit $S = [-1, 1]^k$ with a measure and the corresponding Lebesgue integral, apply the Riesz representation theorem to the bounded linear functional defined by

$$\ell(f) = \int_{-1}^{1} \cdots \int_{-1}^{1} f(x_1, \ldots, x_k) dx_1 \cdots dx_k. \tag{4.3}$$

Appendix B
Topology and Function Spaces

We begin with an important classical result. Let $C[0, 1]$ denote the set of all real-valued continuous functions on $[0, 1]$, endowed with the **sup norm:** $\|f\| = \max\{|f(x)| : x \in [0, 1]\}$. With the distance $d(f, g) = \|f - g\|$, $C[0, 1]$ is a real **Banach space** i.e., it is a vector space (with respect to real scalars) and it is a complete (normed) metric space. Recall that a **norm** $\| \ \| : V \to [0, \infty)$ on a vector space V satisfies: $\|g\| = 0$ iff $g = 0$, $\|\alpha g\| = |\alpha| \cdot \|g\|$ (α scalar, $g \in V$), and $\|f + g\| \le \|f\| + \|g\|$. Also, a subset A of a metric space is **complete** if every Cauchy sequence in A has a convergent subsequence in A.

Theorem 1.2 *(Weierstrass Approximation Theorem)* Polynomials are dense in $C[0, 1]$.

Proof Let $g \in C[0, 1]$. Define a sequence h_n ($n \ge 1$) of polynomials on $[0, 1]$ as

$$h_n(p) = \sum_{i=0}^{n} g\left(\frac{i}{n}\right)\binom{n}{i} p^i (1 - p)^{n-i} \qquad (p \in [0, 1]), \ n \ge 1. \qquad (1.4)$$

Then, for each p one may write $h_n(p) = \mathbb{E}g(X/n)$, where X is a binomial random variable $B(n, p)$. Let $\varepsilon > 0$ be given. There exists $\delta > 0$ such that $|g(p') - g(p'')| \le \varepsilon/2$, if $|p' - p''| \le \delta$ and $p', p'' \in [0, 1]$. Hence

$$|h_n(p) - g(p)| = |\mathbb{E}g(X/n) - g(p)|$$

$$\le \frac{\varepsilon}{2} P\left(\left|\frac{X}{n} - p\right| \le \delta\right) + 2\|g\| P\left(\left|\frac{X}{n} - p\right| > \delta\right)$$

$$\le \frac{\varepsilon}{2} + 2\|g\| \frac{p(1 - p)}{n\delta^2} \le \frac{\varepsilon}{2} + \frac{\varepsilon}{2} = \varepsilon$$

for all p if $n \ge \frac{\|g\|}{\varepsilon\delta^2}$. ∎

Instead of $[0, 1]$, we now consider an arbitrary compact Hausdorff space S. Recall that a topological space S (with a topology $\mathcal{T}$ of open sets) is **Hausdorff** if for every

© Springer International Publishing AG 2016
R. Bhattacharya and E.C. Waymire, *A Basic Course in Probability Theory*,
Universitext, DOI 10.1007/978-3-319-47974-3

pair $x, y \in S$, $x \neq y$, there exist disjoint open sets U, V such that $x \in U$, $y \in V$. A topological space is **compact** if every open cover of S has a finite subcover. That is, if $\{V_\lambda : \lambda \in \Lambda\}$ is a collection of open sets such that $\cup_{\lambda \in \Lambda} V_\lambda = S$, then there exists a finite set $\{\lambda_1, \ldots, \lambda_k\} \subset \Lambda$ such that $\cup\{V_{\lambda_i} : 1 \leq i \leq k\} = S$. By taking complements it is immediately seen that S is compact if and only if it has the **finite intersection property** if $\{C_\lambda : \lambda \in \Lambda\}$ is a collection of closed sets whose intersection is empty, then it has a finite subcollection whose intersection is empty. The following are a few useful related notions. A topological space S is called **locally compact** if every point $x \in S$ has a compact neighborhood. The space S is called σ-**compact** if it is a countable union of compact sets. A *subset D of a topological space $(S, \mathcal{T})$ is compact* if it is **compact** as a topological space with the **relative topology** defined by $D \cap \mathcal{T}$.

It is simple to check, using the finite intersection property, that a real-valued continuous function on a compact space S attains its supremum (and infimum). From this it follows that the space $C(S)$ of real-valued continuous functions on S is a Banach space under the norm (called the **supnorm**) $\|f\| := \max\{|f(x)| : x \in S\}$ It is also an **algebra** i.e., it is a vector space that is also closed under (pointwise) multiplication $(f, g) \mapsto fg \, \forall f, g \in C(S)$. A **subalgebra** of $C(S)$ is a vector subspace that is also closed under multiplication. A subset $\mathcal{H}$ of $C(S)$ is said to **separate points** if for every pair of points $x \neq y$ in S there is a function $f \in \mathcal{H}$ such that $f(x) \neq f(y)$. The following is a far reaching generalization of the Weirstrass approximation theorem 1.2

Theorem 1.3 (*Stone–Weierstrass Theorem*) Let S be a compact Hausdorff space, and $\mathcal{H}$ a subalgebra of $C(S)$. If $\mathcal{H}$ includes constant functions and separates points, then $\mathcal{H}$ is dense in S, i.e., $\overline{\mathcal{H}} = C(S)$.

Proof Step 1. If $f \in \mathcal{H}$, then $|f| \in \overline{\mathcal{H}}$. To prove this use Theorem 1.2 to find a sequence h_n of polynomials converging uniformly to the function $h(p) = \sqrt{p}$ on $[0, 1]$. Now if $f \in \mathcal{H}$ then all polynomial functions of $g \equiv f/\|f\|$ belong to $\mathcal{H}$. In particular, $h_n \circ g^2 \in \mathcal{H}$ ($g^2(x) \equiv (g(x))^2 \in [0, 1]$). But $h_n \circ g^2$ converges uniformly on S to $h \circ g^2 = |f|/\|f\|$, so that the functions $(\|f\|)h_n \circ g^2$ in $\mathcal{H}$ converge uniformly to $|f|$.

Step 2. If $f, g \in \overline{\mathcal{H}}$ then $\max\{f, g\}, \min\{f, g\} \in \overline{\mathcal{H}}$. To see this, write $\max\{f, g\} = \frac{1}{2}(f + g + |f - g|)$, $\min\{f, g\} = \frac{1}{2}(f + g - |f - g|)$ and apply Step 1.

Step 3. Let $x \neq y \in S$, α and β real numbers. Then there exists $f \in \mathcal{H}$ such that $f(x) = \alpha$, $f(y) = \beta$. For this, find $g \in \mathcal{H}$ such that $a \equiv g(x) \neq b \equiv g(y)$. Let $f = \alpha + \frac{\beta - \alpha}{b - a}(g - a)$.

Step 4. Let $f \in C(S)$. Given any $x \in S$ and $\varepsilon > 0$, there exists $g \in \overline{\mathcal{H}}$ such that $g(x) = f(x)$ and $g(y) < f(y) + \varepsilon \, \forall y \in S$. To prove this, fix f, x, ε as above. By Step 3, for each $y \neq x$ there exists $g_y \in \mathcal{H}$ such that $g_y(x) = f(x)$, $g_y(y) = f(y) + \varepsilon/2$. Then y belongs to the open set $\mathcal{O}_y = \{z : g_y(z) < f(y) + \varepsilon\}$, and $S = \cup\{\mathcal{O}_y : y \in S \setminus \{x\}\}$. Let $\{\mathcal{O}_{y_1}, \ldots, \mathcal{O}_{y_k}\}$ be a subcover of S. Define $g = \min\{g_{y_1}, \ldots, g_{y_k}\}$. Then $g \in \overline{\mathcal{H}}$ (by Step 2), $g(x) = f(x)$, and $g(y) < f(y) + \varepsilon$ $\forall y \in S$.

Step 5. To complete the proof of the theorem, fix $f \in C(S)$, $\varepsilon > 0$. For each $x \in S$, let $f_x = g$ be the function obtained in Step 4. Then $V_x := [z \in S : f_x(z) > f(z) - \varepsilon]$, $x \in S$, form an open cover of S (since $x \in V_x$). Let $\{V_{x_1}, \ldots, V_{x_m}\}$ be a finite subcover. Then $\overline{f}_\varepsilon \equiv \max\{f_{x_1}, f_{x_2}, \ldots, f_{x_m}\} \in \mathcal{H}$ (by Step 2), and $f(z) - \varepsilon < \overline{f}_\varepsilon(z) < f(z) + \varepsilon \; \forall z \in S$.   ∎

Among many important applications of Theorem 1.3, let us mention two.

Corollary 1.4 Let S be a compact subset of $\mathbb{R}^m$ ($m \geq 1$). Then the set $\mathcal{P}_m$ of all polynomials in m variables is dense in $C(S)$.

Proof The set $\mathcal{P}_m$ is clearly an algebra that includes all constant functions. Also, let $\mathbf{x} = (x_1, \ldots, x_m) \neq \mathbf{y} = (y_1, \ldots, y_m) \in S$. Define $f(\mathbf{z}) = (z_1 - x_1)^2 + \cdots + (z_m - x_m)^2$. Then $f \in \mathcal{P}_m$, $f(\mathbf{x}) = 0$, $f(\mathbf{y}) > 0$. Hence Theorem 1.3 applies.   ∎

Corollary 1.5 *(Separability of $C(S)$)* Let (S, ρ) be a compact metric space. Then $C(S)$ is a separable metric space.

Proof First observe that S is separable. For there exist finitely many open balls $\{B(x_{j,n} : 1/n) : 1 \leq j \leq k_n\}$ that cover S. Here $B(x : \varepsilon) = \{y \in S : \rho(x, y) < \varepsilon\}$ is a ball with center x and radius ε. Clearly, $\{x_{j,n} : j = 1, \ldots, k_n; n = 1, 2, \ldots\}$ is a countable dense subset of S. To prove separability of $C(S)$, let $\{x_n : n = 1, 2, \ldots\}$ be a countable dense subset of S. Denote by $B_{n,k}$ the ball with center x_n and radius $1/k$ ($k = 1, 2, \ldots$; $n = 1, 2, \ldots$). Also, let $h_m(u) = 1 - mu, 0 \leq u < 1/m, h_m(u) = 0$ for $u \geq 1/m$, define a sequence of continuous functions on $[0, \infty)$ ($m = 1, 2, \ldots$). Define $f_{n,k,m}(x) := h_m(\rho(x, B_{n,k}))$ ($n \geq 1$, $k \geq 1$, $m \geq 1$), and let $\mathcal{M}$ be the set of all (finite) linear combinations of *monomials* of the form $f_{n_1 k_1, m_1}^{j_1} \cdots f_{n_r, k_r, m_r}^{j_r}$ ($r \geq 1$; $j_1, \ldots, j_r$ nonnegative integers). Then $\mathcal{M}$ is a subalgebra of $C(S)$ that includes constant functions and separates points: if $x \neq y$ then there exists $B_{n,k}$ such that $x \in B_{n,k}$ and $y \notin \overline{B}_{n,k}$, implying $f_{n,k,m}(x) = 1$, $f_{n,k,m}(y) < 1$, if m is sufficiently large. By Theorem 1.3, $\mathcal{M}$ is dense in $C(S)$. The countable subset of $\mathcal{M}$ comprising linear combinations of the monomials with rational scalars is dense in $\mathcal{M}$ and therefore in $\overline{\mathcal{M}} = C(S)$.   ∎

Remark 1.1 Let $C(S : \mathbb{C})$ denote the set of all complex-valued continuous functions on a compact metric space S. Under the sup norm $\|f\| := \sup\{|f(x)| : x \in S\}$, $C(S : \mathbb{C})$ is a (complex) Banach space. Letting $\{f_n : n = 1, 2, \ldots\}$ be a dense sequence in the real Banach space $C(S)$, the countable set $\{f_n + if_m : n \geq 1, m \geq 1\}$ is clearly dense in $C(S : \mathbb{C})$. Hence $C(S : \mathbb{C})$ is separable.

The next result concerns the **product topology** of the Cartesian product $S = \times_{\lambda \in \Lambda} S_\lambda$ of an arbitrary collection of compact spaces S_λ ($\lambda \in \Lambda$). This topology comprises all arbitrary unions of sets of the form $V = [\mathbf{x} \equiv (x_\lambda : \lambda \in \Lambda) : x_{\lambda_i} \in V_{\lambda_i}, 1 \leq i \leq k]$, $\lambda_i \in \Lambda$, V_{λ_i} an open subset of S_{λ_i} ($1 \leq i \leq k$), for some $k \geq 1$.

Theorem 1.6 *(Tychonoff's Theorem)* Let S_λ be compact for all $\lambda \in \Lambda$. Then $S = \times_{\lambda \in \Lambda} S_\lambda$ is compact under the product topology.

Proof We will give a proof when Λ is denumerable, say $\Lambda = \{1, 2, \dots\}$, and (S_n, ρ_n) are compact metric spaces, $n \geq 1$. The proof of the general case requires invoking the axiom of choice, and may be found in Folland.[1]

Let $\mathbf{x}^{(n)} = (x_1^{(n)}, x_2^{(n)}, \dots)$, $n \geq 1$, be a sequence in S. We will find a convergent subsequence. Let $\{x_1^{(n(1))} : n \geq 1\}$ be a subsequence of $\{x_1^{(n)} : n \geq 1\}$, converging to some $x_1 \in S_1$, $n(1) > n \; \forall n$. Let $\{x_2^{(n(2))} : n \geq 1\}$ be a subsequence of $\{x_2^{(n(1))} : n \geq 1\}$, converging to some $x_2 \in S_2$, $n(2) > n(1) \; \forall n$. In general, let $\{x_k^{(n(k))} : n \geq 1\}$ be a subsequence of $\{x_k^{(n(k-1))} : n \geq 1\}$, converging to $x_k \in S_k$ ($k = 1, 2, \dots$). Then the *diagonal subsequence* $\mathbf{x}^{(1(1))}, \mathbf{x}^{(2(2))}, \dots, \mathbf{x}^{(k(k))}, \dots$ converges to $\mathbf{x} = (x_1, x_2, \dots, x_k, \dots)$. ∎

Definition 1.2 A family $\mathcal{C}$ of continuous functions defined on a topological space S is said to be **equicontinuous** at a point $x \in S$ if for every $\varepsilon > 0$ there is a neighborhood U of x such that for every $f \in \mathcal{C}$, $|f(y) - f(x)| < \varepsilon$ for all $y \in U$. If $\mathcal{C}$ is equicontinuous at each $x \in S$ then $\mathcal{C}$ is called **equicontinuous**. Also $\mathcal{C}$ is said to be **uniformly bounded** if there is a number $M > 0$ such that $|f(x)| \leq M$ for all $f \in \mathcal{C}$ and all $x \in S$.

The next concept is especially useful in the context of $C(S)$ viewed as a metric space with the uniform metric.

Definition 1.3 A subset A of a metric space is said to be **totally bounded** if, for every $\delta > 0$ there is a covering of A by finitely many balls of radius δ.

Lemma 4 If A is a complete and totally bounded subset of a metric space then A is compact.

Proof Since A may be covered by finitely many balls of radii $1/2$, one of these, denoted by B_1, must contain x_n for infinitely many n, say $n \in N_1$. Next $A \cap B_1$ may be covered by finitely many balls of radii $1/4$. One of these balls, denoted by B_2, contains $\{x_n : n \in N_1\}$ for infinitely many n, say $n \in N_2 \subset N_1$. Continuing in this way, by selecting distinct points $n_1 < n_2 < \cdots$ from $N_1, N_2, \dots$, one may extract a subsequence $\{x_{n_m} : n_m \in N_m\}$ which is Cauchy and, since A is complete, converges in A. Now suppose that $\{U_\lambda : \lambda \in \Lambda\}$ is an open cover of A. In view of the total boundedness of A, if for some $\varepsilon > 0$ one can show that every ball of radius ε which meets A is a subset of some U_λ, then a finite subcover exists. To see that this is indeed the case, suppose not. That is, suppose for every $n \geq 1$ there is a ball B_n of radius at most 2^{-n} which meets A but is not a subset of any U_λ. For each n there is an $x_n \in B_n \cap A$. Since there is a convergent subsequence to $x \in A$, one has $x \in U_\lambda$ for some $\lambda \in \Lambda$. Since U_λ is open and since x is a limit point of the sequence x_n, it follows that $x \in B_n \subset U_\lambda$ for n sufficiently large. This is a contradiction to the construction of B_n, $n \geq 1$. ∎

Theorem 1.7 *(Arzelà-Ascoli)* A collection $\mathcal{C} \subset C[a, b]$ is relatively compact for the uniform metric on $C[a, b]$ if and only if $\mathcal{C}$ is uniformly bounded and equicontinuous.

---

[1]Folland, G.B. (1984), p. 130.

Proof Assume that $\mathcal{C}$ is uniformly bounded and equicontinuous. In view of Lemma 4, it is enough to show the closure of $\mathcal{C}$ is totally bounded and complete to prove relative compactness. The completeness follows from the completeness of $C[a, b]$. For total boundedness it is sufficient to check that $\mathcal{C}$ is totally bounded, since this will be preserved in the closure. Let $\delta > 0$. By equicontinuity, for each $x \in S$ there is an open set U_x containing x such that $|f(y) - f(x)| < \delta/4$ for all $y \in U_x$, and all $f \in \mathcal{C}$. By compactness of $[a, b]$, there are finitely many points $x_1, \ldots, x_n$ in S such that $\cup_{j=1}^{n} U_{x_j} = S$. Now $\{f(x_j) : f \in \mathcal{C}, j = 1, \ldots, n\}$ is a bounded set. Thus there are numbers $y_1, \ldots, y_m$ such that for each $f \in \mathcal{C}$, and each j, $|f(x_j) - y_k| < \delta/4$ for some $1 \leq k \leq m$. Let $X = \{x_1, \ldots, x_n\}$ and $Y = \{y_1, \ldots, y_m\}$. The set Y^X of functions from X into Y is a finite set and $\mathcal{C} = \cup_{g \in Y^X} \mathcal{C}_g$, where $\mathcal{C}_g := \{f \in \mathcal{C} : |f(x_j) - g(x_j)| < \delta/4, 1 \leq j \leq n\}$. Now, to complete the proof of total boundedness, let us see that this covering of $\mathcal{C}$ is by sets $\mathcal{C}_g$ of diameter at most δ. Let $f, h \in \mathcal{C}_g$. Then $|f - h| < \delta/2$ on X. For $x \in [a, b]$, one has $x \in U_{x_j}$ for some j, and therefore $|f(x) - h(x)| \leq |f(x) - f(x_j)| + |f(x_j) - h(x_j)| + |h(x_j) - h(x)| < \delta$.

To prove necessity, let us first observe that if $\overline{\mathcal{C}}$ is compact, then $\overline{\mathcal{C}}$ is totally bounded. For suppose not. Then there is a $\delta > 0$ such that there is no finite cover by balls of radii δ. Thus, for arbitrary but fixed $g_1 \in \overline{\mathcal{C}}$, there is a $g_2 \in \overline{\mathcal{C}}$ such that $\|g_1 - g_2\| := \max_{a \leq x \leq b} |g_1(x) - g_2(x)| > \delta$. This is because otherwise, the ball centered at g_1 would be a cover of $\overline{\mathcal{C}}$. Proceeding by induction, having found $g_1, \ldots, g_n$, there must be a $g_{n+1} \in \overline{\mathcal{C}}$ such that $\|g_k - g_{n+1}\| > \delta$ for $k = 1, \ldots n$. Thus, there is an infinite sequence $g_1, g_2, \ldots$ in $\overline{\mathcal{C}}$ such that $\|g_j - g_k\| > \delta$ for $j \neq k$. Thus $\overline{\mathcal{C}}$ cannot be compact. Now, since $\overline{\mathcal{C}}$ is totally bounded, given any $\varepsilon > 0$ there exist $g_1, \ldots, g_n \in \overline{\mathcal{C}}$ such that for any $f \in \overline{\mathcal{C}}$ one has $\|f - g_k\| < \frac{\varepsilon}{3}$ for some $1 \leq k \leq n$. Since each g_k is a continuous function on the compact interval $[a, b]$, it is bounded. Let $M = \max_{1 \leq k \leq n, a \leq x \leq b} |g_k(x)| + \frac{\varepsilon}{3}$. Then, for $f \in \overline{\mathcal{C}}$, one has $|f(x)| \leq |g_k(x)| + \frac{\varepsilon}{3} \leq M$ for all $a \leq x \leq b$. Thus $\overline{\mathcal{C}}$ is uniformly bounded. Since each g_k is continuous and hence, uniformly continuous on $[a, b]$, there is a $\delta_k > 0$ such that $|g_k(x) - g_k(y)| < \frac{\varepsilon}{3}$ if $|x - y| < \delta_k$. Let $\delta = \min\{\delta_1, \ldots, \delta_n\}$. Then for $f \in \overline{\mathcal{C}}$ one has for suitably chosen g_k, $|f(x) - f(y)| \leq \|f - g_k\| + |g_k(x) - g_k(y)| + \|g_k - f\| < \varepsilon$ if $|x - y| < \delta$. Thus $\overline{\mathcal{C}}$ and hence, $\mathcal{C} \subset \overline{\mathcal{C}}$ is equicontinuous. ∎

Appendix C
Hilbert Spaces and Applications in Measure Theory

C.1 Hilbert Spaces

Let H be a real vector space endowed with an **inner-product** $(x, y) \mapsto \langle x, y \rangle$, i.e.,
(i) $\langle x, y \rangle = \langle y, x \rangle$ *(symmetry)*, (ii) $\langle \alpha x + \beta y, z \rangle = \alpha \langle x, z \rangle + \beta \langle y, z \rangle$ $\forall \alpha, \beta \in \mathbb{R}$
(bilinearity), and (iii) $\langle x, x \rangle \geq 0$ $\forall x$, with equality iff $x = 0$ *(positive definiteness)*.
One writes $\|x\|^2 = \langle x, x \rangle$. to obtain a corresponding *norm* $\| \cdot \|$ on H (see (1.3) below
for the required triangle inequality).

Among the basic inequalities on H is the **parallelogram law**

$$\|x + y\|^2 + \|x - y\|^2 = 2\|x\|^2 + 2\|y\|^2, \tag{1.1}$$

which is easy to check, and the **Cauchy–Schwarz inequality**,

$$|\langle x, y \rangle| \leq \|x\| \cdot \|y\|. \tag{1.2}$$

To prove this, fix x, y. If x or y is 0, this inequality is trivial. Assume then that x, y
are nonzero. Since for all $u \in \mathbb{R}$,

$$0 \leq \|x + uy\|^2 = \|x\|^2 + u^2\|y\|^2 + 2u\langle x, y \rangle,$$

minimizing the right side with $u = -\langle x, y \rangle / \|y\|^2$, one gets $0 \leq \|x\|^2 - \langle x, y \rangle^2 / \|y\|^2$,
from which (1.2) follows. One can now derive the **triangle inequality**

$$\|x + y\| \leq \|x\| + \|y\|, \tag{1.3}$$

by observing that $\|x + y\|^2 = \|x\|^2 + \|y\|^2 + 2\langle x, y \rangle \leq \|x\|^2 + \|y\|^2 + 2\|x\| \cdot \|y\| = (\|x\| + \|y\|)^2$, in view of (1.2). Thus $\| \cdot \|$ is a **norm**: (a) $\|x\| \geq 0$, with
equality iff $x = 0$, (b) $\|\alpha x\| = |\alpha| \cdot \|x\|$ for all $x \in H$ and real scalar α, and
(c) $\|x + y\| \leq \|x\| + \|y\|$ for all $x, y \in H$. If H is a *complete* metric space under
the metric $d(x, y) = \|x - y\|$, then H is said to be a **(real) Hilbert space**.

© Springer International Publishing AG 2016
R. Bhattacharya and E.C. Waymire, *A Basic Course in Probability Theory*,
Universitext, DOI 10.1007/978-3-319-47974-3

Lemma 1 Let M be a closed linear subspace of the Hilbert space H. Then for each $x \in H$, the distance $d \equiv d(x, M) \equiv \inf\{d(x, y) : y \in M\}$ is attained at some $z \in M$ such that $d = d(x, z)$.

Proof Let z_n be such that $d(x, z_n) \to d$ as $n \to \infty$, $z_n \in M \ \forall n$. By (1.1), with $x - z_n, x - z_m$ for x and y, respectively, one has

$$
\begin{aligned}
\|z_n - z_m\|^2 &= 2\|x - z_n\|^2 + 2\|x - z_m\|^2 - \|2x - z_n - z_m\|^2 \\
&= 2\|x - z_n\|^2 + 2\|x - z_m\|^2 - 4\|x - \frac{1}{2}(z_n + z_m)\|^2 \\
&\leq 2\|x - z_n\|^2 + 2\|x - z_m\|^2 - 4d^2 \longrightarrow 2d^2 + 2d^2 - 4d^2 = 0,
\end{aligned}
$$

showing that $\{z_n : n \geq 1\}$ is a Cauchy sequence in M. Letting $z = \lim z_n$, one gets the desired result. ∎

Theorem 1.1 *(Projection Theorem)* Let M be a closed linear subspace of a real Hilbert space H. Then each $x \in H$ has a unique representation: $x = y + z, z \in M$, $y \in M^\perp \equiv \{w \in H : \langle w, v \rangle = 0 \ \forall v \in M\}$.

Proof Let $x \in H$. Let $z \in M$ be such that $d \equiv d(x, M) = d(x, z)$. Define $y = x - z$. Then $x = y + z$. For all $u \in \mathbb{R}$ and $w \in M$, $w \neq 0$, one has

$$ d^2 \leq \|x - (z + uw)\|^2 = \|x - z\|^2 + u^2\|w\|^2 - 2u\langle x - z, w \rangle. \qquad (1.4) $$

If $\langle x - z, w \rangle \neq 0$, one may set $u = \langle x - z, w \rangle / \|w\|^2$ to get $d^2 \leq \|x - z\|^2 - \langle x - z, w \rangle^2 / \|w\|^2 < d^2$, which is impossible, implying $\langle x - z, w \rangle = 0 \ \forall w \in M$. Hence $y \in M^\perp$. To prove uniqueness of the decomposition, suppose $x = w + v, w \in M$, $v \in M^\perp$. Then $w + v = y + z$, and $w - y = z - v$. But $w - z \in M$ and $y - v \in M^\perp$, implying $w - z = 0, y - v = 0$. ∎

The function $x \mapsto z$ in Theorem 1.1 is called the **(orthogonal) projection** onto M, and $x \to y$ is the orthogonal projection onto $M^\perp$. It is simple to check that these projections are linear maps (on H onto M, and on H onto $M^\perp$).

We will denote by H^* the set of all real-valued continuous linear functions (functionals) on H. Note that if $\ell_1, \ell_2 \in H^*$ and $\alpha, \beta \in \mathbb{R}$, then $\alpha \ell_1 + \beta \ell_2 \in H^*$, i.e., H^* is a real vector space. It turns out that H^* is **isomorphic** to H. To see this, note that for each $y \in H$, the functional ℓ_y, defined by $\ell_y(x) = \langle x, y \rangle$, belongs to H^*. Conversely, one has the following result.

Theorem 1.2 *(Riesz Representation Theorem on Hilbert Spaces)* If $\ell \in H^*$, there exists a unique $y \in H$ such that $\ell(x) = \langle x, y \rangle \ \forall x \in H$.

Proof Since $\ell = 0$ is given by $\ell(x) = \langle x, 0 \rangle$, and corresponds to ℓ_0, assume $\ell \neq 0$. Then $M \equiv \{x \in H : \ell(x) = 0\}$ is a closed proper linear subspace of H, $M \neq H$, and therefore $M^\perp \neq \{0\}$. Let $z \in M^\perp$, $\|z\| = 1$. Consider, for any given $x \in H$, the element $w = \ell(x)z - \ell(z)x \in H$, and note that $\ell(w) = 0$. Thus $w \in M$, so that $0 = \langle w, z \rangle = \ell(x) - \ell(z)\langle x, z \rangle$, implying $\ell(x) = \ell(z)\langle x, z \rangle = \langle x, y \rangle$, where $y = \ell(z)z$.

To prove uniqueness of the representation, suppose $\langle x, y_1 \rangle = \langle x, y_2 \rangle \ \forall x \in H$. With $x = y_1 - y_2$ one gets $0 = \langle x, y_1 - y_2 \rangle = \|y_1 - y_2\|^2$, so that $y_1 = y_2$. ∎

A **complex vector space** H is a vector space with the complex scalar field $\mathbb{C}$. An **inner product** on such a space is a function $\langle \, , \, \rangle$ on $H \times H$ into $\mathbb{C}$ satisfying (i) $\langle x, y \rangle = \overline{\langle y, x \rangle}$, ($\overline{\alpha}$ is the complex conjugate of $\alpha \in \mathbb{C}$), (ii) $\langle \alpha x + \beta y, z \rangle = \alpha \langle x, z \rangle + \beta \langle y, z \rangle$, (iii) $\|x\|^2 \equiv \langle x, x \rangle > 0 \ \forall x \neq 0$. If H, with distance $d(x, y) = \|x - y\|$, is a complete metric space, it is called a **complex Hilbert space**. The parallelogram law (1.1) follows easily in this case. For the Cauchy–Schwarz inequality (1.2), take $u = -\frac{\langle x, y \rangle}{\|y\|^2}$ in the relations (for arbitrary $x, y \in H$, and $u \in \mathbb{C}$)

$$0 \leq \|x + uy\|^2 = \|x\|^2 + |u|^2 \cdot \|y\|^2 + \overline{u}\langle x, y \rangle + u\overline{\langle x, y \rangle}$$

to get $0 \leq \|x\|^2 - |\langle x, y \rangle|^2/\|y\|^2$, from which (1.2) follows. The proof of the lemma remains unchanged for complex H. The triangle inequality (1.3) follows as in the case of real H. In the proof of the projection theorem, (1.4) changes (for $u \in \mathbb{C}$) to

$$d^2 \leq \|x - y\|^2 + |u|^2 \cdot \|w\|^2 + \overline{u}\langle x - y, w \rangle + u\overline{\langle x - y, w \rangle},$$

so that taking $u = -\langle x - y, w \rangle/\|w\|^2$, one gets $d^2 \leq \|x - y\|^2 - |\langle x - y, w \rangle|^2/\|w\|^2$, which implies $\langle x - y, w \rangle = 0 \ \forall w \in M$. The rest of the proof remains intact. For the proof of the Riesz representation, the relation $0 = \ell(x) - \ell(z)\langle x, z \rangle$ implies $\ell(x) = \ell(z)\langle x, z \rangle = \langle x, \overline{\ell(z)}z \rangle = \langle x, y \rangle$ with $y = \overline{\ell(z)}z$ (instead of $\ell(z)z$ in the case of real H). Thus Theorems 1.1, 1.2 hold for complex Hilbert spaces H also.

A set $\{x_i : i \in I\} \subset H$ is **orthonormal** if $\langle x_i, x_j \rangle = 0$ for all $i \neq j$, and $\|x_i\| = 1$. An orthonormal set is **complete** if $\langle x, x_i \rangle = 0$ for all $i \in I$ implies $x = 0$. A complete orthonormal subset of a Hilbert space is called an **orthonormal basis** of H. Suppose H is a separable Hilbert space with a dense set $\{y_n : n = 1, 2, \ldots\}$. By the following **Gram–Schmidt procedure** one can construct a countable orthonormal basis for H. Without loss of generality assume $y_n \neq 0$, $n \geq 1$. Let $x_1 = y_1/\|y_1\|$, $u_2 = y_2 - \langle y_2, x_1 \rangle x_1$, $x_2 = u_2/\|u_2\|$, assuming $u_2 \neq 0$. If $u_2 = 0$, replace y_2 by the first y in the sequence such that $y - \langle y, x_1 \rangle x_1 \neq 0$, and relabel $y = y_2$. Having constructed $u_2, x_2, \ldots, u_n, x_n$ in this manner, define $u_{n+1} = y_{n+1} - \sum_{j=1}^{n} \langle y_{n+1}, x_j \rangle x_j$, $x_{n+1} = y_{n+1}/\|y_{n+1}\|$, assuming $u_{n+1} \neq 0$ (if $u_{n+1} = 0$ then find the first y in the sequence that is not linearly dependent on $\{x_1, \ldots, x_n\}$ and relabel it y_{n+1}). The process terminates after a finite number of steps if H is finite dimensional. Otherwise, one obtains a complete orthonormal sequence $\{x_n : n = 1, 2, \ldots\}$. Completeness follows from the fact that if $\langle x, x_n \rangle = 0$ for all n, then $\langle x, y_n \rangle = 0$ for all n, so that $x \in \{y_n : n \geq 1\}^\perp = \{0\}$ (since $\{y_n : n \geq 1\}$ is dense in H, and $A^\perp$ is a closed set for all $A \subset H$). A complete orthonormal set is called a **complete orthonormal basis**, in view of Theorem 1.3 below.

Lemma 2 (*Bessel's Inequality*) Let $\{x_1, x_2, \ldots\}$ be a finite or countable orthonormal subset of a Hilbert space H. Then $\sum_n |\langle x, x_n \rangle|^2 \leq \|x\|^2$ for all $x \in H$.

Proof One has

$$
\begin{aligned}
\left\| x - \sum_n \langle x, x_n \rangle x_n \right\|^2 &= \|x\|^2 - 2\operatorname{Re}\left\langle x, \sum_n \langle x, x_n \rangle x_n \right\rangle + \left\| \sum_n \langle x, x_n \rangle x_n \right\|^2 \\
&= \|x\|^2 - 2\operatorname{Re}\sum_n |\langle x, x_n \rangle|^2 + \sum_n |\langle x, x_n \rangle|^2 \\
&= \|x\|^2 - \sum_n |\langle x, x_n \rangle|^2.
\end{aligned}
$$

The inequality is proven since the expression is nonnegative.   ∎

Theorem 1.3 Let $\{x_1, x_2, \dots\}$ be a complete orthonormal set in a separable Hilbert space. Then for all x one has (a) *(Fourier Expansion)* $x = \sum_n \langle x, x_n \rangle x_n$, and (b) *(Parseval's Equation)* $\|x\|^2 = \sum_n |\langle x, x_n \rangle|^2$, and $\langle x, y \rangle = \sum_n \langle x, x_n \rangle \overline{\langle y, y_n \rangle}$, for all $x, y \in H$.

Proof (a) In view of Bessel's inequality, the series $\sum_n |\langle x, x_n \rangle|^2$ converges, so that $\left\| \sum_M^N \langle x, x_n \rangle x_n \right\|^2 \to 0$ as $M, N \to \infty$. Hence $\sum_n \langle x, x_n \rangle x_n$ converges to $z \in H$, say. Since $\langle x_m, x_n \rangle = 0$, $n \neq m$, and $\langle x_m, x_m \rangle = 1$, it follows that $\langle z, x_m \rangle = \langle x, x_m \rangle$ for all m. Therefore, $\langle z - x, x_m \rangle = 0$ for all m, and hence by completeness of $\{x_n : n \geq 1\}$, $x = z = \sum_n \langle x, x_n \rangle x_n$. Also, the first calculation in (b) follows, since $\|x\|^2 = \|\sum_n \langle x, x_n \rangle x_n\|^2 = \sum_n |\langle x, x_n \rangle|^2$. More generally, one has $x = \sum_n \langle x, x_n \rangle x_n$, $y = \sum_n \langle y, x_n \rangle x_n$, so that $\langle x, y \rangle = \sum_n \langle x, x_n \rangle \overline{\langle y, x_n \rangle}$, using (i) convergence of $\sum_{n=1}^N \langle x, x_n \rangle x_n$ to x and that of $\sum_{n=1}^N \langle y, x_n \rangle x_n$ to y as $N \to \infty$, and (ii) the continuity of the inner product $(u, v) \to \langle u, v \rangle$ as a function on $H \times H$.   ∎

C.2 Lebesgue Decomposition and the Radon–Nikodym Theorem

In the subsection we will give von Neumann's elegant proof of one of the most important results in measure theory.

Let μ, ν be measures on a measurable space $(S, \mathcal{S})$. One says that ν **is absolutely continuous** with respect to μ, $\nu \ll \mu$ in symbols, if $\nu(B) = 0 \; \forall B \in \mathcal{S}$ for which $\mu(B) = 0$. At the other extreme, ν is **singular** with respect to μ if there exists $A \in \mathcal{S}$ such that $\nu(A) = 0$ and $\mu(A^C) = 0$, that is, if μ and ν concentrate their entire masses on disjoint sets: one then writes $\nu \perp \mu$. Note that $\nu \perp \mu$ implies $\mu \perp \nu$. However, $\nu \ll \mu$ does not imply $\mu \ll \nu$.

Theorem 2.1 *(Lebesgue Decomposition and the Radon–Nikodym Theorem)* Let μ, ν be σ-finite measures on $(S, \mathcal{S})$. Then *(Lebesgue decomposition)* (a) there exist unique measures $\nu_a \ll \mu$ and $\nu_s \perp \mu$ such that $\nu = \nu_a + \nu_s$, and there exists a

μ-a.e. unique nonnegative measurable h such that $\nu_a(B) = \int_B h\, d\mu\; \forall B \in \mathcal{S}$. (b) In particular, *(Radon–Nikodym theorem)* if $\nu \ll \mu$, then there exists a μ-a.e. unique $h \geq 0$ such that $\nu(B) = \int_B h\, d\mu\; \forall B \in \mathcal{S}$.

Proof First consider the case of *finite* μ, ν. Write $\lambda = \mu + \nu$. On the real Hilbert space $L^2(\lambda) \equiv L^2(S, \mathcal{S}, \lambda)$, define the linear functional

$$\ell(f) = \int_S f\, d\nu \qquad f \in L^2(\lambda). \tag{2.1}$$

By the Cauchy–Schwarz inequality, writing $\|f\| = (\int |f|^2 d\lambda)^{1/2}$, we have

$$|\ell(f)| \leq \int_S |f| d\lambda \leq \|f\| \cdot (\lambda(S))^{\frac{1}{2}}. \tag{2.2}$$

Thus ℓ is a continuous linear functional on $L^2(\lambda)$. By the Riesz representation theorem (Theorem 1.2), there exists $g \in L^2(\lambda)$ such that

$$\ell(f) \equiv \int_S f\, d\nu = \int_S fg\, d\lambda \qquad (f \in L^2(\lambda)). \tag{2.3}$$

In particular, for $f = \mathbf{1}_B$,

$$\nu(B) = \int_B g\, d\lambda, \qquad \forall\, B \in \mathcal{S}. \tag{2.4}$$

Letting $B = \{x \in S : g(x) > 1\} = E$, say, one gets $\lambda(E) = 0 = \nu(E)$. For if $\lambda(E) > 0$, then (2.4) implies $\nu(E) = \int_E g\, d\lambda > \lambda(E)$, which is impossible. Similarly, letting $F = \{x : g(x) < 0\}$, one shows that $\lambda(F) = 0$. Modifying g on a λ-null set if necessary, we take g to satisfy $0 \leq g \leq 1$ on S. Consider the sets $S_1 = \{x : 0 \leq g(x) < 1\}$ and $S_2 = S_1^c = \{x : g(x) = 1\}$, and define the following measures ν_1, ν_2:

$$\nu_1(B) := \nu(B \cap S_1), \quad \nu_2(B) := \nu(B \cap S_2), \qquad B \in \mathcal{S}. \tag{2.5}$$

Now, using $\lambda = \mu + \nu$, one may rewrite (2.4) as

$$\int_B (1 - g)\, d\nu = \int_B g\, d\mu \qquad (B \in \mathcal{S}). \tag{2.6}$$

For $B = S_2$, the left side is zero, while the right side is $\mu(S_2)$, i.e., $\mu(S_2) = 0$. Since $\nu_2(S_2^c) = 0$ by definition, one has $\nu_2 \perp \mu$. On the other hand, on S_1, $1 - g > 0$, so that $\mu(B) = 0$ implies $\int_{B \cap S_1} (1 - g) d\nu = 0$ implies $\nu(B \cap S_1) = 0$, i.e., $\nu_1(B) = 0$. Hence $\nu_1 \ll \mu$. Thus we have a Lebesgue decomposition $\nu = \nu_a + \nu_s$, with $\nu_a = \nu_1$, $\nu_s = \nu_2$. Its *uniqueness* follows from Corollary 2.3 below. Multiplying both sides of (2.6) by $1, g, g^2, \ldots, g^n$, and adding, we get

$$\int_B (1 - g^{n+1})\, d\nu = \int_B (g + g^2 + \cdots + g^{n+1})\, d\mu \qquad (B \in \mathcal{S}). \qquad (2.7)$$

Since $1 - g^{n+1} \uparrow 1$ (as $n \uparrow \infty$) on S_1, denoting by h the increasing limit of $g + g^2 + \cdots + g^{n+1}$, one gets

$$\nu_a(B) \equiv \nu_1(B) = \nu(B \cap S_1) = \int_{B \cap S_1} h\, d\mu = \int_B h\, d\mu \qquad (B \in \mathcal{S}),$$

completing the proof of (a). Now (b) is a special case of (a). The *uniqueness* of the function h in this case does not require Proposition 2.2 below. For if $\int_B h\, d\mu = \int_B h'\, d\mu\, \forall B \in \mathcal{S}$, then $\int_B (h-h')\, d\mu = 0\, \forall B \in \mathcal{S}$. In particular $\int_{\{h>h'\}} (h-h')\, d\mu = 0$ and $\int_{\{h \le h'\}} (h' - h)\, d\mu = 0$, so that $\int |h - h'| d\mu = 0$.

For the general case of σ-finite measures μ, ν, let $\{A_n : n \ge 1\}$ be a sequence of pairwise disjoint sets in $\mathcal{S}$ such that $\cup_{n=1}^\infty A_n = S$, $\mu(A_n) < \infty$, $\nu(A_n) < \infty\, \forall n$. Applying the above result separately to each A_n and adding up one gets the desired result, using the monotone convergence theorem. ∎

For the next result, call ν a finite **signed measure** if $\nu : \mathcal{S} \to (-\infty, \infty)$ satisfies $\nu(\emptyset) = 0$, and $\nu(\cup_n B_n) = \sum_n \nu(B_n)$ for every pairwise disjoint sequence B_n ($n \ge 1$) in $\mathcal{S}$. If ν takes one of the two values $-\infty$, ∞, but not both, ν is said to be σ-**finite signed measure** if there exists a sequence of pairwise disjoint sets $B_n \in \mathcal{S}$ such that ν is a finite signed measure on each B_n ($n \ge 1$), and $S = \cup_n B_n$.

Proposition 2.2 *(Hahn–Jordan Decomposition)* Suppose ν is a σ-finite signed measure on $(S, \mathcal{S})$. Then *(Hahn decomposition)* (a) there exists a set $C \in \mathcal{S}$ such that $\nu(C \cap B) \ge 0\, \forall B \in \mathcal{S}$, and $\nu(C^c \cap B) \le 0\, \forall B \in \mathcal{S}$ and *(Jordan Decomposition)* (b) defining the measures $\nu^+(B) := \nu(C \cap B)$, $\nu^-(B) := -\nu(C^c \cap B)$, one has $\nu = \nu^+ - \nu^-$.

Proof First assume that ν is finite, and let $u = \sup\{\nu(B) : B \in \mathcal{S}\}$. Let $B_n \in \mathcal{S}$ ($n \ge 1$) be such that $\nu(B_n) \to u$. We will construct a set $C \in \mathcal{S}$ such that $\nu(C) = u$. For each m, consider the partition Γ_m of S by 2^m sets of the form $B_1' \cap B_2' \cap \cdots \cap B_m'$ with $B_i' = B_i$ or B_i^c, $1 \le i \le m$. Let A_m be the union of those among these sets whose ν-measures are nonnegative. Clearly, $\nu(A_m) \ge \nu(B_m)$. Expressing $A_m \cup A_{m+1}$ as a (disjoint) union of certain members of the partition Γ_{m+1} and noting that those sets in Γ_{m+1} that make up $A_{m+1} \backslash A_m$ all have nonnegative ν-measures, one has $\nu(A_m \cup A_{m+1}) \ge \nu(A_m)$. By the same argument, $\nu(A_m \cup A_{m+1} \cup A_{m+2}) \ge \nu(A_m \cup A_{m+1}) \ge \nu(A_m)$, and so on, so that $\nu(\cup_{i=m}^n A_i) \ge \nu(A_m)\, \forall n \ge m$, implying that $C_m \equiv \cup_{i=m}^\infty A_i$ satisfies $\nu(C_m) \ge \nu(A_m) \ge \nu(B_m)$. Hence $\nu(C) = u$, where $C = \lim_{m \to \infty} C_m$. We will now show that $\nu(B \cap C) \ge 0$ and $\nu(B \cap C^c) \le 0\, \forall B \in \mathcal{S}$. First note that $u < \infty$, since ν is finite. Now if $\nu(B \cap C) < 0$ for some B, then $\nu(C \backslash (B \cap C)) > u$, which is impossible. Similarly, if $\nu(B \cap C^c) > 0$ for some B, then $\nu(C \cup (B \cap C^c)) > u$. We have proved the Hahn decomposition (a). The Jordan decomposition (b) follows immediately from this.

If ν is σ-finite, then S is a disjoint union of sets A_n $(n \geq 1)$ such that $\nu_n(B) \equiv \nu(A_n \cap B)$, $B \in \mathcal{S}$, is a finite signed measure for all $n \geq 1$. The Hahn–Jordan decomposition $\nu_n = \nu_n^+ - \nu_n^-$ leads to the corresponding decomposition of $\nu = \nu^+ - \nu^-$, with $\nu^+ = \sum_n \nu_n^+$, $\nu^- = \sum_n \nu_n^-$.   ∎

The measure $|\nu| := \nu^+ + \nu^-$ is called the **total variation** of a σ-finite measure ν.

Corollary 2.3 The Hahn–Jordan decomposition of a σ-finite signed measure ν is the unique decomposition of ν as the difference between two mutually singular σ-finite measures.

Proof It is enough to assume that ν is finite. Let $\nu = \gamma_1 - \gamma_2$ where $\gamma_1 \perp \gamma_2$ are measures, with $\gamma_1(D^c) = 0$, $\gamma_2(D) = 0$ for some $D \in \mathcal{S}$. Clearly, $\gamma_1(D) = \gamma_1(S) = \sup\{\nu(B) : B \in \mathcal{S}\} = \nu(D) = u$, say. As in the proof of Proposition 2.2, it follows that $\gamma_1(B) = \nu(B \cap D)$, $\gamma_2(B) = -\nu(B \cap D^c)$ for all $B \in \mathcal{S}$. If C is as in Proposition 2.2, then $u = \nu(C)$. Suppose, if possible, $\nu^+(B) \equiv \nu(B \cap C) > \gamma_1(B) = \nu(B \cap D)$, i.e., $\nu(B \cap C \cap D^c) + \nu(B \cap C \cap D) > \nu(B \cap D \cap C) + \nu(B \cap D \cap C^c)$, or, $\nu(B \cap C \cap D^c) > \nu(B \cap D \cap C^c) = \gamma_1(B \cap C^c) \geq 0$. But then $\nu(D \cup (B \cap C \cap D^c)) > \nu(D) = \gamma_1(D) = u$, a contradiction. Hence $\nu^+(B) \leq \gamma_1(B) \ \forall B \in \mathcal{S}$. Similarly, $\gamma_1(B) \leq \nu^+(B) \ \forall B \in \mathcal{S}$.   ∎

One may take ν to be a σ-finite signed measure in Theorem 2.1 (and μ a σ-finite measure). Then ν is *absolutely continuous with respect to* μ, $\nu \ll \mu$, if $\mu(B) = 0$ implies $\nu(B) = 0$ ($B \in \mathcal{S}$). Use the Hahn–Jordan decomposition $\nu = \nu^+ - \nu^-$, and apply Theorem 2.1 separately to ν^+ and $\nu^- : \nu^+ = (\nu^+)_a + (\nu^+)_s$, $\nu^- = (\nu^-)_a + (\nu^-)_s$. Then let $\nu_a = (\nu^+)_a - (\nu^-)_a$, $\nu_s = (\nu^+)_s - (\nu^-)_s$.

Corollary 2.4 Theorem 2.1 extends to σ-finite signed measures ν, with $\nu = \nu_a + \nu_s$, where ν_a and ν_s are σ–finite signed measures, $\nu_a \ll \mu$, $\nu_s \perp \mu$. Also, there exists a measurable function h, unique up to a μ–null set, such that $\nu_a(B) = \int_B h \, d\mu \ \forall B \in \mathcal{S}$. If $\nu \ll \mu$, then $\nu(B) = \int_B h \, d\mu \ \forall B \in \mathcal{S}$.

References

The following is a focussed but less than comprehensive list of some supplementary and/or follow-up textbook references. In addition to references cited in the text, it includes resources covering some of the same topics and/or further applications of material introduced in this basic course.

Athreya, K.B., Ney, P.E.: Branching Processes. Springer, NY (1972)

Bhattacharya, R. and Ranga Rao, R. (2010): Normal Approximation and Asymptotic Expansions, SIAM Classics in Applied Mathematics Series, Philadelphia, originally published by Wiley (1976)

Bhattacharya, R., Waymire, E.: Stochastic Processes with Applications. SIAM Classics in Applied Mathematics Series, Philadelphia, originally published by Wiley, NY (1990)

Bhattacharya, R., Waymire, E. (2017): Stochastic Processes: Theory and Applications, vols. 1–4. Springer, NY (in preparation)

Bhattacharya, R., Majumdar, M.: Random Dynamical Systems: Theory and applications. Cambridge University Press, Cambridge (2007)

Billingsley, P.: Convergence of Probability Measures. Wiley, NY (1968)

Billingsley, P.: Probability and Measure, 2nd edn. Wiley, NY (1986)

Breiman, L.: (1992): Probability, Reprinted in SIAM Classics in Applied Mathematics Series, Philadelphia, PA, from Addison Wesley. Reading, MA (1968)

Chung, K.L.: A Course in Probability Theory, 2nd edn. Academic Press, NY (1974)

Davis, M., Etheridge, A.: Louis Bachelier's Theory of Specultation: The origins of modern finance. Princeton University Press, Princeton, NJ (2006)

Doob, J.L.: Stochastic Processes. Wiley, NY (1953)

Durrett, R.: Probability Theory and Examples, 2nd edn. Wadsworth, Brooks & Cole, Pacific, Grove, CA (1995)

Ethier, S.N., Kurtz, T.G.: Markov Processes: Characterization and Convergence. Wiley, NY (1985)

Feller, W.: An Introduction to Probability Theory and Its Applications, vol. 1,3rd ed, vol. 2, 2nd ed. Wiley, NY (1968, 1971)

Folland, G.: Real Analysis. Wiley, NY (1984)

Fürth, R.: Investigations on the Theory of the Brownian Movement. Dover (1954)

Grenander, U.: Probabilities on Algebraic Structures. Wiley, NY (1963)

Halmos, P.: Measure Theory. Springer Graduate Texts in Mathematics, Springer, NY (1974)

Itô, K., McKean, H.P.: Diffusion Processes and Their Sample Paths. Reprint Springer, Berlin (1965)

Kac, M.: Probability, Number Theory and Statistical Physics: Selected Papers, Baclawski, K. Donsker, D., (eds.). Cambridge

Kallenberg, O.: Foundations of Modern Probability, 2nd edn. Springer, NY (2001)

Karlin, S., Taylor, H.M.: A Second Course in Stochastic Processes. Academic Press, NY (1981)

Karatzas, I., Shreve, S.E.: Brownian Motion and Stochastic Calculus, 2nd edn. Springer, NY (1991)

Kesten, H.: Percolation Theory for Mathematicians. Birkhauser (1982)

Kolmogorov, A.N.: Foundations of the Theory of Probability. Chelsea, New York (1950) (English translation of the 1933 German original)

Lamperti, J.: Probability: A survey of the mathematical theory. Wiley, NY (1996)

Laplace, P.-S.: Théorie Analytique des Probabilités, Reprinted in Oeuvres Complète de Laplace 7. Gauthier-Villars, Paris (1812)

Lawler, G., Limic, V.: Random Walk: A Modern Introduction. Cambridge Studies in Advanced Mathematics, no **123**, (2010)

Lévy, P.: Calcul des Probabilités. Gauthier-Villars, Paris (1925)

Lévy, P.: Théorie de l'addition des variables aléatores. Gauthier-Villars, Paris (1937)

Lévy, P.: Processes stochastique et mouvement Brownian. Gauthier-Villars, Paris (1948)

Liggett, T.M.: Continuous Time Markov Processes: An introduction, American Mathematical Society (2009)

Mitzenmacher, M., Upfal, E.: Probability and Computing: Randomized Routing Probabilistic Analysis. Cambridge University Press, Cambridge (2005)

Nelson, E.: Dynamical Theories of Brownian Motion. Princeton Univ. Press, Princeton, N.J. (1967)

Neveu, J.: Mathematical Foundations of the Calculus of Probability. Holden-Day, San Francisco (1965)

Neveu, J.: Discrete Parameter Martingales. North-Holland, Amsterdam (1975)

Parthasarathy, K.R.: Probability Measures on Metric Spaces. Academic Press, NY (1967)

Perrin, J.: 1990. Atoms. Ox Bow Press, French original (1913)

Ramasubramanian, S.: Lectures on Insurance Models. Amer. Math. Soc, Providence (2009)

Revuz, D., Yor, M.: Continuous Martingales and Brownian Motion, 3rd edn. Springer, Berlin (1999)

Rogers, LC.G., Williams, D.: Diffusions, Markov Processes and Martingales, vol. 1 (2nd ed.), vol. 2 Cambridge (2000)

Royden, H.L.: Real Analysis, 3rd edn. MacMillan, NY (1988)

Rudin, W.: Fourier Analysis on Groups, 2nd edn. Wiley, NY (1967)

Shiryaev, A.N.: Probability. Springer, NY (1990)

Skorokhod's A.V.: Studies in the Theory of Markov Processes, (English translation of 1961 Russian edition) (1965)

Spitzer, F.: Principles of Random Walk, 2nd edn. Springer, NY (1976)

Samorodnitsky, G., Taqqu, M.: Stable Non-Gaussian Random Processes: Stochastic Models with Infinite Variance. Chapman and Hall, NY (1994)

Stroock, D.W.: Probability Theory: An Analytic View. Cambridge Univ. Press, Cambridge (1993)

Stroock, D.W., Varadhan, S.R.S.: Multidimensional Diffusion Processes. Springer, Berlin (1979)

Tao, T.: Topics in Random Matrix Theory, Graduate Studies in Mathematics, vol. 132. Amer. Math. Soc, Providence (2012)

Walsh, J.B.: Knowing the Odds: An Introduction to Probability. American Mathematical Society, Providence (2012)

Wax, N.: Selected Papers on Noise and Stochastic Processes. Dover (1954)

Williams, D.: Probability with Martingales. Cambridge Univ. Press, Cambridge (1991)

Symbol Index

(S, ρ) space, 5

$(S, \mathcal{S}, \mu)$, measure space, 1

$(S, \mathcal{T})$ space, 5

$(\Omega, \mathcal{F}, P)$, probability space, 1

$1^\perp$, subspace orthogonal to one, 36

2^Ω, power set of Ω, 2

A^-, closure of set A, 137

A^c, complement of A, 2

A°, interior of set A, 137

B^x, Brownian motion started at x, 190

$C(S)$, space of continuous real-valued functions on S, 237

$C(S : \mathbb{C})$, space of complex-valued continuous functions on S, 243

$C([0, \infty) : \mathbb{R}^k)$, space of continuous functions on $[0, \infty)$ into $\mathbb{R}^k$, 6

$C[0, 1]$, space of real-valued continuous functions on $[0, 1]$, 6

C^y, y-section of the set C, 236

$C_b(S)$ of real-valued bounded functions on S, 11

$C_b^0(S)$, space of continuous functions on S that vanish at infinity, 142

C_x, x-section of the set C, 236

$F(\mathbf{x}) = P(\mathbf{X} \leq \mathbf{x})$, (multivariate) distribution function, 6

$H_{n,k}$, Haar wavelet functions, 173

L^1, abbreviation for $L^1 \equiv L^1(\Omega, \mathcal{F}, P)$, 37

L^2, $L^2(\mathcal{F})$, abbreviations for $L^2(\Omega, \mathcal{F}, P)$, 36

$L^p(\Omega, \mathcal{F}, P)$, L^p space of random variables, 11

$M^\perp$, subspace orthogonal to M, 248

$P \circ X^{-1}$, (induced) probability distribution, 6

$P(A|\mathcal{G})$, conditional probability of A given $\mathcal{G}$, 41

$P^{\mathcal{G}}$ or $Q^{\mathcal{G}}$, regular conditional distribution, 42

$Q(B) := P(X^{-1}(B)) \equiv P(X \in B)$, (induced) probability distribution, 6

$Q_1 * Q_2$, convolution of probability distributions, 26

Q_x, distribution of random walk starting at x, 45

S^x, random walk process (sequence) starting at x, 46

S_n^x, position of random walk starting from x at time n, 45

S_n^{x+}, after-n process, 45

$S_{n,k}$, Schauder functions, 174

$W = P \circ \mathbf{B}^{-1}$, Wiener measure, 148

X^+ part of X, 7

X^-, negative part of X, 7

X_τ^+, after-τ process, 189

X_m^+, after-m process, 187

Z_s^+, after-s process, 190

$[X \in B]$ image notation for random maps, 5

$[x]$, integer part of x, 50

Δ, symmetric difference, 88

$\|X\|_p$, L^p norm of X, 11

δ_x, Dirac point mass measure, 1

$\int f d\mu$, $\int_S f(x)\mu(dx)$, $\int f$, alternative notations for Lebesgue integral of f, 230

$\mathbb{R}^\infty$, infinite real sequence space, 6

$\mathcal{B}(S) := \sigma(\mathcal{T})$, Borel σ-field on S generated by the topology $\mathcal{T}$ for S, 5

$\mathcal{F}_\tau$, pre-τ σ-field, 189

$\mathcal{F}_{t+}$, right-continuous filtration, 191

Cov, covariance, 29

Var, variance, 29

μ^*, outer measure, 226

$\mu_1 \times \mu_2$, product measure, 236

μ_p, order p-moment, 8

$\mu_{t_1, t_2, \ldots, t_n}$, finite-dimensional distribution, 167

$\nu \ll \mu$, absolute continuity of measures, 250

$\nu \perp \mu$, mutual singularity of measures, 250

ν^+, ν^-, positive and negative parts of a signed-measure, respectively, 252

$\nu_\omega(\delta)$, oscillation of function ω at scale δ, 148

$\otimes$, σ-field product, 25

$\otimes_{t\in\Lambda}S_t$, product σ-field, 46

∂A, boundary of set A, 137

∂^α, multi-index partial derivative, 115

π_k, k-th coordinate projection map, 25

π_s, s-th coordinate projection map, 46

$\prec$, subordinate, 238

$\liminf_n A_n$, limit inferior of sequence of sets (events), 2

$\limsup_n A_n$, limit superior of sequence of sets (events), 2

$\{0, 1\}^\infty$, infinite binary sequence space, 3

$\{0, 1\}^m$ binary sequence space, 2

$\{X_{\tau\wedge t} : t \geq 0\}$, stopped process, 59

$\rho(x, F)$, distance from point x to set F, 11

$\sigma(X)$, σ-field generated by X, 5

$\sigma(X_t)$, σ-field generated by X_t, 33

$\sigma(X_t : t \in \Lambda)$, σ-field generated by family of random maps, 33

$\sigma(\{\mathcal{F}_t : t \in \Lambda\})$, σ-field generated by a collection of σ-fields, 34

$\sigma(\mathcal{C})$ generated by $\mathcal{C}$, 4

σ^t, matrix transpose, 191

$\subset$, subset, 1

$\vee$, (lattice) notation for maximum, 59

$\wedge$, (lattice) notation for minimum, 59

d_π, Prohorov metric, 146

d_v, total variation distance, 15

d_{BL}, bounded-Lipschitz metric, 154

$f^{-1}(B)$ image of B under f, 5

i.o., infinitely often, 2

m, λ, dx, alternative notations for Lebesgue measure, 229

$p(x, B)$, homogeneous one-step transition probability from state x to set B at time n, 211

$p(x, B)$, one-step transition probability from a state x to set B, 171

$p(x, dy)$, homogeneous one-step transition probability distribution at time m, 187

$p(x, y)$, one-step transition probability density, 170

$p_m(x, dy)$, one-step transition probability distribution at time m, 187

$p_n(x, B)$, one-step transition probability from state x to set B at time n, 211

$\mathcal{B}^\infty$, infinite product of Borel σ-fields, 45

$\mathcal{F}_\tau$, pre-τ σ-field, 59

$\mathcal{L}(\mathcal{C})$, λ-system generated by $\mathcal{C}$, 4

$\mathcal{P}(S)$, set of probability measures on $(S, \mathcal{S})$, 15

$\mathcal{S}_1 \otimes \cdots \otimes \mathcal{S}_k$, product σ-field, 237

$\mathcal{S}_1 \otimes \cdots \otimes \mathcal{S}_n$, product σ-field, 25

$\mathcal{T}$, tail σ-field, 87

$\hat{\mu}$, Fourier transform of finite-signed measure μ, 115

$\hat{f}(\xi)$, Fourier transform, 112

$\mathbb{Z}^+$, set of nonnegative integers, 54, 59

$\mathbb{C}$ of complex numbers, 23

$\mathbb{E}(X|\mathcal{G})$, conditional expectation of X given $\mathcal{G}$, 36

$\mathbb{E}X$, expected value of X, 7

$\mathbb{R}$, set of real numbers, 7

$\mathbb{R}^\Lambda$, the indicated product space, 172

$\mathbb{Z}$, set of integers, 46

$\mathbf{1}_A$, indicator function of A, 7

$\mathbf{X}^{n+}$, after-n process, 211

$\tilde{\mathcal{F}}_t$, trace σ-field, 60

a.e., almost everywhere, 233

Arrows $\downarrow, \uparrow$, limits from above (right-sided), and from below (left-sided), respectively, 11

i.i.d., independent and identically distributed, 31

Re(z), the real part of z, 123

Widearrow $\Rightarrow$, weak convergence, convergence in distribution, 75

Index

A

Adapted, 55
  σ-field, 55
Adjustment coefficient, Lundberg
    coefficient, 72
After-τ process, 189
After-m process, 187
After-n process, 45
After-s process, 190
Alexandrov theorem, 137
Algebra, 225, 242
Almost everywhere (a.e.), 233
Almost sure (a.s.) convergence, 10
Arzelà-Ascoli theorem, 244
Asymmetric random walk, 221

B

Banach space, 11, 241
Bernstein polynomials, 92
Berry–Esseen bound, 83, 97, 126
Bessel's inequality, 249
Binomial distribution, 22
Birth–death chain
  reflecting boundary, 216
Bochner Theorem, 119
Boltzmann paradox, 215
Borel σ-field, 5, 228
Borel–Cantelli lemma I, 3
Borel–Cantelli lemma II, 34
Boundary value (hitting) distribution for
    Brownian motion, 196
Bounded-Lipschitz metric, 154
Branching rate, 70
Brownian motion, 148, 194
  boundary value (hitting) distribution for
    Brownian motion, 196
  diffusion coefficient, 191
  drift coefficient, 191

homogeneity of Brownian motion
    increments, 179
independent increments of Brownian
    motion, 179
k-dimensional, 176
k-dimensional Brownian motion with
    respect to filtration, 191
Kolmogorov model, 172
law of the iterated logarithm, 183
Lévy modulus of continuity, 186
maximal inequality for Brownian
    motion, 183
nowhere differentiable paths, 186
scale invariance of Brownian motion,
    179
set of zeros, 197
standard, 191
strong Markov property, 193
symmetry transformation of Brownian
    motion, 179
time inversion invariance of Brownian
    motion, 179
two-point boundary hitting probabilities,
    198
upper class, 198
wavelet construction, 173
zeros of Brownian motion, 186, 197
Brownian sheet, 181

C

Canonical model, canonical construction, 7
Canonical process, 178
Cantor function, 230
Cantor measure, 230
Cantor set, 229
Carathéodory extension theorem, 226
Cartesian product, 237
Cauchy distribution, 132
Cauchy–Schwarz inequality, 247

© Springer International Publishing AG 2016
R. Bhattacharya and E.C. Waymire, *A Basic Course in Probability Theory*,
Universitext, DOI 10.1007/978-3-319-47974-3

Central limit theorem
   classical, 81, 125
   Lindeberg, 78
   Lyapounov, 81
   multivariate, 84, 143
Cesàro limit, 163
Change of variables, 8
Characteristic function, 115
   Taylor expansion, 125
Chebyshev estimation of distribution
   function, 14
Chebyshev estimation of sample size, 30,
   82
Chebyshev sampling design, 82
Chernoff estimation of sample size, 100
Chernoff inequality, 96
Chung–Fuchs lemma, 120
   recurrence criterion, 123, 134
Coin toss, 2, 3, 28
Compact
   finite intersection property, 242
Compact space, 242
Complete metric space, 241
Complete orthonormal set, 249
Complex vector space, 249
Concentration inequalities, 98
Conditional expectation
   Cauchy-Schwarz inequality, 51
   Chebyshev inequality, 51
   contraction, 38
   convergence theorems, 38
   dominated convergence, 38
   first definition, L^2-projection, 36
   linearity, 38
   monotone convergence, 38
   ordering, 38
   second definition on L^1, 37
   smoothing, 38
   substitution property, 38
Conditional probability, 41
   regular conditional distribution, 41
Conditional probability density function, 44
Conjugate exponents, 13
Convergence
   almost sure (a.s.), 9
   in distribution, 75, 137
   in measure, 233
   in probability, 10
   vague, 142
   weak, 75
Convergence-determining, 139
Convex function, 12
   line of support property, 12

Convolution, 26, 47, 115, 116
   smoothing property, 132
Coordinate projections, 25, 46
Covariance, 29
Cramér–Chernoff Theorem, 94
Cramér–Lèvy Continuity Theorem, 117
Cramér–Lundberg model, 71
Cramér–Wold device, 143
Critical probability
   percolation, 29
Cumulant-generating function, 92
Cylinder sets, 45

D
Density
   with respect to Lebesgue measure, 6
Diffusion coefficient, 148, 191, 209
Dilogarithmic random walk, 222
Disintegration formula, 42
Disorder parameter, 70
Distribution
   absolutely continuous, 250
   singular, 250
Distribution function, 228
   absolutely continuous, 6
Doeblin minorization, 213
Donsker's invariance principle, 201
Doob–Blackwell theorem, 42
Doob maximal inequality, 56
Doob maximal inequality for moments, 57
Doob's upcrossing inequality, 66
Drift coefficient, 148, 191
Dynkin's π-λ theorem, 4

E
Ehrenfest model, 215, 222
Equicontinuity lemma, 119
Equicontinuous, 244
Esscher transform, 93
Events, 1
   increasing, decreasing, 2
   infinitely often, i.o., 2
   limsup, liminf, 2
   tail, 87
Eventually for all, 34
Expected value
   mean, first moment, 7
Exponential distribution, 48, 140

F
Fatou's lemma, 234

Fejér average, 104
Fejér kernel, 105
Feller property, 221
Feller's Gaussian tail probability estimates,
    82
Field, 225
Filtration, 55
    right-continuous, 191
Finite-dimensional cylinders, 167
Finite-dimensional distribution, 21, 167
Finite-dimensional rectangles, 46
Finite intersection property, 242
First return time for random walk, 195
Fluctuation-dissipation, 214
Fourier coefficients
    multidimensional, 112
    of a function, 104
    of a measure, 109
Fourier inversion formula, 112
Fourier series, 104
    multidimensional, 112
Fourier transform
    inversion, 112, 113, 130
    inversion for lattice random variables,
        133
    location-scale change, 131
    multidimensional, 112
    of a function, 113
    of a measure, 115
    range, 119
    uniqueness theorem, 116
Fubini–Tonelli Theorem, 237
Functional central limit theorem (FCLT),
    147, 201

G
Gamma distribution, 48
Gaussian distribution, 48, 82, 130, 133
Geometric distribution, 140
Gram–Schmidt procedure, 249

H
Haar wavelet functions, 173
Hahn decomposition, 252
Hahn–Jordan decomposition, 252
Hausdorff estimates, 100
Helly selection principle, 142
Herglotz theorem, 110
Hewitt–Savage zero–one law, 88
Hilbert cube embedding, 143
Hilbert space, 11, 247
    complex, 249

Hitting time, 196, 198
    birth–death chain, 216
Hoeffding estimation of sample size, 100
Hoeffding lemma, 98
Holtzmark problem, 129

I
Inclusion-exclusion formula, 18
Independence
    independent and identically distributed,
        i.i.d., 31
    of σ-fields, 32, 34
    of events, 26
    of families of random maps, 34
    of random maps, 30
Independent
    pairwise, 91
Independent increments, 54, 218
Inequality
    Bonferroni, 19
    Cauchy–Schwarz, 13
    Chebyshev, 13
    Chernoff, 96
    conditional Jensen inequality, 38
    Hoeffding, 99
    Hölder, 13
    Jensen, 13
    Lundberg, 71
    Lyapounov, 13
    Markov, 13, 14
    Minkowski triangle, 13
    Neveu-Chauvin, 40
    upcrossing, 66
Infinitely divisible, 84
Infinitely divisible distribution, 128
Initial distribution, 211
Inner product, 247, 249
Inner product space norm, 247
Integrable, 232
Integral, 230
    complex, 23
    Lebesgue, 9
Integral limit theorem, 139
Integration
    complex functions, 23
Integration by parts, 9
Invariance principle, 147, 148, 201
Invariant distribution, 212
Irreducible
    Markov chain, 214
Itô calculus, 221

J
Jordan decomposition, 252

K
Kolmogorov-Chentsov theorem, 180
Kolmogorov consistency theorem, 168
Kolmogorov existence theorem, 168
Kolmogorov extension theorem, 31, 47, 168
Kolmogorov maximal inequality, 56, 160
Kolmogorov three-series theorem, 161, 162
Kolmogorov zero–one law, 87
Kronecker's Lemma, 163

L
λ-system, 4
Langevin equation, 219
Laplace transform, 92
Large deviation rate, 96
Large deviations
    Bahadur–Ranga Rao, 96
Law of large numbers
    Etemadi proof, 89
    Marcinkiewicz–Zygmund, 162
    reverse martingale proof, 88
    strong, 163, 166
    weak law, 92
Law of rare events, 118
L^1-convergence criterion, 17
Lebesgue Decomposition
    Radon–Nikodym Theorem, 250
Lebesgue measure, 229, 237
Lebesgue's dominated convergence
    theorem, 234
Lebesgue's monotone convergence theorem,
    234
Lebesgue–Stieltjes Measures, 228
Legendre transform, 93, 101
Levy–Ciesielski construction
    Brownian motion, 173
Light-tail distribution, 71
LIL, law of the iterated logarithm, 202
LIL, law of the iterated logarithm for
    Brownian motion, 183, 202
Line of support, 12
Linear functional
    bounded, 237
    continuous, 237
    positive, 237
Local limit theorem, 139, 140
Locally compact, 242
L^p-convergence criterion, 18
L^p-spaces, 11

Lundberg coefficient, adjustment
    coefficient, 72

M
Machine learning
    concentration inequalities, 100
Mann–Wald theorem, 140
Markov chain
    feller property, 221
    irreducible, 214
Markov process
    discrete parameter, 170
    homogeneous, 218
Markov property, 178, 187, 189–191, 211
    general random walk, 45
    homogeneous transition probabilities,
        187, 189
    stationary transition probabilities, 187
Markov property of Brownian motion, 190
Martingale
    Brownian motion, 198
    first definition, 53
    reverse, 68
    second general definition, 55
Martingale differences, 54
Mathematical expectation
    expected value, 7
Matrix square root, 84
Maximal inequality
    Doob, 56
    Kolmogorov, 56, 160
    Skorokhod, 160
Maxwell–Boltzmann distribution, 222
Measurable
    Borel, 5
    function, 5, 230
    map, 5
    rectangles, 25, 230, 235
    set, 1
    space, 1, 230
Measure, 1
    absolutely continuous, 250
    complete, 225
    completion of, 225
    continuity properties, 2
    countably additive, 1, 225
    dirac point mass, 1
    distribution function, 228
    finite, 1
    Lebesgue, 1, 229
    Lebesgue–Stieltjes, 228
    on a field, algebra, 225

outer, 226
probability, 1
σ-finite, 1
signed, 252
singular, 250
subadditivity, 225
Measure space, 1
completion, 225
Measure-determining
events, 5
functions, 11
Metric
bounded-Lipschitz, 154
Metrizable topology, 6
Minorization
Doeblin, 213
Modulus of continuity for Brownian
motion, 186
Moment
formula, 8
higher order, 8
Moment-generating function, 92
Monomials, 243
Multiplicative cascade measure, 69
Multivariate normal distribution, 84
μ^*-measurable, 226

N
Natural scale, 217
Net profit condition, 70
Nonatomic distribution, 230
Norm, 241
Normal distribution, 48
Normed vector space
complete, 247

O
One-dimensional simple symmetric random
walk
recurrence, 46
Optional sampling theorem, 61
Brownian motion, 198
Optional stopping theorem, 61
Optional time, 58, 59
Ornstein–Uhlenbeck process, 185, 210, 218
Maxwell–Boltzmann invariant
distribution, 222
stationary, 222
time change of Brownian motion, 222
Orthogonal projection, 36, 248
Orthogonality
conditional expectation, 36

Orthonormal, 104, 249
basis, 249
complete, 249
set, 104
Oscillation of a function on a set, 154

P
Parallelogram law, 247
Parseval relation, 117, 133
multidimensional, 117
Partition, 43
P-continuity, 137
Percolation, 49
binary tree, 28
critical probability, 29
π-λ theorem, 4
Picard iteration, 219
π-system, 4
Plancheral identity, 112–114
Poisson approximation, 118
Poisson distribution, 84
Poisson process
interarrival times, 222
Polish space, 42
Polya's recurrence theorem, 122
Polya's theorem
uniform weak convergence, 83
Positive-definite, 110, 119
Pre-τ σ-field, 189, 192
Probability
monotonicity, 18
sub-additivity, 18
Probability space, 1
Product σ-field, 167, 235, 237
Product measure, 236, 237
Product measure space, 236
Product space, 46, 167
Product topology, 243
Progressive measurability, 59
Prohorov metric, 146
Prohorov theorem, 145
Projection
orthogonal, 248
Projection theorem, 248
Pseudo-metric, 102

R
Radon-Nikodym theorem
Lebesgue decomposition, 250
Random dynamical system, 215
Random field, 172

Gaussian, 172
Random matrices
    i.i.d. product, 215
Random power series, 35
Random series, mean-square-summability
        criterion, 161
Random signs problem, 165
Random variable
    continuous, 6
    discrete, 7
    integrable, 7
    random map, 5
    random vector, 6
    simple, 7
Random walk, 45, 120
    rth passage times, 195
    asymmetric, 65, 221
    first return time, 195
    Gaussian, 124
    general k-dimensional, 45, 212
    lazy, 221
    Markov property, 45, 187
    recurrence, 120, 196
    simple, 183
    symmetric, 46
    with absorption, 222
    with reflection, 221
Recurrence
    Chung–Fuchs criteria, 123, 134
    one-dimensional simple symmetric
        random walk, 46, 195
    random walk, 65, 196
    simple symmetric random walk on
        $\mathbb{Z}^k$, $k \leq 2$, 122
Recurrent
    neighborhood, 120
Reflection principle, 183
    Browian motion, 183
    simple random walk, 183
Regular conditional distribution, 41
Regular measure, 238
Relative topology, 242
Reverse martingale, 68
Reverse submartingale, 68
    convergence theorem, 68
Riemann–Lebesgue Lemma, 113
Riesz Representation Theorem on $C(S)$,
        237
Ruin probability
    insurance risk, 70

S
Sample points, 1

Sample space, 1
Sampling design, 100
Schauder basis, 173
Schauder functions, 174
Scheffé theorem, 14
Second moment bound, 29
Second moment inequality, 29
Section
    of a measurable function, 236
    of a measurable set, 236
Separability of $C(S)$, 243
Set
    boundary, 137
    closure, 137
    P-continuity, 137
σ-compact, 145, 242
σ-field, 1
    adapted, 55
    Borel, 5
    complete, 225
    filtration, 55
    generated, 5
    generated by collection of events, 4
    join, 4
    P-complete, 60
    pre-τ, 59
    product, 230
    tail, 87
    trace, 60
    trivial, 36
σ-finite, 225
σ-finite signed measure, 252
Signed measure, 252
Simple function, 230
    standard approximation, 231
Simple symmetric random walk
    reflection principle, 183
    unrestricted, 221
Singular distribution, 230
Size-bias
    exponential size-bias, 93
Skorokhod embedding, 199
Skorokhod Embedding Theorem, 200
Skorokhod maximal inequality, 160
Smoothing
    conditional expectation, 38
Sparre Andersen model, 71
Stable law, symmetric, 128, 129
Stationary process, 212, 223
Stochastic differential equation, 221
Stochastic process, 46
    continuous-parameter, 46
    discrete-parameter, 46

Stokes law, 219
Stone–Weierstrass Theorem, 169, 242
Stopped process, 59, 192
Stopping time, 59, 189
    approximation technique, 63
Strong law of large numbers, 163, 166
    reverse martingale proof, 88
Strong Markov property
    Brownian motion, 193
    discrete-parameter, 189
Submartingale, 54, 55
Submartingale convergence theorem, 67
Subordinate, 238
Substitution property
    conditional expectation, 38
Supermartingale, 54, 55
Supnorm, 242
Support of function, 238
Symmetric difference, 88
Symmetric stable distribution, 129
Symmetrization, 132

T
Tail event, 87
Taylor expansion, 125
Tight, 145
Tilting
    size-bias, 93
Topological space, 5
    compact, 242
    Hausdorff, 241
    metrizable, 6
    pointwise convergence, 6
    uniform convergence, 6
    uniform convergence on compacts, 6
Total variation distance, 15, 136
Total variation of a signed measure, 253
Totally bounded, 244
Transience
    simple asymmetric random walk, 72
    simple symmetric random walk on
        $\mathbb{Z}^k, k \geq 3$, 122
Transition probabilities, 171
    homogeneous, 211
    one-step, 211
    stationary, 211
Triangle inequality, 247
Triangular array, 81
Trigonometric polynomials, 103
Truncation method

Kolmogorov, 165
Tulcea extension theorem, 47, 169
Tychonoff's Theorem, 243

U
$UC_b(S)$, space of bounded uniformly
    continuous real-valued functions on,
    11
Uncorrelated, 29, 91
Uniformly asymptotic negligible (u.a.n.),
    81
Uniformly bounded, 244
Uniformly integrable, 16
Unrestricted simple symmetric random
    walk, 221
Upcrossing inequality, 66
Upper class functions, 198
Urysohn lemma, 238

V
Vague convergence, 142
Variance, 29
Variance of sums, 29
Vector space
    complex, 249

W
Wald's identity, 72
Wavelet construction
    Brownian motion, 173
Weak convergence, 75, 136
    finite-dimensional, 76
    Polya's theorem, 83
    uniform, 83
Weak law of large numbers (WLLN), 92
Weak topology, 136
Weierstrass approximation theorem, 92,
    103, 241, 242
    periodic functions, 103
Wiener measure, 147, 173, 190, 209

Z
Zero–one law
    Blumenthal, 198
    Hewitt–Savage, 88
    Kolmogorov, 87

Printed in the United States
By Bookmasters